ORGANIC ACIDS
and FOOD
PRESERVATION

ORGANIC ACIDS and FOOD PRESERVATION

Maria M. Theron

J. F. Rykers Lues

CRC Press
Taylor & Francis Group
Boca Raton London New York

CRC Press is an imprint of the
Taylor & Francis Group, an **informa** business

CRC Press
Taylor & Francis Group
6000 Broken Sound Parkway NW, Suite 300
Boca Raton, FL 33487-2742

First issued in paperback 2019

ISBN-13: 978-0-367-38363-3

Library of Congress Cataloging-in-Publication Data

Theron, Maria M.
    Organic acids and food preservation / Maria M. Theron, J.F. Rykers Lues.
      p. cm.
    Includes bibliographical references and index.

    1. Food--Preservation. 2. Organic acids. I. Lues, J. F. Rykers (Jan Frederick Rykers) II. Title.

TP371.2.T446 2011
664'.028--dc22
                                         2010027909

Visit the Taylor & Francis Web site at
http://www.taylorandfrancis.com

and the CRC Press Web site at
http://www.crcpress.com

Contents

List of Figures xiii
List of Tables xv
Preface xvii
Acknowledgments xix
Author biographies xxi

Chapter 1   Introduction 1
1.1   The evolution of preservation with organic acids: From stone
        age to space age 1
        1.1.1   More than a century of preservation with
                  organic acids 1
        1.1.2   Toward preservative-free food 2
1.2   Unrivaled advantages 3
1.3   Economic implications: "Safer food, better business" 3
1.4   Legislative issues in food production 6
1.5   Problems in an "organic world" 7
        1.5.1   Nutrition and consumer perceptions 9
        1.5.2   Pesticides 9
        1.5.3   Mycotoxins 10
        1.5.4   Food safety control measures 10
        1.5.5   Seeking alternatives 10
1.6   New and emerging pathogens 11
        1.6.1   Introduction 11
        1.6.2   Foodstuffs implicated 11
        1.6.3   Laboratory methodologies 12
        1.6.4   Epidemiology 12
        1.6.5   A review of emerging organisms 13
                  1.6.5.1   Bacteria 13
                  1.6.5.2   Fungi 14
                  1.6.5.3   Viruses 15
                  1.6.5.4   Protozoa 15
References 15

Chapter 2   Nature and composition of organic acids **21**
2.1   General characterization 21
2.2   Structural description 23
2.3   An overview of individual organic acids and
       their applications 25
       2.3.1   Acetic acid...25
       2.3.2   Ascorbic acid 26
       2.3.3   Benzoic acid..26
       2.3.4   Cinnamic acid 28
       2.3.5   Citric acid..29
       2.3.6   Formic acid 30
       2.3.7   Fumaric acid 31
       2.3.8   Gluconic acid 32
       2.3.9   Lactic acid 33
       2.3.10   Malic acid..35
       2.3.11   Propionic acid...36
       2.3.12   Sorbic acid 37
       2.3.13   Succinic acid 39
       2.3.14   Tartaric acid..39
       2.3.15   Other acids..40
2.4   General applications...41
2.5   Food products naturally containing organic acids.........................42
       2.5.1   Fruit 42
       2.5.2   Juices..43
       2.5.3   Wine and vinegar 43
       2.5.4   Dairy..44
       2.5.5   Coffee...44
       2.5.6   Bakery products..44
       2.5.7   Honey 44
References 44

Chapter 3   Application of organic acids in food preservation **51**
3.1   Introduction..51
3.2   Foodstuffs 51
       3.2.1   Meat 51
               3.2.1.1   Cured meat...52
               3.2.1.2   Poultry 52
               3.2.1.3   Seafood 52
       3.2.2   Acidic foods..53
       3.2.3   Confectionery...53
       3.2.4   Fruits and vegetables 53
       3.2.5   Fruit juices..53
       3.2.6   Salads...54
       3.2.7   Vegetables 54

Contents   vii

	3.2.8	Dairy	54
	3.2.9	Soft drinks	55
	3.2.10	Sport drinks	55
	3.2.11	Animal feed	55
3.3	Industrial applications		55
	3.3.1	Labeling	55
	3.3.2	Vacuum packaging	56
	3.3.3	Meat	56
	3.3.4	Processed meats	58
	3.3.5	Seafood	59
	3.3.6	Poultry	59
	3.3.7	Dipping/spraying	60
	3.3.8	Acidified foods	60
3.4	Salts of organic acids		60
	3.4.1	Potassium sorbate	61
	3.4.2	Sodium benzoate	63
	3.4.3	Sodium lactate	64
	3.4.4	Other	64
3.5	Organic acid combinations		64
	3.5.1	Combinations in general	65
	3.5.2	Salt combinations	66
		3.5.2.1 Possible adverse effects	66
	3.5.3	Aromatic compounds	67
	3.5.4	Ethanol	67
	3.5.5	Irradiation	67
	3.5.6	Emulsifiers	68
	3.5.7	Spices	68
	3.5.8	Liquid smoke	68
3.6	Considerations in the selection of organic acids		69
	3.6.1	Sensory properties	69
	3.6.2	Color stability	69
	3.6.3	Flavor	69
	3.6.4	Carcass decontamination	70
	3.6.5	Chemical stability	70
3.7	Organic acids in antimicrobial packaging		70
	3.7.1	Antimicrobial films	70
	3.7.2	Active packaging	71
	3.7.3	Edible films	72
	3.7.4	Modified atmosphere packaging (MAP)	73
3.8	Organic acids in animal feed preservation		73
	3.8.1	The essence of preserving feed	73
	3.8.2	The postantibiotic era	74
	3.8.3	Chicken feed	75
	3.8.4	In combination with heat treatment	75

3.8.5   Propionic acid in feed 76
3.8.6   Organic acids in animal nutrition.. 77
3.9   Concentrations 77
3.9.1   Pressure toward decreased concentrations 77
3.9.2   Concentrations effective against common pathogens........ 78
3.9.3   Daily consumption of organic acids 78
3.9.4   Legislation 79
3.10   A review of current methodologies 79
3.11   Recommended applications 80
3.11.1   Carcasses... 80
3.11.2   Processed meats.. 80
3.12   Control of common pathogens 81
3.12.1   Chickens.. 81
3.12.2   Fruit 82
3.12.3   Vegetables 82
3.13   Organic acids as additives in chilled foods 82
3.14   Marinating... 83
References 84

Chapter 4   Microbial organic acid producers.. **97**
4.1   Introduction... 97
4.2   Predominant antimicrobial substances produced by LAB 99
4.2.1   Lactic acid 99
4.2.2   Bacteriocins.. 101
4.3   Principles of lactic acid fermentation... 101
4.4   Other applications of LAB 105
4.5   Genetic and bioinformatic characterization of LAB....................... 105
4.6   Acetic acid bacteria (AAB).. 106
4.6.1   Acetic acid (vinegar) production 106
4.6.2   Microorganisms involved in the production
of vinegar 106
4.6.2.1   Acetobacter and yeasts.................................... 106
4.6.3   Industrial importance—essential versus undesirable 108
4.6.4   Glucose, acid, and ethanol tolerance................................ 108
4.7   Susceptibility of and resistance to organic acids 109
4.8   Other organisms 110
4.8.1   Fungi.. 110
4.8.2   Other bacteria.. 111
References 112

Chapter 5   Mechanisms of microbial inhibition..................................... **117**
5.1   Introduction.. 117
5.2   Activity of organic acids... 118
5.3   Physiological actions of organic acids 119

        5.3.1   Introduction...119
        5.3.2   Bacterial membrane disruption............................. 121
        5.3.3   Accumulation of toxic anions 121
        5.3.4   Inhibition of essential metabolic reactions 122
        5.3.5   Stress on intracellular pH homeostasis.............. 122
5.4   Factors that influence organic acid activity 123
        5.4.1   Number of undissociated organic acids that enter the
                    bacterial cell... 124
        5.4.2   Acidity constant (pK$_a$ value)................................. 124
        5.4.3   Water activity (a$_w$) 125
        5.4.4   Temperature 125
        5.4.5   Production of H$_2$O$_2$.. 126
5.5   The role of pH.. 126
5.6   Antibacterial action 128
5.7   Antifungal action... 129
5.8   Antiviral action 131
5.9   Acidified foods... 131
5.10   Comparing effectiveness of organic acids with
          inorganic acids 133
5.11   Spectra of inhibition... 134
5.12   Improving effectiveness 134
5.13   (Physical) factors that will enhance effectiveness............ 136
        5.13.1   Ozone... 136
        5.13.2   Ultrasound.. 136
        5.13.3   Ionizing radiation 136
        5.13.4   Heat treatment 137
        5.13.5   Steam washing... 137
        5.13.6   Vacuum 138
        5.13.7   Freezing... 138
        5.13.8   Storage temperature 138
        5.13.9   Do interactions exist? 138
        5.13.10 Buffering 139
5.14   Comparisons among organic acids..................................... 139
References 142

Chapter 6   Problems associated with organic acid preservation 151
6.1   Adverse effects on humans and animals............................ 151
        6.1.1   Chemical reactions in humans ("allergies") 151
        6.1.2   Organic acids as pro-oxidants 152
6.2   Adverse effects on foodstuffs 153
6.3   Protective effects on microorganisms 153
6.4   Sensorial effects and consumer perception....................... 154
6.5   Recommended daily intake 154
6.6   Odors and palatability 155

6.7   Cost 156
6.8   Application methods... 156
6.9   Oxidation 158
6.10   Ineffectiveness.. 158
6.11   Influence on tolerance to other stresses...................................... 159
References 159

Chapter 7   Large-scale organic acid production **165**
7.1   Introduction... 165
7.2   Naturally occurring weak organic acids...................................... 165
7.3   Microbial physiology and organic acids 165
7.4   Substrates and yields... 168
7.5   Industrial fermentation... 170
         7.5.1   Monopolar 171
         7.5.2   Bipolar 171
7.6   Organic acid demand... 173
7.7   Lactic acid production..174
         7.7.1   Factors affecting production of lactic acid 176
7.8   Citric acid production 177
References 180

Chapter 8   Resistance to organic acids **185**
8.1   Introduction... 185
8.2   Intrinsic (natural) resistance 185
         8.2.1   Bacteria.. 186
         8.2.2   Fungi.. 187
8.3   Development of resistance 188
8.4   Inducible resistance 189
8.5   Mechanisms of resistance.. 191
         8.5.1   Bacteria.. 191
         8.5.2   Fungi.. 193
8.6   Transmission of resistance 196
8.7   Extent of the situation 196
8.8   *E. coli* O157:H7 197
8.9   Protective effects of organic acids... 198
8.10   Possible advantages of organic acid resistance 198
8.11   Industry strategies... 198
         8.11.1   Targets 199
References 199

Chapter 9   Acid tolerance **205**
9.1   Introduction... 205
9.2   Delineating the difference among acid adaptation, acid
         tolerance, and acid resistance.. 205

9.3   Role of organic acids in tolerance.. 206
9.4   Acid tolerance of gastrointestinal pathogens.............................. 208
9.5   Cross-resistance to secondary stresses.. 210
9.6   Mechanisms of acid tolerance development.................................. 210
        9.6.1   Passive homeostasis 212
        9.6.2   Active pH homeostasis 212
9.7   Known acid-tolerant organisms... 213
9.8   Development of acid tolerance.. 215
9.9   Implications of acid tolerance 216
9.10   Contribution of acidic foodstuffs 217
9.11   Analytical procedures.. 217
9.12   Interacting mechanisms 218
9.13   Control strategies... 218
References 219

Chapter 10 Modeling organic acid activity 225
10.1   Introduction... 225
10.2   Genomics 227
10.3   Growth models in defined systems 229
10.4   Different predictive models 230
        10.4.1   Partial least squares regression (PLS)............................... 231
        10.4.2   Stoichiometric models.. 231
10.5   Predictive indices for organic acids 232
10.6   Toward improving on existing models 233
10.7   Significance of modeling 236
References 238

Chapter 11 Legislative aspects...243
11.1   Introduction... 243
11.2   Differences in regulatory authorities.. 243
11.3   Application guidelines for organic acid preservation.................. 245
11.4   The role of general food safety regulations 248
11.5   Codex Alimentarius Commission.. 250
11.6   Proposed amendments 251
11.7   Role of government and parastatals 253
11.8   Feed preservation 254
11.9   Commercial trials 255
References 255

Chapter 12 Incidental and natural organic acid occurence.................261
12.1   Introduction... 261
12.2   Honey 261
12.3   Sourdough 261
12.4   Berries... 262

12.5 Wine..262
12.6 Coffee..264
12.7 Vinegar...264
12.8 Acid foods..264
12.9 Kombucha...264
12.10 Edible films...265
12.11 Summary 265
References 268

Chapter 13 Biopreservation 271
13.1 Introduction..271
13.2 LAB and biopreservation..272
13.3 Other organisms implicated in biopreservation.......................274
13.4 New technologies and applications..275
13.5 Consumer acceptance 276
13.6 Organic acids and probiotics 276
References 277

Chapter 14 Novel applications for organic acids.............................281
14.1 Emerging challenges..281
14.2 Consumer satisfaction...281
14.3 Optimizing organic acid application in animal feed 282
14.4 Preservative combinations 283
14.5 Antimicrobial packaging...283
        14.5.1   Factors for the design of antimicrobial film
                     or packaging...284
14.6 Optimizing commercial trials 285
14.7 New possibilities in minimally processed foods.........................285
14.8 Alternatives to washing techniques 286
14.9 Alternative application regimes 286
14.10 Recognizing the need in RTE foods..288
References 289

Chapter 15 Detection of organic acids..293
15.1 Introduction..293
15.2 Traditional detection methods...293
15.3 Contemporary methods...295
15.4 The importance of effective detection 299
15.5 Detection in specific foodstuffs...302
15.6 Characteristics of detected organic acids.....................................303
15.7 Comparing sample preparation techniques 303
References 304

Index 307

List of figures

Figure 2.1   General structure of a few organic acids.   22

Figure 2.2   Structure of acetic acid.   25

Figure 2.3   Structure of ʟ-ascorbic acid.   26

Figure 2.4   Structure of benzoic acid.   27

Figure 2.5   Structure of cinnamic acid.   28

Figure 2.6   Structure of citric acid.   29

Figure 2.7   Structure of formic acid.   31

Figure 2.8   Structure of fumaric acid.   32

Figure 2.9   Structure of ᴅ-gluconic acid.   32

Figure 2.10 Structure of ᴅ-lactic acid.   33

Figure 2.11 Structure of malic acid.   35

Figure 2.12 Structure of propionic acid.   36

Figure 2.13 Structure of sorbic acid.   37

Figure 2.14 Structure of succinic acid.   39

Figure 2.15 Structure of tartaric acid.   40

Figure 2.16 Structure of phenyllactic acid.   40

Figure 2.17 Structure of gallic acid.   41

Figure 4.1 Homolactic fermentation.   103

Figure 4.2 Heterolactic fermentation.   104

Figure 5.1 A schematic illustration of the transport system
            and consequent proposed action of organic acids on
            microbial cells.   118

Figure 7.1 Electrodialytic production of fermented organic acids.   172

Figure 7.2 Conventional production of lactic acid.   175

Figure 7.3 Production of lactic acid by *Rhizopus* and
              simultaneous saccharification and fermentation (SSF).   176

Figure 7.4 Conventional production of sodium citrate by
              utilization of the fungus *Aspergillus niger*.   179

Figure 8.1 The content of undissociated acid declines with an
              increase in pH value.   186

Figure 13.1 Schematic representation of the processes involved
              in biopreservation.   272

List of tables

Table 2.1 Organic Acids Frequently Assayed in Foods and
Dissociation Constant   24

Table 2.2 Organic Acids Naturally Found in Foodstuffs   42

Table 3.1 Application of Decontaminants as Proposed by the
Scientific Veterinary Committee of the European Union (1996) 57

Table 3.2 Application of Organic Acid Sprays   58

Table 3.3 Factors That Determine the Selection of Organic Acids
in Meat Decontamination   59

Table 3.4 Compliance Guidelines for Chilling of Thermally
Processed Meat and Poultry Products   62

Table 3.5 Example of a Pathway Implemented in the Production
of Feed   76

Table 4.1 Some Commonly Known Lactic Acid Bacteria
Important in Food Processing, Preservation, and Spoilage   99

Table 4.2 Main Antifungal Compounds Produced by Lactic
Acid Bacteria   100

Table 4.3 Importance of the Various Lactic Acid Bacteria and
Their Products in Food Production and Preservation   102

Table 5.1 Growth pH Limits for Some Foodborne Microorganisms   127

Table 5.2 Some Common Foodstuffs and pH   132

Table 6.1 Some Effects Detected on Sensory Properties of
Subprimal or Retail Cut Red Meat (Beef) after Application
of Organic Acids   157

Table 7.1 Natural Occurrence of Organic Acids   166

Table 9.1 Genes Involved in Increased Acid Tolerance in
          Salmonella enterica sv. *typhimurium* Cells Not Adapted to
          Low pH   212

Table 10.1 Differences between Mathematical Characterization
           of Bacterial Growth and Mathematical Modeling
           Techniques Used in Biotechnology   226

Table 11.1 Foodstuff Commonly Associated with Foodborne
           Illness Outbreaks   244

Table 11.2 Organic Acids Included in the FDA's List of GRAS
           Food Additives   246

Table 12.1 Organic Acids Prevalent in Various Other Frequently
           Consumed Complementary Foodstuffs Considered as
           Potential Alternative Preservatives   266

Table 13.1 Factors Involved with Antimicrobial Activity of LAB
           in Biopreservation   274

Table 13.2 Antimicrobial Action Performed by the Presence of LAB   274

Table 14.1 Washing of Minimally Processed Fruits and
           Vegetables with Chemical-Based Treatments   287

Table 15.1 Advantages and Disadvantages of Various Traditional
           Detection Methods for Organic Acids   296

Table 15.2 Advantages and Disadvantages of Various Modern
           Detection Methods for Organic Acids   300

Preface

The nutritional demands of the world for the availability of tasty nutritious food that is free of pathogens, but also free of harmful additives, are growing more intense. The demand, therefore, is for foods produced with milder treatments (e.g., less heat, salt, sugar, and chemicals), and new technologies for preventing the growth of pathogenic bacteria are increasingly required. Despite the wide range of potential antimicrobials, relatively few are suitable for use in practice, in particular food products, and in response to these modern consumer trends and also to food legislation, the food industry is faced with a serious challenge to successfully attain their objective. Organic acids have for many years been used in food preservation, but because of their natural occurrence in food and because they inhibit the growth of most microorganisms, they are ever increasingly being used.

This informational book is aimed at providing as much as possible of the available information that may be needed on the organic acids in food preservation. This information is relevant for hands-on use in the food industry, various academic disciplines, research, education, and food technologies. Problems identified thus far with regard to the organic acids have been approached and discussed and possible solutions investigated. This was made possible by studying the newest relevant research findings reported from industries and laboratories around the globe. In a quest for providing the ultimate reference book, the outcome of this book project is a combination of information, application regimes, and future prospects of the organic acids as food preservatives. This may provide the food industry as well as the food science and academic worlds with a book that contains most of the factual information on organic acids in food preservation, in combination with the latest and greatest research data (the "past, present, and future of organic acids"). In other words, the authors endeavored to represent the actual situation to the reader.

This is a unique book by itself, as to our knowledge there is no other book available that focuses solely on the subject of the organic acids as food preservatives. The book not only provides information on these acids, but also highlights the problems and provides relevant solutions where possible. Information available on the organic acids is expanding daily and this book should be a decidedly relevant source of information in both the food industry, which is, after all, one of the strongest industries in the world, as well as several disciplines in the academic world, including microbiology, food science, food technology, biochemistry, and biotechnology. Audiences from both the formal and the informal sectors from countries all over the world will be able to benefit from this book, even developing countries. Food-related problems are similar all over the world. Numerous problems in food preservation are encountered even in developed countries, who also claim to be leaders in the fields of food production, food processing, and food safety.

Numerous books have, over the last century (the authors found one as early as 1905!) and longer, been published on food preservation, which include methods of preservation, as well as the variety of food additives. However, it is surprising to find that there appears to be no book, older or recently published, that fully addresses the organic acids, even though it may be assumed that researchers, scientists, industrial processors, and even food technologists have recognized the need for such a single volume that covers most of the information (the broad and vast spectrum) available on the organic acids as natural food preservatives. Because many of these acids are naturally produced by various microorganisms, occur naturally in foods, and inhibit the growth of most microorganisms, they have been of great value as preservatives. Although organic acids have for many years been used in food, especially processed food, as additives, either as flavorants or preservatives, have they really been assessed as to their continuing effectiveness and sustainability? The concept of compiling all the relevant important information on the organic acids into one descriptive book may just be the ultimate *vade mecum* for any scientist, researcher, food processor, or food technologist.

Acknowledgments

The authors wish to thank the personnel of the Library and Information Centre at the CUT, and in particular, Anita du Toit for the absolutely amazing task of collecting all the data.

A very special thank you also to Kaylene Maasdorp and Marelize van Rooyen for taking care of the referencing.

Author biographies

Maria Theron is a full-time researcher in the Unit for Applied Food Science and Biotechnology (UAFSB) at the Central University of Technology (CUT), Free State in Bloemfontein, South Africa. She holds a PhD in medical microbiology from the University of the Free State (UFS). She worked as medical scientist and lecturer in the Faculty of Health Sciences at the UFS from 1984–2004, her research focusing on antimicrobial resistance development in anaerobic bacteria. In 2005, she entered the food arena, but her research focus remained on antimicrobial activity and efficacy ("different song, same tune"). She is currently steering a project on "Organic Acids as Antimicrobials in Food Preservation" within the research niche area of the UAFSB, funded by the National Research Foundation (NRF) of South Africa.

Ryk Lues is currently full professor and head of the Programme Environmental Health at the Central University of Technology (CUT), Free State, South Africa. He also heads the Unit for Applied Food Safety and Biotechnology at the CUT. Lues holds an MSc (microbiology) and a PhD (food science) from the University of the Free State and his field of specialization comprises organic acid biotechnology and social–behavioral aspects affecting food microbiology and hygiene. He has, to date, authored numerous articles in ISI-accredited journals and contributed to a number of books and book chapters on topics related to food microbiology, research methodology, and food hygiene systems auditing and management. He also supervised several masters, doctorate, and postdoctorate students and often acts as an external examiner, referee, and auditor.

chapter one

Introduction

1.1 The evolution of preservation with organic acids: From stone age to space age

1.1.1 More than a century of preservation with organic acids

The ability to store large quantities of food has played a pivotal role in mankind's development. Evidence of food preservation dates back to the postglacial era, from 15,000 to 10,000 BC, and the first use of biological methods has been traced back to 6000 to 1000 BC, with evidence of fermentation processes used in producing beer, wine, vinegar, bread, cheese, butter, and yogurt (Soomro, Masud, and Anwaar, 2002).

Evidence also exists that points to a notable award that was offered by the Emperor Napoleon of France in 1787 for anyone who could propose a way to preserve the food intended for the soldiers in the French armies. This award was won in 1809 by Nicholas Appert for his idea to put food in bottles, stopper them, and heat the contents, killing bacteria and preventing further contamination. However, he was apparently not aware at the time of the existence of bacteria and that they were the cause of the spoilage. In 1819 Peter Dunrand used metal containers to apply the same principle. These containers were called cannisters, which is the origin of the word "can" (Saint Xavier University, 2000).

The cause of food spoilage was affirmed in 1864 by Louis Pasteur who showed that microorganisms in foods were the cause of such contamination. He subsequently found that microbes were killed by heating and that sealed containers could be used to preserve food by preventing recontamination from atmospheric air.

In 1940, a major development in the distribution and storage of foods was achieved, when low-cost home refrigerators and freezers became available. Other developments followed and included artificial drying, vacuum packaging, ionizing, radiation, and chemical preservation (Soomro, Masud, and Anwaar, 2002). A combination of techniques in which acids have played an important role have emanated as part of consequent preservation protocols for various foodstuffs (Hsiao and Siebert, 1999). One of the fundamental principles of food safety has been based on the division of foods into low-acid and high-acid categories and the according application of preservation treatments (Nakai and Siebert, 2003). Similarly to the entire food preservation process that can be traced back centuries, the use of weak acids as antimicrobials also has a long history. The first reported chemical preservative was sulfite which was employed for centuries for the sterilization of wine vessels. Nowadays the use of sulfite as a preservative has been largely curtailed because of reports of adverse effects on human health, especially in steroid-dependent asthmatics, and consequently in preservation protocols the sulfites have been predominantly replaced by weak organic acids (Piper, 1999).

A contributing factor to the organic acids being of considerable value as food preservatives is that they are also food ingredients, often naturally produced by microorganisms and widely distributed in nature as normal constituents of plants or animal tissues (Gauthier, 2005). Consequently, in more recent times it has become commonplace to include organic acids in chemical additives to control microbial contamination in foods and feeds (Ricke et al., 2005). Vinegar is another preservative that was applied by ancient civilizations and used as a seasoning or preservative agent (Tesfaye et al., 2002). The present-day definition of vinegar as a food constituent clearly hints at its being a pivotal component due to its preservation ability: "A liquid fit for human consumption, produced from a suitable raw material of agricultural origin, containing starch, sugars, or starch and sugars by the process of double fermentation, alcoholic and acetous, and contains a specified amount of acetic acid" (FAO/WHO, 1987; Tesfaye et al., 2002).

Benzoic acid is the oldest and most commonly used preservative (Barbosa-Cánovas et al., 2003). In 1970 it was reported that sorbic acid had been proposed for extensive use as a preservative, but its safety-in-use had to be established before being added to the existing list of permitted food additives (Shtenberg and Ignat'ev, 1970). A study done in 1956 reported on the use of sorbic acid as a preservative in fresh apple juice (Ferguson and Powrie, 1957). Worldwide, approximately 40% of food grown for human consumption is lost to pests and microorganisms (Saint Xavier University, 2000). The preservation process itself was of little public interest until a decade ago. Today it is one of the key issues addressed by every food processor (Marz, 2000).

1.1.2   Toward preservative-free food

Preservation of food requires the control of microorganisms throughout the food chain, and proper monitoring to ensure the safety and stability of the product during its shelf life (Prange et al., 2005). It is said that except for the food grown in one's own garden, all food products contain preservatives, and that every manufacturer adds preservatives to the food during processing (*Food Additives World*, 2008). However, there is an increasing demand for foods without chemical preservatives (Bégin and Van Calsteren, 1999). Therefore, in attempting to satisfy consumer demands for foods that are ready to eat, yet fresh-tasting, rich in nutrients and vitamins, and also minimally processed and preserved, the food industry, worldwide, is increasingly faced with numerous challenges (Leguerinel and Mafart, 2001) and forced to respond by prioritizing the development of minimally processed food products that are at the same time nutritious and with high organoleptic quality. This places even greater strain on the general market in commercialization as minimally processed products generally present a very short shelf life, which may often be less than a week (Valero et al., 2000).

It has also become evident that microorganisms are continuously adapting to survive in the presence of previously effective control methods (Rico et al., 2008). In response to safety concerns from these dilemmas, research intensified to address new techniques demanded in all steps of the production and distribution chain for maintaining required food quality that will at the same time inhibit undesired microbial growth (Rico et al., 2008; Samelis et al., 2002). Research has also shifted the focus to the potential implications of new decontamination techniques on pathogen behavior as well as on the microbial ecology of food products (Samelis et al., 2002).

1.2   Unrivaled advantages

Organic acids have a long history of being utilized as food additives and preservatives in preventing food deterioration and extending the shelf life of perishable food ingredients, and have been demonstrated to be effective under a wide variety of food processing conditions. They have been applied in pre- and postharvest food production and processing and various organic acids are approved or listed in FDA regulations for various technical purposes, in addition to preservation (Ricke, 2003). These include the application of organic acids as acidulants, antioxidants, flavoring agents, pH adjusters, and even nutrients (Smulders and Greer, 1998). Traditionally organic acids have, therefore, been extensively used in the food and pharmaceutical industries as preservatives, chemical intermediates, and also buffer media (Bailly, 2002). Much research has been devoted to the application of organic acids as preservatives because of their bactericidal activity, and their "generally recognized as safe" (GRAS) status (Tamblyn and Conner, 1997).

1.3   Economic implications: "Safer food, better business"

Sustainable economic growth requires safe sustainable resources for industrial production (Kamm and Kamm, 2004). However, microbial contamination of foods plays a major role in economic losses to food processing companies worldwide and these problems are being experienced despite intensified control measures, extensive research on food safety, and modern food preservation regimes. Entire industries may be affected, experiencing significant losses as a result of food spoilage as well as food poisoning (Brul et al., 2002a). Contamination of foods with pathogenic organisms may even have devastating effects on the foreign trade of a country. This was evident in the early 1990s in Peru during the initial stages of a cholera epidemic. Such economic losses may be devastating even when the hazard is recognized before any consumers have been affected (Molins, Motarjemi, and Käferstein, 2001). To make matters even worse, public health officials have predicted that a number of underlying forces may make foodborne illnesses even more of a problem in years to come. Factors such as emerging pathogens, improper food preparation and storage practices among consumers, insufficient training of retail employees, increasing global food supply, and an increase in the number of people at risk because of aging and compromised capacity to fight foodborne illnesses will have to be contended with in the combat against the effects of microbial contamination on food safety (Medeiros et al., 2001).

Several studies have demonstrated the high cost to society of foodborne illness in different countries, including the United States (1996) and Canada (1990), and there is a growing literature on the importance of reducing foodborne illness in developing countries (Motarjemi et al., 1996; Moy, Hazzard, and Käferstein, 1997; Unnevehr and Jensen, 1999). From these studies the immense benefits in improving food safety are apparent, yet these benefits have not been fully internalized by market mechanisms to reward the companies involved (Unnevehr and Jensen, 1999). Food preservation requires the control of microorganisms throughout the food chain and in this regard there is a need to predict the behavior of undesirable microorganisms and thus to get an insight to their cellular functioning under conditions generally encountered under relevant food manufacturing conditions (Brul et al., 2002a). Producers or retailers may, however, not be able to ascertain or certify the safety of foods, as a result of the wide spectrum of potential microbial contaminating agents and the hazards posed by them when encountered in food or food environments (Unnevehr and Jensen, 1999). It is essential that thorough risk assessment be the very first step taken by all leading food industries in optimizing the quality and safety of food products (Brul et al., 2002b). Although markets may undergo changes simultaneously with changes in the marketing environment as well as political frameworks, direct contact with the industry provides reliable and precise quantitative information with regard to food additives and food safety (Marz, 2002).

It is worth mentioning in particular the spoilage of foods and beverages characterized by high sugar contents, low pH, and low water activities (a_w) by yeasts as posing a major economic problem (Hazan, Levine, and Abeliovich, 2004). In the summer months fungal spoilage of bakery products before their expiration date is an important concern for manufacturers. Adding to the problem is that preservatives added to bakery products with relatively high pH (Juneja and Thippareddi, 2004) may be ineffective, as they do not have any impact on their shelf life (Marín et al., 2003). Although many of these products are carrying useless preservatives that could just as well be excluded from their recipes, reduction in preservatives to subinhibitory levels has been shown to stimulate growth of fungi or to stimulate mycotoxin production (Marín et al., 2003; Suhr and Nielsen, 2004). However, it is possible that these products have to rely on a suitably low a_w for preservation (Marín et al., 2003).

In addition to economic losses as a result of spoilage by fungal growth on bread products, production of mycotoxins and the health hazards associated with these are also a serious concern (Arroyo, Aldred, and Magan, 2005). A mycotoxin commonly associated with bread products is ochratoxin A, the presence of which mainly comes from the wheat flour used for manufacturing. Ochratoxin A is only partly destroyed in wheat grain or in flour during the making of bread and little information is available on the effect of preservatives in bread products on ochratoxin A production by spoilage molds, especially in moderate climatic conditions (Arroyo, Aldred, and Magan, 2005).

The evidence is clear that measures should be taken in the food safety arena in addressing all the factors implicated in such serious economic problems caused as a result of microbial contamination of food. The advantages of a better understanding of the physiological actions of preservative systems to the food manufacturing industry are endless. It is envisaged, among others, to improve food quality and wholesomeness through lower thermal treatment, provide possibilities for new products (mildly preserved, organic foods), save energy through better controlled processes, use less waste and cooling water, reduce the use of cleaning and disinfecting agents, and produce less waste through better process control (Brul et al., 2002a).

It is, however, believed that food poisoning and in particular full control of its morbidity is well within reach, speculated from a consistent decline in the United States, among others, in the incidence of food-transmitted salmonellosis, as a result of new developments such as the introduction of pathogen reduction steps in food processing that have been implemented in the United States by regulation (Olsen, 2001; Struijk, Mossel, and Moreno Garcia, 2003). Similarly improved protection of the consumer population is envisaged to be achievable in all areas of food-transmitted infections. However, infection pressure on the food environment (from raw food, in particular from animal origin) will require the implementation of a decontamination step which should be introduced in production and distribution at a very specific point. This should be in addition to the usual hygienic practices already in place (Struijk, Mossel, and Moreno Garcia, 2003).

1.4   Legislative issues in food production

In developed economies significant changes have been made to food safety regulations over the last two decades, where many countries have increased their standards for ensuring safety and quality of food intended for domestic consumption as well as for the export market. However, developing countries are reported to experience difficulties in meeting all the requirements associated with implementing high-level sanitary measures involved in technical regulations and public and private standards, as well as conformity tests (Martinez and Bañados, 2004). Furthermore, public authorities have been encouraging both the food and feed industries in developing extensive quality management systems in attempting to improve food safety, to restructure a system for inspection, and also to try to boost consumers' trust by promoting adequate information (Rohr et al., 2005). Another important factor is that continuous monitoring of proposed regulations is achievable by online government resources in updating food safety recommendations (Cabe-Sellers and Beattie, 2004). Both the U.S. Food and Drug Administration (FDA) and U.S. Department of Agriculture (USDA) are involved in the provision of extensive oversight in regulating measures implemented in ensuring the safety of food. These also include compulsory and meticulous studies on safety, with regard to issues such as toxicology and pharmacokinetics, prior to the approval of an antimicrobial agent or preservative for use and monitoring of the food supply to ensure the agent is being used correctly (Donoghue, 2003).

Of specific importance are regulations with regard to safe food production in the meat industry. In the list of feed additives authorized by European Union (E.U.) legislation, organic acids are recommended as preservatives, and their use is currently allowed in all livestock species. It is recognized that the organic acids produce no detectable abnormal residues in meat and are, therefore, considered safe substances (Castillo et al., 2004). In the United States organic acids such as lactic acid and acetic acid have GRAS status, and are approved for use. Differences in regulatory authorities are evident, particularly in the meat industry. The European Union requires strictly hygienic processing sufficient to assure product safety and does not allow any additional antimicrobial decontaminant, as it regards decontamination strategies as a means of concealing poor hygiene. Contrary to the E.U. requirements, it may be generally accepted that U.S. abattoirs initiated the incorporation of organic acid sprays as a component of the carcass dressing process. In the adaptation of E.U. regulations with regard to meat treatment, it should be kept in mind that even with the best and most current hygienic practices, contamination of carcasses is inevitable. The USDA, therefore, regards such interventions as an important addition to hazard analysis critical control point (HACCP) in meeting the standards required for pathogen control (Smulders and Greer, 1998).

1.5   Problems in an "organic world"

Consumers are increasingly, and often without grounds, questioning the ability of modern food systems to provide safe and nutritious food of high quality. In response to these consumer conceptions authorities and the industry have to contend with frustration with these attitudes of the public toward food risks, as they believe the public worries about the wrong things. This then often results in twisted agendas and specific efforts by food producers being unjustified and often wrongly allocated (Macfarlane, 2002). The demand for safer food is also constantly growing as consumers, specifically in the developed countries, are becoming more affluent, live longer, and with modern technology and available information, have a better understanding of the relationship between diet and health (Unnevehr and Jensen, 1999). As a result consumers today are sceptical about any additives as preservatives and various industries are compelled to reduce quantities or totally remove additives that are often essential in ensuring quality and safety of a specific foodstuff (Suhr and Nielsen, 2004). Contrary to consumer demands, studies have shown that the use of suboptimal concentrations of preservatives such as the organic acids may stimulate the growth of spoilage fungi (including *Hyphopichia burtonii*, *Candida guillermondii*, and *Aspergillus flavus*) which may lead to mycotoxin production (Arroyo, Aldred, and Magan, 2005). However, as individuals may be sensitive to varying preservatives and may present allergic reactions, there are circumstances when the type and amount of preservatives should be controlled (Tang and Wu, 2007).

When considering consumer prerequisites in a typical decision-making process, safety is usually a nonnegotiable attribute. Evidence suggests that consumers expect all food to be intrinsically safe. However, consumer perceptions also relate to human subjectivity which ultimately determines their purchasing and consumption decisions. The meat crisis and subsequent decline in beef consumption, particularly in Europe, have forced governments and the meat industry into action in restoring consumer confidence in meat safety and wholesomeness (Tang and Wu, 2007; Verbeke et al., 2007). Consumers also judge the quality of fresh foodstuffs such as fresh-cut fruit and vegetables on appearance and freshness at the time of purchase. However, satisfaction and subsequent purchases will depend upon the consumer's experience in terms of texture and flavor of the product. Then again, everyone uses a different set of criteria to interpret the quality of product and the term "acceptability" is a practical approach to quality when compared to a criterion, the quality limit (Rico et al., 2007).

Consumers are also very concerned about pesticides used in food production, because of possible residues in the food product. In surveys it has been found that both organic and conventional food buyers believe organic produce to be better (Macfarlane, 2002). Consumers, and more specifically young people, are showing increased interest in more education about food safety (Cabe-Sellers and Beattie, 2004). However, notwithstanding the fact that the consumer is gaining more knowledge with regard to food safety and quality, there is the gap between the experts and the public that continues to exist (Macfarlane, 2002). The public perception of risk often incorporates many factors in addition to the quantitative measure of risk. It is, however, interesting to acknowledge that from constant research findings, it is evident that what the consumers say they want, and what they actually buy, are often not the same. The desires and realities of the public are also often difficult to understand and impractical to achieve. People want easy-to-prepare food (such as microwave foods) to look, smell, and taste as good as those made in an oven. In attempting to satisfy these desires, more not fewer additives are then needed. The nature of foods also continues to change, more specifically the increasingly popular prepared foods (Forman, 2002).

Foods are produced or processed by many technologies and this poses problems and challenges for researchers who are concerned with the factors responsible for consumer choice, acceptance, and purchase behavior. Consumer perceptions about safety, risks, benefits, and cost associated with new and modern technologies can influence their purchase decisions in a negative way. Much research has therefore been devoted to the evaluation of consumers' concerns regarding specific food safety issues such as irradiated foods, bioengineered foods, foods containing pesticides, processed foods that use laser light sources, and of course, microbial contaminated foods (Cardello, Schutz, and Lesher, 2007). Consumers' attitudes toward food safety are not an independent issue and have been shown to influence and predict their behavior. They are shown to be linked to consumers' demographic and socioeconomic status, culture, personal preferences, and experience (Cheng, Yu, and Chou, 2003).

As a result of various food safety incidents consumers are increasingly asking questions about the food they consume and how it is produced, which has resulted in increasing demand for organic products (Martinez and Bañados, 2004). Surveys have indicated that many consumers choose organic foods because of a belief in nutritional benefits of such organic products. Findings of one survey demonstrated that the main reasons consumers purchased organic foods were avoidance of pesticides (70%), freshness (68%), health and nutrition (67%), and to avoid genetically modified foods (55%; Winter and Davis, 2006). It is, however, often very costly to test for the safety of products, causing some reason for concern, as private markets often fail to provide for adequate food safety, a safety that is not readily apparent to consumers (Unnevehr and Jensen, 1999).

Another concern in the food safety arena is consumers who may prefer consuming raw or undercooked foods, despite the risks known to them. Although consumers need to be educated about avoiding such foods it is very difficult to change customs and cultures of a population and it may be more effective to consider other more cost-effective interventions (Molins, Motarjemi, and Käferstein, 2001). However, increased public scrutiny about the use of antibiotics and other additives in the animal feed industry has now directed research toward alternative means for manipulating gastrointestinal microflora in livestock (Castillo et al., 2004).

1.5.1 Nutrition and consumer perceptions

Some studies have concluded that organic production methods would mean increased nutrient content, particularly organic acids and polyphenolic compounds. Although many of these substances are considered to have potential human health benefits as antioxidants, their impact on human health when consuming greater levels of organic acids and polyphenolics has not yet been determined (Winter and Davis, 2006).

1.5.2 Pesticides

In agricultural development pesticides have become indispensable as plant protection agents and essential in enhancing food production. Pesticides also play a significant role, apart from their role in food production, in keeping society free from deadly diseases such as malaria, dengue, encephalitis, and filiariasis among others. Moreover, pesticides are also important in rural health programs, controlling biting or irritating insects and other pests that may plague humans and animals (Rekha, Naik, and Prasad, 2006).

However, indiscriminate use has led to occupational hazards in the developing world and is posing a severe risk to human health. The overall trend is in developing a food production system that is socially, ecologically, and economically acceptable as well as sustainable. The ideal would be the development of an alternative for economic crop production without the use of agrochemicals. One such alternative to synthetic pesticide use is organic agriculture and food processes, which are already widely in practice in developed countries. The aims of organic food production is to encourage and enhance biological cycles within the farming systems in attempting to maintain and increase soil fertility, to minimize negative effects caused by fertilizers and pesticides, and ultimately to produce adequate quantities of high-quality food. However, this process is costly, labor intensive, and not always effective (Rekha, Naik, and Prasad, 2006).

1.5.3 Mycotoxins

Mycotoxins are secondary metabolites produced by filamentous fungi. These mycotoxins may cause a toxic response, referred to as mycotoxicosis, when ingested by humans or animals. *Aspergillus, Fusarium,* and *Penicillium* are the most abundant mycotoxin producers, contaminating foods through fungal growth before harvest, during harvest, or often during improper storage. Most common mycotoxins are the aflatoxins, which also represent one of the most potentially carcinogenic substances known. Aflatoxins are rated as Class I human carcinogens by the International Agency for Research on Cancer (IARC) (Binder, 2007). It is, therefore, one of the more serious problems to deal with in an organic food production setup, where common preservatives and antifungal agents are almost nonexistent.

1.5.4 Food safety control measures

Effective control measures are the cornerstone for ensuring food safety. These include methods available in preventing food contamination and also decontamination, in instances where this may be necessary. It has been shown, over and over again, that present production methods are inadequate to totally prevent food contamination. Moreover, the complexity of food handling and processing provides ample opportunity for contamination, as well as survival and growth of pathogenic organisms. It also cannot be foreseen that in the near future, production methods would be able to ensure contamination-free foods, as many pathogens are part of the normal flora of the environment, and to make things even worse, public health authorities all over the world are faced with newly recognized types of foodborne illnesses emerging and reemerging (Molins, Motarjemi, and Käferstein, 2001).

1.5.5 Seeking alternatives

The food safety problem has reached a stage where the presence of harmful bacteria has become an even greater concern, as consumers remain worried about new technologies and ingredients in their food (Smith DeWaal, 2003). A real need has developed to find alternatives for preservation in an ever-demanding consumer world. One area where various alternatives have been suggested, is the preservation of fresh-cut fruit and vegetables, to improve the efficacy of washing treatments. Alternative methods have been proposed and include (1) antioxidants, (2) irradiation, (3) ozone, (4) organic acids, (5) modified atmosphere packaging, and (6) whey permeate among others. These methods are actually modified methods, but have not yet gained widespread acceptance by the industry. Their acceptance limit is also primarily defined by economic and physiological factors, whereas the quality of a product is largely defined by its intrinsic properties. For example, for fruits and vegetables, the product properties such as color, firmness, and taste are known to change over time (Rico et al., 2007).

1.6 New and emerging pathogens

1.6.1 Introduction

It has become evident that a need exists for better understanding of the effectiveness of classical preservatives such as the organic acids as well as naturally occurring antimicrobial biomolecules in combination with the other common components of food preservation systems. This would entail understanding the concept of microorganisms dying, surviving, adapting, or growing as well as the physiological and molecular mechanisms within cells that result in these phenotypes (Brul and Coote, 1999). Microbiological problems and the implications of this in food safety have been elucidated at the WHO food strategic planning meeting (WHO, 2003; Galvez et al., 2007). Of major concern is the emergence of new pathogens together with problems with pathogens not previously associated with food consumption. There are constant changes taking place in food production, preservation, and packaging and because microorganisms, and specifically pathogenic organisms, are not limited to cardinal ranges of temperature, pH, and water activity, but can adapt to survive at values outside those given in textbooks, altered food safety hazards have emerged (Galvez et al., 2007). New organisms continue to be added to the list of potential diseases and increased attention is focused on foods not previously considered to be common carriers of hazardous contaminants. For example, in 2004 a widespread outbreak of *Yersinia pseudotuberculosis* was reported that involved 47 cases, with iceberg lettuce identified as the carrier (Cabe-Sellers and Beattie, 2004).

1.6.2 Foodstuffs implicated

Meat products are among the leading foods associated with listeriosis, in addition to salmonellosis (Mbandi and Shelef, 2002). Although the meat from a healthy animal is sterile, it is often contaminated from various origins, such as dirty skin, hooves, hair, cutting tools, infected personnel, polluted water, improper slaughtering procedure, postslaughter handling, and also during storage (Durango, Soares, and Andrade, 2006). Pathogens that often contaminate freshly slaughtered beef carcasses typically originate from the bovine gastrointestinal tract, and contaminate via fecal matter or via contents from intestinal organs that are accidentally damaged during removal (Berry and Cutter, 2000). Due to the ubiquitous nature and increasing presence of the pathogen *Listeria monocytogenes* in the slaughterhouse and meat packaging environments, the incidence and behavior of this pathogen in meat products are receiving increasing attention. Contamination of ready-to-eat (RTE) meats with *Listeria monocytogenes* and *Salmonella* spp. occurs mainly at postprocessing, and it is common practice for these products to be consumed without further heating. *L. monocytogenes* can, therefore, be hazardous to the consumer in both refrigerated and temperature-abused RTE meats (Mbandi and Shelef, 2002).

An example of foodstuffs where hazardous microbial contaminants are increasingly found, are salad bars, offering freshly cut lettuces. Such salad bars have become increasingly popular, and have introduced new environments that support the growth of foodborne pathogens, including *L. monocytogenes*. These salad bars are found in restaurants, supermarkets, and convenience stores and *L. monocytogenes* has been isolated from fresh and minimally processed vegetables. However, washing and sanitizing treatments should be able to reduce the presence or at least prevent proliferation of *L. monocytogenes* during storage of vegetable foods (Allende et al., 2007).

1.6.3   Laboratory methodologies

Bacteria are the only pathogens routinely tested for in food samples, whereas viruses and parasites are typically tested in stool specimens of persons infected after food consumption. The food processor tests for specific types of bacteria and can choose among culture tests, gene probes, manual immunoassays, and instruments (automated immunoassays) (Kroll, 2001).

1.6.4   Epidemiology

Food poisoning as well as food spoilage outbreaks are the ultimate cause for concern for any food producer or health authority. It is, however, worth mentioning a few very important cases. Several processed meat products, such as roast beef, turkey, and meat-containing Mexican foods have been implicated in outbreaks of *Clostridium perfringens*. *C. perfringens* is an important foodborne pathogen that is estimated to cause, in the United States alone, 248,000 cases of foodborne illness annually. Spores of *C. perfringens* are widely distributed in soil and water and often contaminate raw meat and poultry during slaughter operations. Outbreaks occur primarily as a result of consumption of foods that are improperly handled after cooking. However, a serious problem is that spores present in raw materials that are used in the preparation of meat products can survive the traditional heat processing schedules employed by the meat industry. These heat-activated spores surviving in cooked foodstuffs may then germinate, outgrow, and multiply during subsequent chilling operations, especially if chill rates are not properly followed, if products are not properly refrigerated, or if products have been temperature abused (Juneja and Thippareddi, 2004).

1.6.5   A review of emerging organisms

1.6.5.1   Bacteria

Since the first recognized outbreak in 1982, *Escherichia coli* O157:H7 has emerged as a serious, potentially life-threatening, human foodborne pathogen, implicated in causing diarrhea, hemorrhagic colitis, and hemolytic uremic syndrome (Jordan, Oxford, and O'Byrne, 1999; Cheng, Yu, and Chou, 2003). *E. coli* O157:H7 is a member of the enterohemorrhagic *E. coli* that causes devastating bloody diarrhea in its victims (Audia, Webb, and Foster, 2001). It has become clear in recent years that much more attention is necessary to prevent contamination by *E. coli* O157:H7 in foods, especially low pH foods (Cheng, Yu, and Chou, 2003). *E. coli* O157:H7 is associated with eating contaminated undercooked beef and less frequently with drinking raw milk, unpasteurized apple cider, water, or even person-to-person contact (Ryu and Beuchat, 1999). The apparent source of these pathogens is the bovine gastrointestinal tract from where the contents can contaminate various processed meat products (Audia, Webb, and Foster, 2001). Safety hazards of unpasteurized fruit and vegetable beverages have also been noted. *E. coli* O157:H7 is known to be acid tolerant and has a low infectious dose (10–2000 cfu/g) (Chikthimmah, LaBorde, and Beelman, 2003).

   L. monocytogenes is another major concern to manufacturers worldwide. This is due to the high mortality rate of listeriosis in susceptible populations and due to the pathogen being resistant to a number of food preservation practices. The ability of the organism to grow at refrigeration temperatures, in particular, and the ability to grow on dry surfaces, which would normally inhibit bacterial growth, make it well adapted to food environments (Barker and Park, 2001). Between 2000 and 2003, there were at least 111 reported recalls of RTE meat and poultry products suspected of being contaminated with *L. monocytogenes* (Geornaras et al., 2005).

   Campylobacter gastroenteritis is one of the important diseases that has a serious impact on public health in developing countries. Poultry has initially been infected by these bacteria and the consumption of contaminated chicken meat has frequently been implicated as the source of

Campylobacter infection. It has been shown that water is an important route of horizontal transmission on broiler farms (Chaveerach et al., 2002).

Bacterial strains that reside under the lactic acid bacteria (LAB) have been reported to survive disinfection. These organisms then occur as spoilage organisms and can also affect fermentation processes (Sidhu, Langsrud, and Holck, 2001). In addition to this, there is increasing evidence that LAB can act as opportunistic pathogens (Sims, 1964; Adams, 1999; Sidhu, Langsrud, and Holck, 2001). Several lactobacilli species have been associated with human diseases such as septicemia, rheumatic diseases, vascular diseases, meningitis, lung abscesses, infective endocarditis, peritonitis, and urinary tract infections (Aguirre and Collins, 1993; Harty et al., 1994; Sidhu, Langsrud, and Holck, 2001). Some of the LAB strains known to be implicated in foodborne diseases include *Lactobacillus casei, Lactobacillus plantarum, Lactobacillus rhamnosus,* and the *Lactobacillus acidophilus* group (Sidhu, Langsrud, and Holck, 2001).

1.6.5.2 Fungi

Yeasts and fungi are a major spoilage threat for many food products that are preserved at low pH, low a_w, or high levels of preservatives (Piper et al., 2001). Of importance in the spoilage of bakery products are species of *Eurotium, Aspergillus,* and *Penicillium* (Marén et al., 2003). Another important spoilage fungus is *Fusarium,* often found in cereal grains where they might produce a number of mycotoxins (Schnürer and Magnusson, 2005). Adding to these is *Zygosaccharomyces,* a genus associated with the more extreme spoilage yeasts. These spoilage yeasts are osmotolerant, highly fermentative, and extremely resistant to most preservatives (Steels et al., 2000). *Zygosaccharomyces bailii* originally attracted attention as a spoilage agent of wine, with its high ethanol tolerance and ability to metabolize acetic acid in the presence of the complex mixtures of sugars present in wine fermentations. It is now known that *Z. bailii* was the first *Zygosaccharomyces* gene of resistance to food preservatives that was characterized genetically (Mollapour and Piper, 2001). *Zygosaccharomyces lentus* is another important food spoilage organism able to grow in a wide range of food products, particularly at a low pH as well as foods and drinks containing high levels of sugar. It is likely to be more significant than *Z. bailii* in the spoilage of chilled products (Steels et al., 1999).

In the past few years, interest has been increased in the yeast *Yarrowia lipolytica,* due to its numerous biotechnological applications. Although it has often been reported in incidents from food environments, sometimes with serious consequences, this organism is, however, not considered to be among the more dangerous food spoilage yeasts. *Y. lipolytica* is usually isolated from lipid-rich environments such as mayonnaise and salad dressing, but can be found in dairy products, specifically butter, cheese, and yogurt. The presence of this organism in these environments could be attributed to their being tolerant or even resistant toward stress environments, such as the organic acids. Utilization and toxicity depend on their capacity to use (as sole carbon and energy source) most carboxylic acids often used as preservatives in the food industry (Rodrigues and Pais, 2000).

1.6.5.3   Viruses

Reports on Norovirus outbreaks have increased in the past few years. In 2007 the existence of a new Norovirus strain of genogroup II/4 was confirmed by the CDC Enteric Virus Branch in Atlanta, Georgia. This virus is acid-stable and can pass through the stomach undamaged. There is an obvious problem emerging with regard to Norovirus infections (Leuenberger et al., 2007).

1.6.5.4   Protozoa

Protozoa, such as *Cyclospora cayetanensis*, and also various foodborne trematodes have been associated with infections, often with severe, chronic, or fatal health consequences (Molins, Motarjemi, and Käferstein, 2001).

References

Adams, M.R. 1999. Safety of industrial lactic acid bacteria. *Journal of Biotechnology* 68:171–178.

Aguirre, M. and Collins, M.D. 1993. Lactic acid bacteria and human clinical infection. *Journal of Applied Microbiology* 75:95–107.

Allende, A., Martinez, B., Selma, V., Gil, M.I., Suarez, J.E., and Rodriguez, A. 2007. Growth and bacteriocin production by lactic acid bacteria in vegetable broth and their effectiveness at reducing *Listeria monocytogenes* in vitro and in fresh-cut lettuce. *Food Microbiology* 24:759–766.

Arroyo, M., Aldred, D., and Magan, N. 2005. Environmental factors and weak organic acid interactions have differential effects on control of growth and ochratoxin A production by *Penicillium verrucosum* isolates in bread. *International Journal of Food Microbiology* 98:223–231.

Audia, J.P., Webb, C.C., and Foster, J.W. 2001. Breaking through the acid barrier: An orchestrated response to proton stress by enteric bacteria. *International Journal of Medical Microbiology* 291:97–106.

Bailly, M. 2002. Production of organic acids by bipolar electrodialysis: Realizations and perspectives. *Desalination* 144:157–162.

Barbosa-Cánovas, G.V., Fernández-Molina, J.J., Alzamora, S.M., Tapia, M.S., López-Malo, A., and Chanes, J.W. 2003. General considerations for preservation of fruits and vegetables. In: *Handling and Preservation of Fruits and Vegetables by Combined Methods for Rural Areas*. Rome: Food and Agriculture Organization of the United Nations.

Barker, C. and Park, S.F. 2001. Sensitization of *Listeria monocytogenes* to low pH, organic acids, and osmotic stress by ethanol. *Applied and Environmental Microbiology* 67:1594–1600.

Bégin, A. and Van Calsteren, M.R. 1999. Antimicrobial films produced from chitosan. *International Journal of Food Microbiology* 26:63–67.

Berry, E.D. and Cutter, C.N. 2000. Effects of acid adaptation of *Escherichia coli* O157:H7 on efficacy of acetic acid spray washes to decontaminate beef carcass tissue. *Applied and Environmental Microbiology* 66:1493–1498.

Binder, E.M. 2007. Managing the risk of mycotoxins in modern feed production. *Animal Feed Science and Technology* 133:149–166.

Brul, S. and Coote, P. 1999. Preservative agents in foods. Mode of action and microbial resistance mechanisms. *International Journal of Food Microbiology* 50:1–17.

Brul, S., Coote, P., Oomes, S., Mensonides, F., Hellingwerf, K., and Klis, F. 2002a. Physiological actions of preservative agents: Prospective of use of modern microbiological techniques in assessing microbial behavior in food preservation. *International Journal of Food Microbiology* 79:55–64.

Brul, S., Klis, F.M., Oomes, S.J.C.M., et al. 2002b. Detailed process design based on genomics of survivors of food preservation processes. *Trends in Food Science and Technology* 13:325–333.

Cabe-Sellers, B.J. and Beattie, S.E. 2004. Food safety: Emerging trends in foodborne illness surveillance and prevention. *Journal of the American Dietetic Association* 104:1708–1717.

Cardello, A.V., Schutz, H.G., and Lesher, L.L. 2007. Consumer perceptions of foods processed by innovative and emerging technologies: A conjoint analytic study. *Innovative Food Science & Emerging Technologies* 8:73–83.

Castillo, C., Benedito, J.L., Mendez, J., et al. 2004. Organic acids as a substitute for monensin in diets for beef cattle. *Animal Feed Science and Technology* 115:101–116.

Chaveerach, P., Keuzenkamp, D.A., Urlings, H.A., Lipman, L.J., and Van, K.F. 2002. In vitro study on the effect of organic acids on *Campylobacter jejuni/coli* populations in mixtures of water and feed. *Poultry Science* 81:621–628.

Cheng, H.Y., Yu, R.C., and Chou, C.C. 2003. Increased acid tolerance of *Escherichia coli* O157:H7 as affected by acid adaptation time and conditions of acid challenge. *Food Research International* 36:49–56.

Chikthimmah, N., LaBorde, L.F., and Beelman, R.B. 2003. Critical factors affecting the destruction of *Escherichia coli* O157:H7 in apple cider treated with fumaric acid and sodium benzoate. *Journal of Food Science* 68:1438–1442.

Donoghue, D.J. 2003. Antibiotic residues in poultry tissues and eggs: Human health concerns? *Poultry Science* 82:618–621.

Durango, A.M., Soares, N.F.F., and Andrade, N.J. 2006. Microbiological evaluation of an edible antimicrobial coating on minimally processed carrots. *Food Control* 17:336–341.

FAO/WHO Food Standards Programme. 1987. Codex standards for sugars, cocoa products and chocolate and miscellaneous. Codex standard for vinegar. In Codex alimentarius. Regional European standard, Codex Stan 162. Geneva.

Ferguson, W.E. and Powrie, W.D. 1957. Studies on the preservation of fresh apple juice with sorbic acid. *Applied Microbiology* 5:41–43.

Food Additives World. 2008. Preservatives. *http://www.foodadditivesworld.com/* (Accessed 15 June 2009).

Forman, C. 2002. The food additives business. *Food and Beverage*. Report Code: FOD 009D. Published July 2002.

Galvez, A., Abriouel, H., Lopez, R.L., and Ben, O.N. 2007. Bacteriocin-based strategies for food biopreservation. *International Journal of Food Microbiology* 120:51–70.

Gauthier, R. 2005. Organic acids and essential oils, a realistic alternative to antibiotic growth promoters in poultry. I *Forum Internacional de avicultura* 17–19 August 2005, pp. 148–157.

Geornaras, I., Belk, K.E., Scanga, J.A., Kendall, P.A., Smith, G.C., and Sofos, J.N. 2005. Post-processing antimicrobial treatments to control *Listeria monocytogenes* in commercial vacuum-packaged bologna and ham stored at 10 degrees C. *Journal of Food Protection* 68:991–998.

Harty, D.W.S., Oakey, H.J., Patrikakis, M., et al. 1994. Pathogenic potential of lactobacilli. *International Journal of Food Microbiology* 24:179–189.

Hazan, R., Levine, A., and Abeliovich, H. 2004. Benzoic acid, a weak organic acid food preservative, exerts specific effects on intracellular membrane trafficking pathways in *Saccharomyces cerevisiae. Applied and Environmental Microbiology* 70:4449–4457.

Hsiao, C.P. and Siebert, K.J. 1999. Modeling the inhibitory effects of organic acids on bacteria. *International Journal of Food Microbiology* 47:189–201.

Jordan, K.N., Oxford, L., and O'Byrne, C.P. 1999. Survival of low-pH stress by *Escherichia coli* O157:H7: Correlation between alterations in the cell envelope and increased acid tolerance. *Applied and Environmental Microbiology* 65:3048–3055.

Juneja, V.K. and Thippareddi, H. 2004. Inhibitory effects of organic acid salts on growth of *Clostridium perfringens* from spore inocula during chilling of marinated ground turkey breast. *International Journal of Food Microbiology* 93:155–163.

Kamm, B. and Kamm, M. 2004. Principles of biorefineries. *Applied Microbiology and Biotechnology* 64:137–145.

Kroll, D. 2001. The growing food testing business: Highlighting pathogens, pesticides and GMOs. *Food and Beverage*. Report Code FOD011C, Published October 2001. http://www.bccresearch.com. (Accessed November 15, 2006).

Leguerinel, I. and Mafart, P. 2001. Modeling the influence of pH and organic acid types on thermal inactivation of *Bacillus cereus* spores. *International Journal of Food Microbiology* 63:29–34.

Leuenberger, S., Widdowson, M.A., Feilchenfeldt, J., Egger, R., and Streuli, R.A. 2007. Norovirus outbreak in a district general hospital—New strain identified. *Swiss Medical Weekly* 137:57–61.

Macfarlane, R. 2002. Integrating the consumer interest in food safety: The role of science and other factors. *Food Policy* 27:65–80.

Marín, S., Abellana, M., Rubinat, M., Sanchis, V., and Ramos, A.J. 2003. Efficacy of sorbates on the control of the growth of *Eurotium* species in bakery products with near neutral pH. *International Journal of Food Microbiology* 87:251–258.

Martinez, G.M. and Bañados, F. 2004. Impact of EU organic product certification legislation on Chile organic exports. *Food Policy* 29:1–14.

Marz, U. 2000. World markets for fermentation ingredients. *Food and Beverage Publications*. Report Code FOD020A, Published February 2000. http://www.bccresearch.com. (Accessed July 29, 2006).

Marz, U. 2002. World markets for citric, ascorbic, isoascorbic acids: Highlighting antioxidants in food. *Food and Beverage*. Report Code FOD017B, Published June 2002. http://www.bccresearch.com. (Accessed July 29, 2006).

Mbandi, E. and Shelef, L.A. 2002. Enhanced antimicrobial effects of combination of lactate and diacetate on *Listeria monocytogenes* and *Salmonella* spp. in beef bologna. *International Journal of Food Microbiology* 76:191–198.

Medeiros, L.C., Kendall, P., Hillers, V., Chen, G., and DiMascola, S. 2001. Identification and classification of consumer food-handling behaviors for food safety education. *Journal of the American Dietetic Association* 101:1326–1339.

Molins, R.A., Motarjemi, Y., and Käferstein, F.K. 2001. Irradiation: A critical control point in ensuring the microbiological safety of raw foods. *Food Control* 12:347–356.

Mollapour, M. and Piper, P.W. 2001. The ZbYME2 gene from the food spoilage yeast *Zygosaccharomyces bailii* confers not only YME2 functions in *Saccharomyces cerevisiae*, but also the capacity for catabolism of sorbate and benzoate, two major weak organic acid preservatives. *Molecular Microbiology* 42:919–930.

Motarjemi, Y., Käferstein, F., Moy, G., et al. 1996. Importance of HACCP for public health and development. The role of the World Health Organization. *Food Control* 7:77–85.

Moy, G., Hazzard, A., and Käferstein, F. 1997. Improving the safety of street-vended food. *World Health Statistics Quarterley* 50:124–131.

Nakai, S.A. and Siebert, K.J. 2003. Validation of bacterial growth inhibition models based on molecular properties of organic acids. *International Journal of Food Microbiology* 86:249–255.

Olsen, S.J., Bishop, R., Brenner, F.W., Roels, T.H., et al. 2001. The changing epidemiology of *Salmonella*: Trends in serotypes isolated from humans in the United States, 1987–1997. *Journal of Infectious Diseases* 183:753–761.

Piper, P., Calderon, C.O., Hatzixanthis, K., and Mollapour, M. 2001. Weak acid adaptation: The stress response that confers yeasts with resistance to organic acid food preservatives. *Microbiology* 147:2635–2642.

Piper, P.W. 1999. Yeast superoxide dismutase mutants reveal a pro-oxidant action of weak organic acid food preservatives. *Free Radical Biology and Medicine* 27:11–12.

Prange, A., Birzele, B., Hormes, J., and Modrow, H. 2005. Investigation of different human pathogenic and food contaminating bacteria and moulds grown on selenite/selenate and tellurite/tellurate by X-ray absorption spectroscopy. *Food Control* 16:723–728.

Rekha, B., Naik, S.N., and Prasad, R. 2006. Pesticide residue in organic and conventional food-risk analysis. *Journal of Chemical Health and Safety* 13:12–19.

Ricke, S.C. 2003. Perspectives on the use of organic acids and short chain fatty acids as antimicrobials. *Poultry Science* 82:632–639.

Ricke, S.C., Kundinger, M.M., Miller, D.R., and Keeton, J.T. 2005. Alternatives to antibiotics: Chemical and physical antimicrobial interventions and foodborne pathogen response. *Poultry Science* 84:667–675.

Rico, D., Martin-Diana, A.B., Barat, J.M., and Barry-Ryan, C. 2007. Extending and measuring the quality of fresh-cut fruit and vegetables: A review. *Trends in Food Science & Technology* 18:373–386.

Rico, D., Martin-Diana, A.B., Barry-Ryan, C., Frias, J.M., Henehan, G.T.M., and Barat, J.M. 2008. Use of neutral electrolysed water (EW) for quality maintenance and shelf life extension of minimally processed lettuce. *Innovative Food Science & Emerging Technologies* 9:37–48.

Rodrigues, G. and Pais, C. 2000. The influence of acetic and other weak carboxylic acids on growth and cellular death of the yeast *Yarrowia lipolytica*. *Food Technology and Biotechnology* 38:27–32.

Rohr, A., Luddecke, K., Drusch, S., Muller, M.J., and Alvensleben, R. 2005. Food quality and safety—Consumer perception and public health concern. *Food Control* 16:649–655.

Ryu, J.-H. and Beuchat, L.R. 1999. Changes in heat tolerance of *Escherichia coli* O157:H7 after exposure to acidic environments. *Food Microbiology* 16:317–324.

Saint Xavier University. 2000. *Food Preservation*.

Samelis, J., Sofos, J.N., Kendall, P.A., and Smith, G.C. 2002. Effect of acid adaptation on survival of *Escherichia coli* O157:H7 in meat decontamination washing fluids and potential effects of organic acid interventions on the microbial ecology of the meat plant environment. *Journal of Food Protection* 65:33–40.

Schnürer, J. and Magnusson, J. 2005. Antifungal lactic acid bacteria as bio-preservatives. *Trends in Food Science and Technology* 16:70–78.

Shtenberg, A.J. and Ignat'ev, A.D. 1970. Toxicological evaluation of some combination of food preservatives. *Food and Cosmetics Toxicology* 8:369–380.

Sidhu, M.S., Langsrud, S., and Holck, A. 2001. Disinfectant and antibiotic resistance of lactic acid bacteria isolated from the food industry. *Microbial Drug Resistance* 7:73–83.

Sims, W. 1964. A pathogenic *Lactobacillus*. *Journal of Pathologic Bacteriology* 87:99–105.

Smith DeWaal, C. 2003. Safe food from a consumer perspective. *Food Control* 14:75–79.

Smulders, F.J. and Greer, G.G. 1998. Integrating microbial decontamination with organic acids in HACCP programmes for muscle foods: Prospects and controversies. *International Journal of Food Microbiology* 44:149–169.

Soomro, A.H., Masud, T., and Anwaar, K. 2002. Role of lactic acid bacteria (LAB) in food preservation and human health – A review. *Pakistan Journal of Nutrition* 1:20–24.

Steels, H., James, S.A., Roberts, I.N., and Stratford, M. 1999. *Zygosaccharomyces lentus*: A significant new osmophilic, preservative-resistant spoilage yeast, capable of growth at low temperature. *Journal of Applied Microbiology* 87:520–527.

Steels, H., James, S.A., Roberts, I.N., and Stratford, M. 2000. Sorbic acid resistance: The inoculum effect. *Yeast* 16:1173–1183.

Struijk, C.B., Mossel, D.A.A., and Moreno Garcia, B. 2003. Improved protection of the consumer community against food-transmitted diseases with a microbial aetiology: A pivotal food safety issue calling for a precautionary approach. *Food Control* 14:501–506.

Suhr, K.I. and Nielsen, P.V. 2004. Effect of weak acid preservatives on growth of bakery product spoilage fungi at different water activities and pH values. *International Journal of Food Microbiology* 95:67–78.

Tamblyn, K.C. and Conner, D.E. 1997. Bactericidal activity of organic acids in combination with transdermal compounds against *Salmonella typhimurium* attached to broiler skin. *Food Microbiology* 14:477–484.

Tang, Y. and Wu, M. 2007. The simultaneous separation and determination of five organic acids in food by capillary electrophoresis. *Food Chemistry* 103:243–248.

Tesfaye, W., Morales, M.L., Garcia-Parrilla, M.C., and Troncoso, A.M. 2002. Wine vinegar: Technology, authenticity and quality evaluation. *Trends in Food Science & Technology* 13:12–21.

Unnevehr, L.J. and Jensen, H.H. 1999. The economic implications of using HACCP as a food safety regulatory standard. *Food Policy* 24:625–635.

Valero, M., Leontidis, S., Fernandez, P.S., Martinez, A., and Salmeron, M.C. 2000. Growth of *Bacillus cereus* in natural and acidified carrot substrates over the temperature range 5–30°C. *Food Microbiology* 17:605–612.

Verbeke, W., Frewer, L.J., Scholderer, J., and De Brabander, H.F. 2007. Why consumers behave as they do with respect to food safety and risk information. *Analytica Chimica Acta* 586:2–7.

WHO. 2003. Food safety strategic planning meeting: Report of a WHO strategic planning meeting, WHO headquarters, Geneva, 20–22 February 2001. Geneva: World Health Organization.

Winter, C.K. and Davis, S.F. 2006. Organic foods. *Journal of Food Science* 71:R117–R124.

chapter two

Nature and composition of organic acids

2.1   General characterization

Organic acids are typical products of microbial metabolism. All organic acids occur naturally in a variety of vegetable and animal substrates and can, therefore, be either naturally present as constituents of foods as a result of normal biochemical metabolic processes, direct addition as acidulants, hydrolysis, or bacterial growth, or can later be added directly or indirectly to the products (Gomis, 1992).

An organic acid is an organic compound with acidic properties and containing carbon, as do all organic compounds. The most common organic acids are the carboxylic acids whose acidity is associated with their carboxyl group –COOH. These are generally weak acids, whereas sulfonic acids, containing the group OSO_3H, are relatively stronger acids. The acidity of an acid is determined by the relative stability of the conjugate base of the acid, although other groups can also confer acidity. This acidity is, however, usually weak and is the result of the –OH, –SH, enol group, and the phenol group (Figure 2.1) (Theron and Lues, 2007).

Organic acids do not dissociate completely in water as opposed to the strong mineral acids. Although the lower molecular weight acids such as formic and acetic acids are water soluble, the higher molecular weight organic acids such as benzoic acid, are insoluble in the molecular (neutral) form. Most organic acids are, however, very soluble in organic solvents. Exceptions to this solubility may be prevalent in the presence of other substituents that may have an effect on the polarity of the compound (Theron and Lues, 2007).

Organic acids have the common denominator of having carbon in their structure, and although less reactive than inorganic acids, they are able to dissolve minerals in the mineral reserve in the soil. Organic acids that have 10 or fewer carbons in their structure, are distinguished from fatty acids that, on the other hand, have straight carbon even-number chains of 12 to 24. Organic acids exist in two basic forms: pure acids or buffered acids. Included in the pure acids are lactic acid, propionic acid, acetic acid, citric acid, and benzoic acid, whereas the calcium and sodium salts of propionic, acetic, citric, and benzoic acids are buffered organic

phenol:

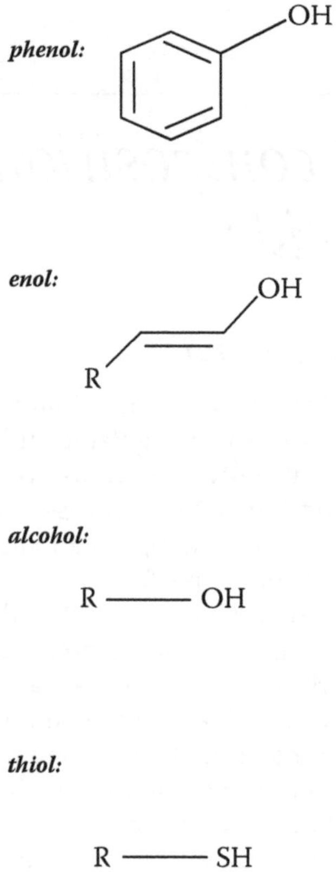

enol:

alcohol:

R ——— OH

thiol:

R ——— SH

Figure 2.1 General structure of a few organic acids.

acids. These buffered organic acids are safer to handle and less caustic to machinery (Hoffman and Possin, 2000). Most organic acids have an advantage because of their simple structure and small molecular size or mass, which allows them to move freely throughout cells. Organic acids also have many vital roles and functions in plants, including contributing to growth and production of fruit, and are therefore applied in many sectors of agriculture (AMCA, 2005). The organic acids commonly found in foods differ considerably in structure and in their inhibitory effects on different bacteria (Nakai and Siebert, 2003).

Various organic acids or organic acid compounds are used as additives that are directly incorporated into human food. Alternatively these acids can be the indirect result of fermentation activity of starter cultures often added to foods such as dairy, vegetable, or meat products. Organic acids may also originate from raw materials or be produced by fermentation during processing and storage (Ricke, 2003). One of the most important characteristics of organic acids is that they have a direct influence on the flavor and quality of some natural and many processed foods to which they are added as stabilizers or preservatives (Gomis, 1992). On the other hand, although organic acids are generally recognized as safe (GRAS) they may also produce adverse sensory changes (Gomis, 1992; Min et al., 2007). Dilute solutions of organic acids (1–3%) are, however, generally without effect on the desirable sensory properties of, for example, meat when used as a carcass decontaminant (Smulders and Greer, 1998; Min et al., 2007).

Weak organic acids have an extended history as food preservatives due to their general antimicrobial activity. They are, in fact, the most commonly used chemical preservatives of food and are GRAS, broad-spectrum, antimicrobial agents (Plumridge et al., 2004; Steiner and Sauer, 2003). Organic acids are such effective food preservatives because, apart from their antimicrobial inhibitory activities, they also act as acidulants and in so doing reduce bacterial growth by lowering the pH of food products to levels that inhibit bacterial growth (Dziezak, 1990; Hinton, Jr., 2006).

2.2   Structural description

Organic acids primarily include the saturated straight-chain monocarboxylic acids. Derivatives of such organic acids may consist of unsaturated, hydroxylic, phenolic, and multicarboxylic acids. Often organic acids are also referred to as fatty acids, volatile fatty acids, weak acids, or carboxylic acids (Ricke, 2003). Organic acids are not members of a homologous series as they differ in the number of carboxy groups, hydroxy groups, and carbon–carbon double bonds in their molecules (Hsiao and Siebert, 1999; Permprasert and Devahastin, 2005) and can be classified according to (1) the type of carbon chain (aliphatic, alicyclic, aromatic, and heterocyclic), (2) being saturated or nonsaturated, (3) substituted or nonsubstituted, or (4) the number of functional groups (mono-, di-, tricarboxylic, etc.) (Gomis, 1992). It is, therefore, not typically possible to predict the effect on food systems containing more than one acid (Hsiao and Siebert, 1999). However, it is known that these short-chain volatile organic acids with carbon numbers ranging from 2 to 12 significantly affect the flavor and quality of food (Hsiao and Siebert, 1999; Yang and Choong, 2001). The lowest monocarboxylic (C_1–C_4) aliphatic acids are rather volatile liquids with a distinct pungency, whereas those acids containing more carbon atoms are of a relatively oily substance and slightly water soluble.

In comparison, dicarboxylic acids are colorless solids with melting points at about 100°C. All these acids form somewhat soluble metal salts and esters, of which the latter are adequately volatile for gas chromatography analysis. They also have spectral absorption properties that make them suitable for HPLC analysis (Gomis, 1992; Hsiao and Siebert, 1999).

Table 2.1 Organic Acids Frequently Assayed in Foods and
Dissociation Constant[a]

Monocarboxylic		
Organic acid	Formula	pK_a
Acetic acid	CH_3COOH	4.76
Benzoic acid	$C_7H_6O_2$	4.20
Butyric acid	$C_4H_8O_2$	4.83
Cinnamic acid	$C_9H_8O_2$	4.44
Formic acid	CH_2O_2	3.75
Gallic acid	$C_7H_6O_5$	4.41
Lactic acid	$C_3H_6O_3$	3.83
Propionic acid	$C_3H_6O_2$	4.87
Pyruvic acid	$C_3H_4O_2$	2.39
Sorbic acid	$C_6H_8O_2$	4.76

Dicarboxylic			
Organic acid	Formula	pK_1	pK_2
Ascorbic acid	$C_6H_8O_6$	4.10	11.79
Fumaric acid	$C_4H_4O_4$	3.09	4.60
Malic acid	$C_4H_6O_5$	3.46	5.21
Succinic acid	$C_4H_6O_4$	4.22	5.70

Tricarboxylic				
Organic acid	Formula	pK_1	pK_2	pK_3
Citric acid	$C_6H_8O_7$	2.79	4.30	5.65
Isocitric acid	$C_6H_8O_7$	3.28	4.71	6.39

[a] A dissociation constant K_a is sometimes expressed by its pK_a, where $pK_a = -\log_{10} K_a$. These pK_as are mainly used for covalent dissociations because these dissociation constants can vary greatly.

The strengths of weak acids are measured on the pK_a scale. The pK_a value is the pH at which the acid and anion concentrations are equal (Table 2.1). The smaller the number on this scale is, the stronger the acid. In solution, weak acids do not fully dissociate into ions, but form equilibria between unchanged acid molecules and their respective charged anions and protons:

$$HA \quad \leftrightarrow \quad A^- \quad + \quad H^+$$
(acid molecule)   (anion)   (proton)

2.3 An overview of individual organic acids and their applications

2.3.1 Acetic acid

Acetic acid (Figure 2.2) is one of the oldest chemicals known to humanity and is produced naturally during spoilage of fruit and certain other foods by the presence and activity of, among others, the acetic acid bacteria (AAB), one of which is *Acetobacterium*, commonly found in foodstuffs, water, and soil. Acetic acid is one of the main products of AAB metabolism (Gonzalez et al., 2005). The acid has GRAS status and is approved worldwide for use as a food additive and preservative (Smulders and Greer, 1998). It is known to give a strong flavor profile and also increases the effectiveness of other flavor additives and substances present in food, particularly in bread (Şimşek, Çon, and Tulumoğlu, 2006).

Acetic acid is known to be a stable substance and is also referred to as "ethanoic acid" or "glacial acetic acid" (Ricke, 2003). Acetic acid has only one COOH group, which implies that a higher concentration is needed than for an organic acid with more than one COOH group, to achieve the same pH. It is a weak acid and cannot dissociate as completely as a strong acid. A lower ionization constant (K_a) is characteristic of a weaker acid, as it yields a lower amount of hydrogen ions in solution. The ionization amount of weak acid in a solution must be determined from the ionization constant (K_a). In comparison to citric acid with 3 K_a (7.447×10^{-4}, 1.733×10^{-5}, and 4.018×10^{-6}), acetic acid with 1 K_a (1.753×10^{-5}) is weaker than citric acid (Permprasert and Devahastin, 2005). However, acetic acid is more inhibitory than lactic acid, because of its higher pK_a value (Farkas, 1998). On the other hand, acetate is far less inhibitory to yeasts than the more lipophilic sorbate (trans, trans hexane dienoate), although they have the same pK_a (4.76). This may be due to the sorbate's much higher capacity to dissolve in membranes (Bauer et al., 2003). In culture media acetic acid is

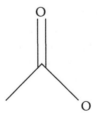

Chemical formula: CH_3COOH

Figure 2.2 Structure of acetic acid.

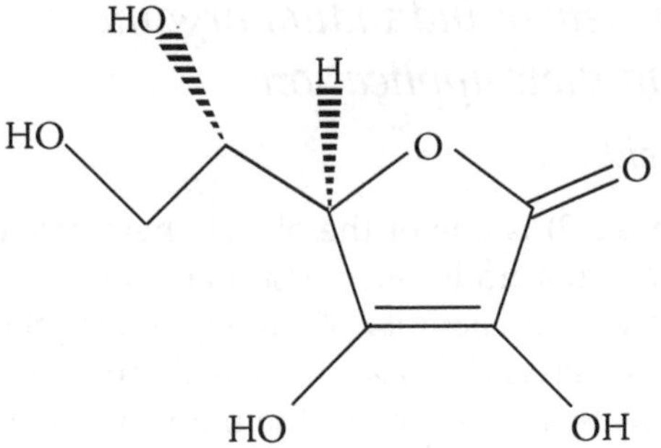

Chemical formula: $C_6H_8O_6$

Figure 2.3 Structure of L-ascorbic acid.

also much more inhibitory against *Listeria* than lactic acid and addition of acetic acid usually results in greater cell destruction, but similar concentrations of citric acid and lactic acid will reduce the pH of a broth such as tryptic soy broth more than acetic acid (Doyle, 1999).

Acetic acid is lipid soluble and able to rapidly diffuse through the plasma membrane, a factor that has a dramatic effect on the pH_i of a cell (Greenacre et al., 2003). Intracellular acidification may, however, play a role in acids with short aliphatic chains (such as acetic acid), and much higher concentrations (20–80 mM) are needed for growth inhibition (Hazan, Levine, and Abeliovich, 2004).

2.3.2 Ascorbic acid

L-ascorbic acid (Figure 2.3) and its salts have, for many decades, been leading antioxidants used on fruits and vegetables in fruit juices. This application method is effective in preventing browning and other oxidative reactions. Ascorbic acid also acts as an oxygen scavenger by removing molecular oxygen in polyphenol oxidase reactions (Rico et al., 2007). L-ascorbic acid and its various derivatives have GRAS status and are common preservatives in the production of canned foods (Perez-Ruiz, Martinez-Lozano, and Garcia, 2007).

2.3.3 Benzoic acid

Benzoic acid (Figure 2.4) is one of the oldest and most commonly and widely used chemical preservatives. However, the use of sodium benzoate as a food preservative has been optimal in products that are acidic in nature, specifically suitable for foods and beverages with pH < 4.5. The

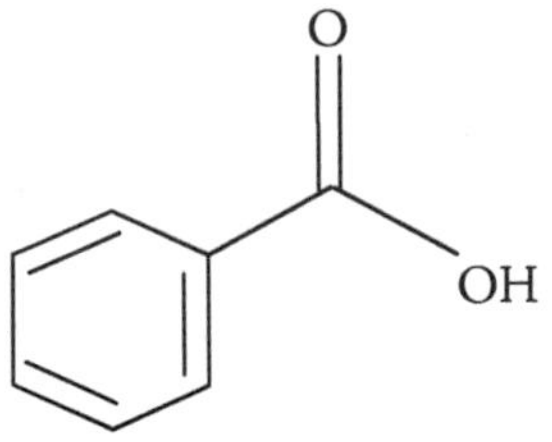

Chemical formula: $C_7H_6O_2$

Figure 2.4 Structure of benzoic acid.

salt sodium benzoate is also more effective in food systems where pH ≤ 4 (Barbosa-Cánovas et al., 2003), although it is mainly associated with the preservation of fruit and fruit juices (Suhr and Nielsen, 2004). Here it is primarily applied as an antifungal agent (Barbosa-Cánovas et al., 2003). Benzoic acid is also often used in combination with sorbic acid in many types of products. This combination is particularly popular in confectionery (Suhr and Nielsen, 2004). It has been detected as a natural constituent of cranberries, raspberries, plums, prunes, cinnamon, and cloves (Barbosa-Cánovas et al., 2003). In other applications, benzoic acid has also been used as a mild antiseptic in cough medicines, mouthwashes, toothpastes, and as an antifungal in ointments (Otero-Losada, 2003). Adverse effects of sodium benzoate have been reported to be hyperactivity and asthma (Poulter, 2007).

Derivative benzoates and parabenzoates have been used primarily in fruit juices, chocolate syrup, pie fillings, pickled vegetables, relishes, horseradish, and cheese (Barbosa-Cánovas et al., 2003). Other foodstuffs where sodium benzoate is used include soft drinks, baked goods, and lollipops (Poulter, 2007). Benzaldehyde and benzoic alcohol are better known to be yeast inhibitors. Benzoic acid has been found to release fewer protons than sulphite, nitrite, or acetic acid and it may be speculated that benzoic acid is not a classic weak-acid preservative. However, due to a lower pK_a value, benzoic acid releases three to four times more protons than sorbic acid. This is a sizable concentration of protons although not as much as other weak-acid preservatives (Stratford and Anslow, 1998). Inhibition of growth is strongly pH-dependent and most effective under acidic conditions. Under these conditions the protonated form of the acid is predominantly found (Visti, Viljakainen, and Laakso, 2003). Another unexpected discovery was that benzoic acid appears to be a pro-oxidant. This was unexpected as it is a well-known fact 2-hydroxybenzoic acid (or salicylic acid) acts as a scavenger of free radicals *in vivo* (Piper, 1999).

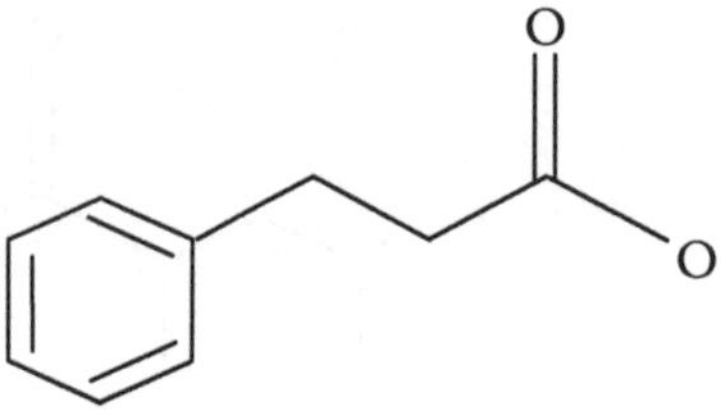

Chemical formula: $C_9H_8O_2$

Figure 2.5 Structure of cinnamic acid.

Benzoic acid possesses GRAS status, and is conjugated in the liver to produce benzoylglycine (or hippuric acid), which is then excreted in urine (Piper, 1999). It is permitted for use in nondairy dips where pH ≤ 4.5, where appropriate amounts of benzoic acid (or sorbic acid) appear to be safe as the acid has been tested by labs throughout the world (Tang and Wu, 2007). Although sensory characteristics of benzoic acid are unknown, the acid could be regarded as a classical mild pungent stimulus. The taste of sodium benzoate cannot be detected by one in every four of the population, but for those who can taste it, it is perceived as sweet, sour, salty, or sometimes bitter. Benzoic acid has also been found to be superior in bactericidal activity against coliform and lactic acid bacteria in both the stomach and the intestinal content (Otero-Losada, 1999).

2.3.4 Cinnamic acid

Cinnamic acid (Figure 2.5) is a phenolic component (a derivative of phenylalanine) of several spices, including cinnamon (Roller, 1995). It consists of a relatively large family of organic acid isomers extracted from plants or chemically synthesized. Cinnamic acid is also GRAS and used as a component in several food flavorings and has been found to exhibit antibacterial, antifungal, and antiparasitic activities. However, it has been found to be particularly effective at acidic pH against fungi. Although high concentrations of cinnamic acid have been reported to cause browning in some fruits (e.g., nectarines, limes, and pears), it has been found to be very effective in prolonging shelf life of several important fruit products, such as fresh tomato slices stored at 4°C or 25°C. The Japanese are known to use cinnamic acid as an antimicrobial agent in fish paste (Roller, 1995).

Cinnamic acid is well known for its role as a phenolic compound that gives the oil cinnamon its characteristic odor and flavor. It is soluble in water or ethanol and in nature cinnamic acid derivatives are known to

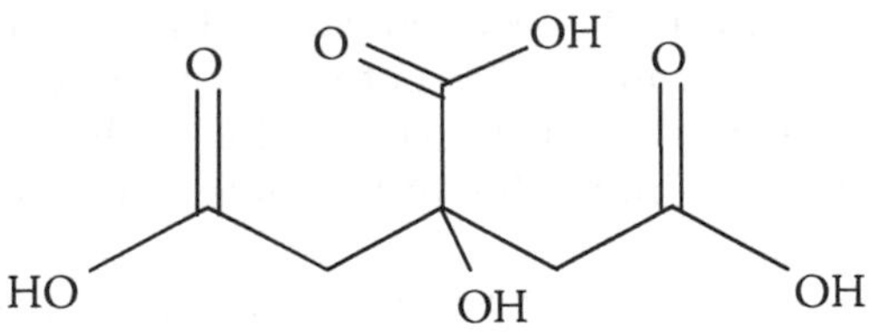

Chemical formula: $C_6H_8O_7$

Figure 2.6 Structure of citric acid.

be important metabolic building blocks in the production of lignins for higher plants. Cinnamic acid is classified as a skin, eye, and respiratory tract irritant (Roller, 1995).

2.3.5   Citric acid

Citric acid (Figure 2.6) is a popular acidulant and, due to its flexibility, its use is standard in virtually every preserved food (Marz, 2002). Together with its salts, citric acid is, in fact, one of the most commonly used organic acids in the food as well as pharmaceutical industries (Couto and Sanroman, 2006). It also has an important role in the metal processing and chemical industries (Nikbakht, Sadrzadeh, and Mohammadi, 2007). Citric acid does not fit under the description of the so-called classic weak organic acid preservatives, which would be "a lipophilic, undissociated acid that inhibits microbial growth at low pH, by causing intracellular acidification" (Nielsen and Arneborg, 2007). Citric acid is more of a "chelator," which may be defined as "a lipophobic, dissociated acid, inhibiting growth of microorganisms by chelating divalent metal ions from the medium" (Brul and Coote, 1999; Stratford, 1999; Nielsen and Arneborg, 2007). However, this has been found to be true, more specifically in bacteria (Rammel, 1962; Imai, Banno, and Iijima, 1970; Graham and Lund, 1986; Nielsen and Arneborg, 2007).

Citric acid is a typical product of fermentation and was originally produced by large fermentation companies in Western countries. More recently the technological applications of citric acid (e.g., in detergents) have contributed to the production and marketing of more environmentally friendly products (Marz, 2002). Although citric acid has a fresh acidic flavor (which is somewhat different from that of malic acid), it has a pleasant taste and this, together with its flavor-enhancing characteristics and also its being highly soluble, have rendered the substance a dominating position in the market. The food industry is the largest consumer of citric acid, estimated to use almost 70% of the total production. This is followed by the pharmaceutical industry (about 12%) with various other applications using the remaining 18% of the citric acid produced (Couto and Sanroman, 2006).

Citric acid has three COOH groups and is an hydroxy acid (Fite et al., 2004). It contains twice as many hydrophilic groups as hydrophobic groups, has a strong affinity with water, and cannot be adsorbed at the surface (Permprasert and Devahastin, 2005). It is one of the largest carboxylic organic acids produced by *Aspergillus niger* through fermentation of low molecular weight carbohydrate solutions (Nikbakht, Sadrzadeh, and Mohammadi, 2007). Citric acid is the most important organic acid produced: global production is estimated to be around 620,000 metric tons per year and represents 60% of all food acidulants utilized worldwide (Krishnakumar, 1994; Moresi and Sappino, 1998; Nikbakht, Sadrzadeh, and Mohammadi, 2007). The entire production of citric acid is carried out by fermentation. In studies done on organic acid production by various yeast species, citrate was most often produced in the greatest amounts by all the isolates tested (Guerzoni, Lanciotti, and Marchetti, 1993). Citric acid can also be produced by chemical synthesis, although the cost is much higher than with fermentation (Couto and Sanroman, 2006).

Despite the popularity of citric acid as a food additive and preservative, its sharp and overwhelming acidic taste is, however, inclined to overpower the flavor of sweeteners or other flavorants that may be present in a foodstuff. Citric acid also presents a short-lived tartness flavor. Other disadvantages have also been noted when using citric acid as a food additive. For example, due to its hygroscopic nature, it tends to cake when used in dry powders such as soft drink or beverage mixes. It also has an uneven particle size and is not always free-flowing. However, despite these few drawbacks, citric acid is by far the most commonly used food acidulant. The initial sharpness associated with citric acid is preferable in some end products such as citrus-flavored drinks (Fowlds, 2002).

The inhibitory effect of citric acid on *Listeria monocytogenes* growth has been found to be much more evident than the inhibitory effect of ascorbic acid. The difference in inhibition between both acids is most obvious when the temperature is higher. Because of the pK values of citric acid (pK_1 = 3.14, pK_2 = 4.77, and pK_3 = 6.39), it dissociates strongly inside microbial cells (Carrasco, Garcia-Gimeno, and Seselovsky, 2006).

2.3.6 Formic acid

Formic acid (Figure 2.7) is the shortest chain organic acid (HCOOH). This characteristic may be beneficial in enabling it to diffuse into a microbial

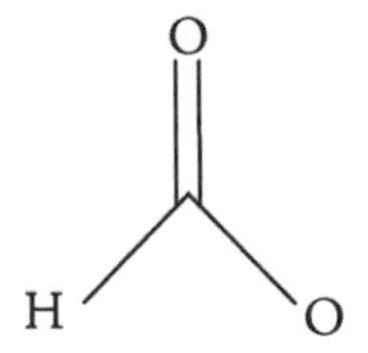

Chemical formula: CH_2O_2

Figure 2.7 Structure of formic acid.

cell and to accumulate very quickly at the expense of the microbial membrane pH, causing acidification of the cytoplasm (El-Ziney, De, and Debevere, 1997; Chaveerach et al., 2002). The principal use of formic acid is as a preservative in livestock feed. As such it is sprayed on fresh hay or other silage to arrest the decaying processes and may also cause the feed to retain its nutritive value longer. It is, therefore, widely used in the preservation of winter feed for cattle.

A small excess of undissociated formic acid has a stronger bactericidal effect on the culturability of *Campylobacter jejuni* and *Campylobacter coli* in comparison with a large excess of undissociated propionic or acetic acid at the same pH levels. It may be possible that the structure of these acids is the most important factor in the inhibition activity of *C. jejuni* and *C. coli* viability (Chaveerach et al., 2002).

2.3.7   Fumaric acid

Fumaric acid (Figure 2.8) is one of the strongest of the weak organic acids used as a food acidulant, primarily due to its low pK_a values (3.03 and 4.54), and is widely found in nature (Comes and Beelman, 2002). It is an essential part of plant life and also the main by-product in L-lactic acid production by *Rhizopus oryzae*, resulting from a specific metabolic pathway (Wang et al., 2005). Fumaric acid is one of the cheaper food acids, but has a low solubility in water and may also tend to dissolve slowly. However, it is sometimes supplied as a very fine powder, which includes a wetting agent. This is referred to as water-soluble fumaric acid. Although fumaric acid is primarily added to foodstuffs as an acidulant, it also has a bacteriostatic function (Fowlds, 2002).

Fumaric acid would then appear to be a "stronger" acid in food and as a result of this, less fumaric acid would be required to achieve a specific level of tartness compared to other food acids. It is known to improve the quality of food and beverages and may also reduce the cost of such products. Fumaric acid may also be used in animal feed. Since 1946 fumaric

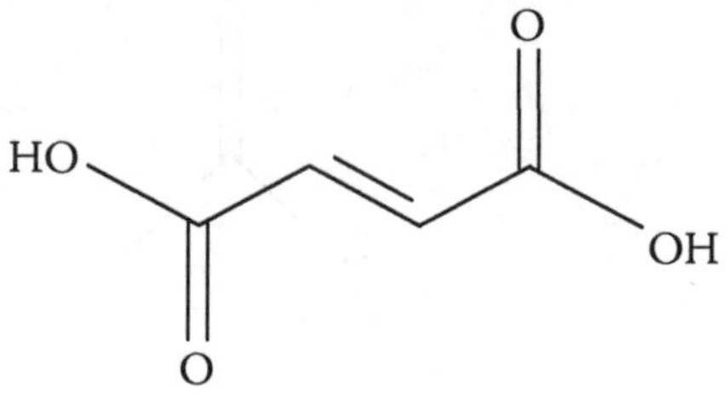

Chemical formula: $C_4H_4O_4$

Figure 2.8 Structure of fumaric acid.

acid has been used in food and beverage products and it is still used in wheat and corn tortillas, sour dough and rye breads, biscuit dough, fruit juices, gelatin desserts, gelling aids, pie fillings, and wine. However, this relatively strong acid taste and low solubility have made it less desirable for certain food applications. The use of fumaric acid has, therefore, declined since the 1980s, when in 1985 nearly 50 million pounds were consumed. Now fumaric acid has to compete with other acidulants such as citric acid, tartaric acid, and malic acid, although it is less expensive than some of these substitutes (Fowlds, 2002).

2.3.8 Gluconic acid

Gluconic acid (Figure 2.9) is a mild organic acid derived from glucose by a simple oxidation reaction, facilitated by the enzyme glucose oxidase in fungi and glucose dehydrogenase in bacteria (such as *Gluconobacter*). Microbial production of gluconic acid dates back several decades and has

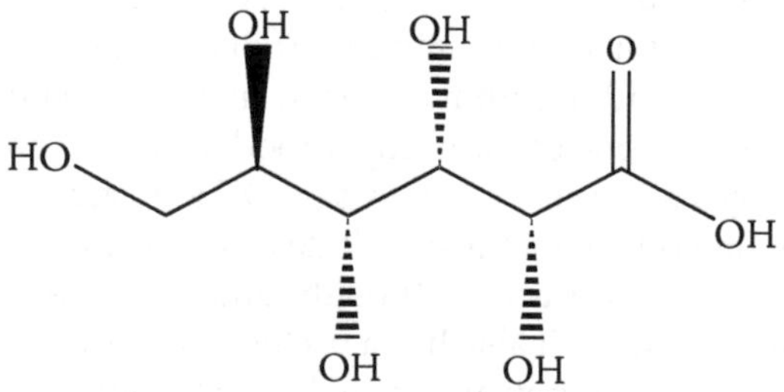

Chemical formula: $C_6H_{12}O_7$

Figure 2.9 Structure of D-gluconic acid.

since been the preferred method for production. *Aspergillus niger*, in particular, is most widely used in the fermentation process. Gluconic acid and its derivatives (such as sodium gluconate) are widely applied in the food and pharmaceutical industries. Apart from being a mild organic acid, gluconic acid is also noncorrosive, nonvolatile, and nontoxic. It is popular for causing a refreshing sour taste in various foodstuffs, including wine and fruit juices and is used for prevention of milkstone in the dairy industry. The derivative glucono-δ-lactone functions as a slow-acting acidulant in processing meat such as sausage, whereas sodium gluconate is known to be highly overpowering (Ramanchandran et al., 2006).

2.3.9 Lactic acid

Lactic acid (Figure 2.10) is an organic acid with a wide range of industrial applications. It is also an hydroxyacid (2-hydroxypropionic, $CH_3CHOHCOOH$) (Fite et al., 2004), classified as GRAS by the FDA and very often used in foods as an acidulant, flavoring agent, pH buffering agent, and of course as a preservative (Valli et al., 2006). Although lactic acid also has wide applications in the pharmaceutical, leather, and textile industries and as a chemical feedstock (Couto and Sanroman, 2006), a review published in 1995 reported that in the United States 85% of the lactic acid was used in food or food-related applications (Zhang, Jin, and Kelly, 2007). Lactic acid is not naturally present in foods, but is produced during fermentation of foods by lactic acid bacteria. These foods include sauerkraut, pickles, olives, and some meats and cheeses (Barbosa-Cánovas et al., 2003). Lactic acid is used in foodstuffs for acidification, to increase flavor and aroma, and is a potent microbial inhibitor. Lactic acid is applied to a diverse range of foodstuffs, including meats, fish, vegetables, cereals, and cake products. In fermented drinks it is used to contribute specifically to aroma and preservation (De la et al., 2005).

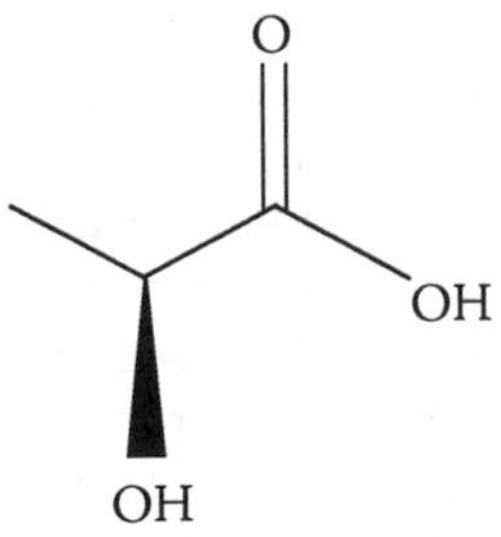

Chemical formula: $C_3H_6O_3$

Figure 2.10 Structure of D-lactic acid.

Lactic acid exists naturally in two optical isomers: D(–)-lactic acid and L(+)-lactic acid of which L(+) is preferred in food applications (Zhang, Jin, and Kelly, 2007). This is the form used by the human metabolism and the enzyme L-lactate dehydrogenase (Couto and Sanroman, 2006). Elevated levels of the D-isomer are harmful to humans (Zhang, Jin, and Kelly, 2007). The inhibitory activity of lactic acid is also stereospecific. L-lactic acid is much more inhibitory for *E. coli* strains than for the D-isomer, whereas *L. monocytogenes* is most sensitive to D-lactic acid (Gravesen et al., 2004). D- and L-lactic acid have an identical effect on the pH_i of *L. monocytogenes* as well as *E. coli* (McWilliam Leitch and Stewart, 2002). Lactic acid penetration across the cell membrane appears to be similar for the two isomers. An additional antimicrobial mechanism has been described for lactic acid that involves specific interaction with a chiral molecule (Gravesen et al., 2004). *E. coli* and *L. delbrueckii* subsp. *bulgaricus* produce D-lactic acid (Benthin and Villadsen, 1995; Bunch et al., 1997), whereas *L. monocytogenes* produces L-lactic acid. Bacterial strains are least sensitive to the isomer that they intrinsically produce. This implies an additional resistance mechanism, involving (1) an enzymatic reaction, or (2) a stereospecific efflux system (Gravesen et al., 2004).

Lactic acid is a major end product from fermentation of a carbohydrate by lactic acid bacteria (Tormo and Izco, 2004). However, lactic acid can be produced commercially by either chemical synthesis or fermentation. The chemical synthesis results in a racemic mixture of the two isomers whereas during fermentation an optically pure form of lactic acid is produced. However, this may depend on the microorganisms, fermentation substrates, and fermentation conditions. Lactic acid can be produced from renewable materials by various species of the fungus *Rhizopus*. This has many advantages as opposed to bacterial production because of amylolytic characteristics, low nutrient requirements, and the fungal biomass, which is a valuable fermentation by-product (Zhan, Jin, and Kelly, 2007).

Sodium lactate is the sodium salt of natural lactic acid (L+) and is a normal component of muscle tissue (Choi and Chin, 2003). It is primarily used in meat products to extend cold storage and the effects are more bacteriostatic than bactericidal (Bolton et al., 2002). Sodium lactate is also produced by microbial fermentation and has been used to control microbial growth during storage, and for improving the flavor and production of processed meat (Choi and Chin, 2003). The spectrum of activity of sodium lactate is actually quite narrow and may even be specific for a certain group of organisms (Lemay et al., 2002). It also has an antibacterial effect on pathogens such as *Clostridium botulinum* and *L. monocytogenes,* the latter being notorious for causing severe problems in meat products (Choi and Chin, 2003). Depending on the specific conditions, the inhibitory activity of lactic acid against *L. monocytogenes* can be either bacteriocidic or bacteriostatic (Gravesen et al., 2004).

L. *monocytogenes* can utilize lactate as a carbohydrate source (Kouassi and Shelef, 1996). Stereospecific antimicrobial activity of lactic acid is not really applicable to *L. monocytogenes*. This also suggests considerable species-to-species variation in relative sensitivity to the two isomers, which may differentially influence the development of beneficial and unwanted bacteria, and in so doing, alter the composition of complex microflora. Clearer insight into the mechanisms involved could enhance the optimal use of lactic acid or lactic acid producers (Gravesen et al., 2004).

Lactic acid is commonly found, which contributes to its wide use in food and food-related industries. It also has the potential for production of biodegradable and biocompatible polymers. These products have been proven to be environmentally friendly alternatives to biodegradable plastics derived from petrochemical materials (Zhang, Jin, and Kelly, 2007). Lactic acid is slightly lipid soluble and diffuses slowly through the cell membrane. As a result of this, the disruption of the cell pH_i is not its main mode of inhibition (Gravesen et al., 2004).

Lactic acid is the major LAB metabolite, causing pH reductions that inhibit many microorganisms. However, both acetic and propionic acids have higher pK_a values than lactic acid and, therefore, have a higher proportion of undissociated acid at a certain pH (Schnürer and Magnusson, 2005).

2.3.10 Malic acid

Malic acid (Figure 2.11) is an alpha hydroxy organic acid, widely found in apples and other fruits and is, therefore, sometimes referred to as a fruit acid or, more specifically, an apple acid. It is also found in plants and animals, including humans. Malic acid is also known as hydroxybutaneoic acid or hydroxysuccinic acid. It is a chiral molecule, the naturally occurring

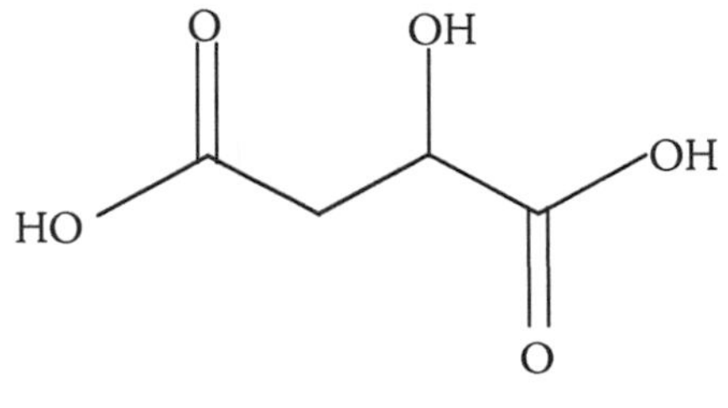

Chemical formula: $C_4H_6O_5$

Figure 2.11 Structure of malic acid.

stereoisomer being the L-form. The L-form is also the biologically active one. Malic acid is also widely found in vegetables and is a potent inhibitor of the growth of yeasts and some bacteria (Barbosa-Cánovas et al., 2003). Malic acid is more effective than acetic acid and lactic acid in inhibiting growth of thermophilic bacteria, but is not as effective as lactic acid in suppressing growth of *L. monocytogenes* (Doyle, 1999).

Malic acid has a smooth lingering taste. Although it also has a tart taste, this is not as sharp as that of citric acid, yet it is longer lasting. It has also been found to mask the bitter aftertaste of synthetic sweeteners, yet when using sweeteners such as aspartame, it was found that the slower increase to peak tartness does not overpower the sweeteners and as such, less sweetener is needed. However, in some end products such as citrus-flavored drinks, the initial sharpness associated with citric acid is preferable (Fowlds, 2002).

Water used in many manufacturing processes often contains small amounts of hard water salts. However, malic acid is highly soluble, and the solubility of its calcium salt prevents cloudiness of end products containing malic acid as an acidulant. Malic acid and its salts have been confirmed by the FDA as GRAS to be applied as a flavor enhancer and acidulant as well as a pH control agent at levels varying from 6.9% for hard candy to 0.7% for miscellaneous uses (Fowlds, 2002).

2.3.11   Propionic acid

Propionic acid (Figure 2.12) was first described in 1844. Propionic acid occurs in foods by natural processing. It has been found in Swiss cheese at concentrations of up to 1% and is produced by the bacterium *Propionibacterium shermanii*. Antimicrobial activity of propionic acid is reported to be primarily against molds and bacteria (Barbosa-Cánovas et al., 2003). Commercial products containing propionic acid include

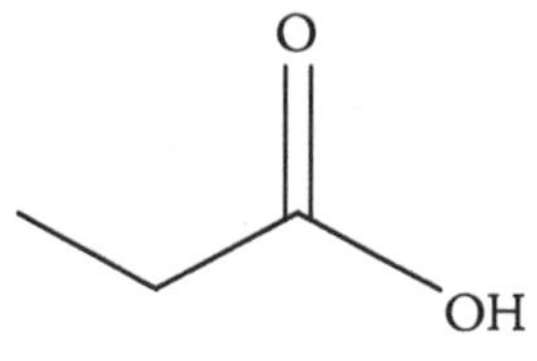

Chemical formula: $C_3H_6O_2$

Figure 2.12 Structure of propionic acid.

Microguard® and Bioprofit®, where the use of a *Propionibacterium freuden-reichii* strain, along with *Lactobacillus rhamnosus* increases inhibitory activity against fungi and some Gram-positive bacteria (Caplice and Fitzgerald, 1999). The esters of propionic acid have fruitlike odors and are sometimes used as solvents or artificial flavorings. Propionic acid can also be used as an intermediate in the production of other chemicals such as polymers. Cellulose acetate propionate is a useful thermoplastic, and vinyl propionate is used, in more specialized applications, to make pesticides and pharmaceuticals (Schnürer and Magnusson, 2005).

Propionic acid is effective at reducing fungal growth, especially at lower pH, by affecting fungal membranes at pH values below 4.5. Similar to acetic acid, propionic acid also inhibits amino acid uptake. The salts of propionic acid, such as sodium propionate and ammonium propionate show a similar effect against yeasts and filamentous molds at a low pH (Schnürer and Magnusson, 2005).

2.3.12   Sorbic acid

Sorbic acid (Figure 2.13) is the weak organic acid most extensively used in food preservation. It has GRAS status and it is used in any food, animal feed, or pharmaceutical needing preservation (Piper, 1999). Together with the organic acids, benzoic acid and acetic acid, sorbic acids (or *trans-trans*-2,4-hexadienoic acids) are the most commonly used chemical preservatives of food and are GRAS, broad-spectrum, antimicrobial agents (Plumridge et al., 2004). Sorbic acid is also often used in combination with benzoic acid for various types of products and confectionery (Suhr and Nielsen, 2004).

Sorbic acid is preferred to other organic acids because of its physiological harmlessness and organoleptic neutrality. In the United States, sorbic acid is a GRAS substance and its use is permitted in any food product to which preservatives may be added (Banerjee and Sarkar, 2004). Sorbic

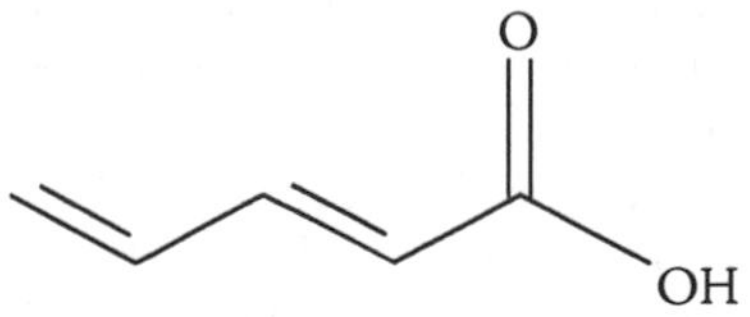

Chemical formula: $C_6H_8O_2$

Figure 2.13 Structure of sorbic acid.

acid is a six-carbon monocarboxylic acid, unsaturated at positions 2 and 4, known to be a membrane-active compound (Stratford and Anslow, 1998).

Sorbic acid and its salts have several advantages as food preservatives. They are highly active against fungi as well as a wide range of bacteria, in particular the catalase-positive organisms, their effective concentrations normally do not alter the taste or odor of a food product, and they are also considered harmless. However, potassium salt is more commonly used because of its stable nature. Potassium sorbate is also much more soluble and very suitable for use in dipping and spraying decontamination practices (Gonzalez-Fandos and Dominguez, 2007). Solubility is 58.2% (139 g/100 ml) at 20°C. It may also be used for spraying or metering fruit and vegetable products. The inhibitory activity of sorbates is similar to sorbic acid, but usually 25% more potassium sorbate is needed to have the same effect. However, sorbates have many uses because of their milder taste, greater effectiveness, and broader pH range (up to 6.5), when compared to either benzoate or propionate. This would effectively mean that in foods with a very low pH, sorbate levels as low as 200 ppm may give more than adequate protection. Sorbate can, therefore, be applied to a wide range of food products and beverages, including among others, syrups, fruit juices, wines, jellies, jams, salads, and pickles (Barbosa-Cánovas et al., 2003). The higher solubility of potassium sorbate, therefore, renders it the preferred form of sorbic acid as a preservative in foods. However, sorbic acid is more soluble in oils than the potassium salt. Sorbic acid and potassium sorbate are not very effective at controlling spoilage in bakery products, where the pH may exceed pH 7. Potassium sorbate may only be useful at pH 6 but not at pH 7.5 (Marín et al., 2003).

Sorbic acid has a pK_a value of 4.76 and in a pH 4.2 solution about 75% will be in its undissociated form. At pH 7 > 1% will be undissociated. Although acetic acid has the same pK_a value as sorbic acid, a tenfold higher concentration of acetic acid is required to produce the same extent of growth inhibition as sorbic acid (Papadimitriou et al., 2007).

Until recently, the use of sorbic acid was a standard approach for preservation in the food industry. However, much attention has now focused on the development of resistance against this weak organic acid preservative (Brul et al., 2002).

As benzoic acid is generally assumed to be safe because it is conjugated in the liver to produce benzoylglycine (hippuric acid), sorbate is also largely excreted in this case as the oxidation product 2,4-hexadienedioxic acid (Piper, 1999). Sorbate has been found to have a bacteriostatic effect on *L. monocytogenes*, but conflicting reports on the efficacy of sorbate against *L. monocytogenes* may be due to variations in media or food, pH, or sorbate concentrations (Gonzalez-Fandos and Dominguez, 2007). Potassium sorbate has been found to be much less effective than sorbic acid at inhibiting fungal growth, specifically mainly at lower temperatures (15 and 20°C;

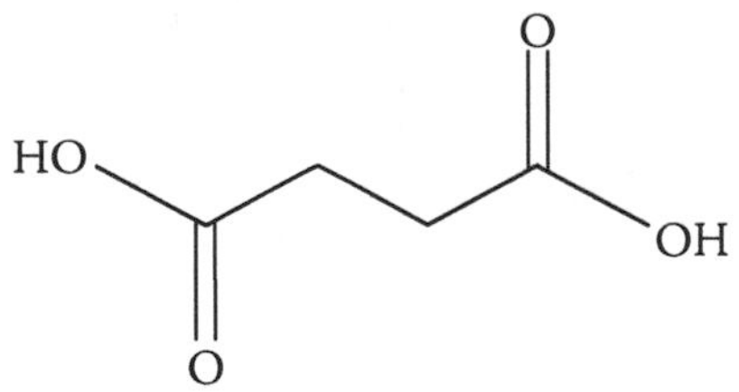

Chemical formula: $C_4H_6O_4$

Figure 2.14 Structure of succinic acid.

Marín et al., 2003), whereas sorbic acid is less linked to pH (Shtenberg and Ignat'ev, 1970).

2.3.13   Succinic acid

Succinic acid (also called butanedoic acid) (Figure 2.14) is a dicarboxylic acid of four carbon atoms and occurs naturally in plant and animal tissues. It plays a significant role in the intermediary metabolism (Krebs cycle) in the body. Succinic acid is a colorless crystalline solid medium with a salty bitter taste. It is soluble in water, slightly soluble in ethanol, ether, acetone, and glycerine, but not in benzene, carbon sulfide, carbon tetrachloride, and oil ether (ChemicalLand21.com, 2008). Succinic acid is used in foods as a sequestrant, buffer, and neutralizing agent. It is also used as a chemical intermediate in medicine, and in the manufacturing of lacquers and perfume esters (*Hawley*, 1988).

2.3.14   Tartaric acid

Tartaric acid (Figure 2.15) is also one of the hydroxy acids (Fite et al., 2004). It is naturally present in fruits such as pineapples and grapes and is consequently also the major acid in wine and wine vinegars (Barbosa-Cánovas et al., 2003; Morales, Gonzalez, and Troncoso, 1998). Primarily due to its low pK_a values (3.03 and 4.54), tartaric acid is often used as an acidulant (Comes and Beelman, 2002). The tartness profile of tartaric acid is smoother than that of citric acid and it has a higher peak acid taste than citric acid and this is also longer lasting. Tartaric acid may well be the most expensive among the acids commonly used in food (Fowlds, 2002).

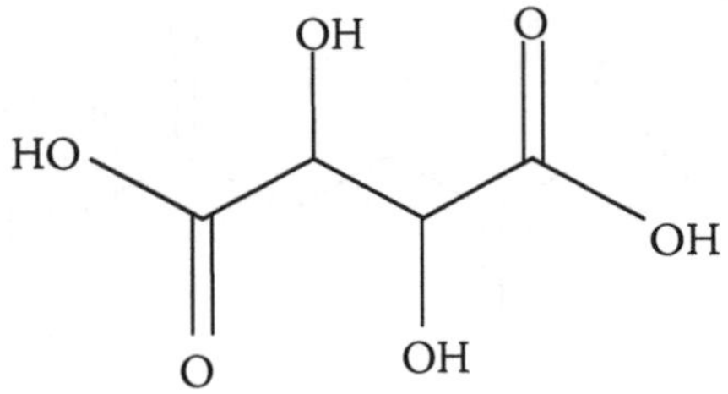

Chemical formula: $C_4H_6O_6$

Figure 2.15 Structure of tartaric acid.

2.3.15 Other acids

Phenyllactic acid (Figure 2.16) is another organic acid with an interesting potential for application as an antimicrobial agent in food. It has a broad spectrum of inhibition, specifically against a variety of foodborne fungi. Concentrations required for antifungal activity, particularly against molds from bakery products are generally lower than concentrations required against bacterial food contaminants. Phenyllactic solutions are odorless (Lavermicocca, Valerio, and Visconti, 2003).

Gallic acid (also called 3,4,5-trihydroxybenzoic acid) (Figure 2.17) is a natural plant phenol. It is an organic acid found in walnuts, sumac, witch hazel, tea leaves, oak bark, and other plants. Gallic acid has been found to

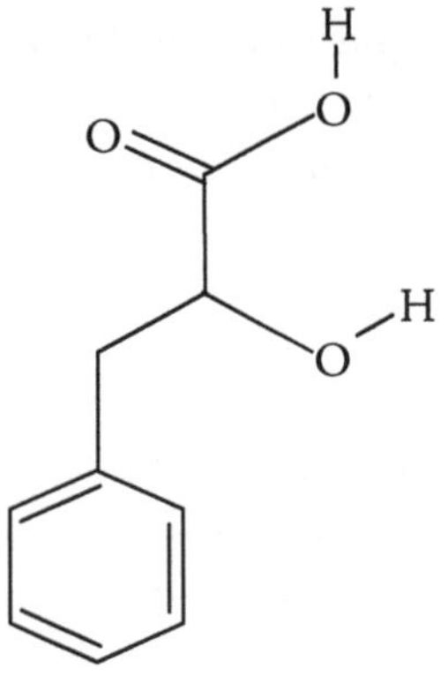

Chemical formula: $C_9H_{10}O_3$

Figure 2.16 Structure of phenyllactic acid.

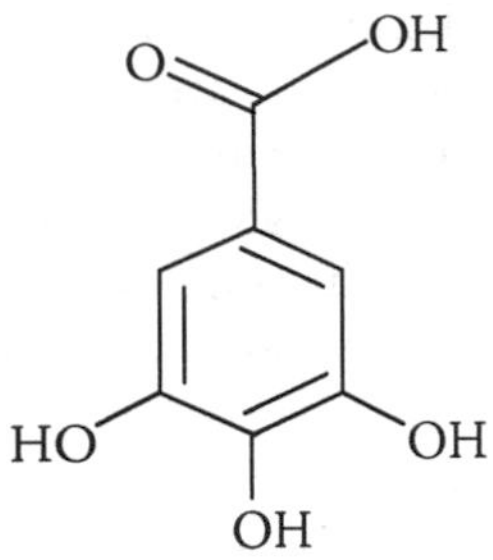

Chemical formula: $C_7H_6O_5$

Figure 2.17 Structure of gallic acid.

possess antifungal and antiviral properties, and derivatives such as propyl gallate, octyl gallate, lauryl gallate, and dodecyl gallate are also widely used as food additives in reducing rancidity and as preservatives (Van der Heijden, Janssen, and Strik, 1986; Serrano et al., 1998; Niho et al., 2001).

2.4   General applications

Organic acids, inorganic anions, amino acids, and carbohydrates are important compounds in many fields, such as chemistry, biochemistry, pharmaceuticals, agriculture, and food science (Cai et al., 1999). Organic acids have traditionally been applied to a wide variety of foods and are currently the most commonly used food preservatives (Gould and Jones, 1989; Wen, Wang, and Feng, 2007). These compounds and their salt derivatives have for many decades been used to effectively control the growth of select microorganisms by lowering the pH of foods. They are added deliberately to food products or can be functional as by-products as a result of fermentation to acidify the food to control microbial contamination and more important, foodborne pathogens (Murphy et al., 2006). Organic acids have also been found to be beneficial in extending cold storage of meat products (Bolton et al., 2002). Some organic acids, such as ascorbic acid and citric acids are most generally used as preservatives for canned foods or even fresh salad vegetables (Perez-Ruiz, Martinez-Lozano, and Garcia, 2007; Rico et al., 2007). Salts of organic acids are used extensively in meat and poultry products to enhance microbiological safety of these products by controlling pathogens such as *Listeria monocytogenes*. In processed meat, buffered salts, alone or in combination, are used as ingredients to provide an additional measure of safety during chilling (Juneja and Thippareddi, 2004).

Some organic acids, such as sorbic acid and benzoic acid, are widely regarded as the most active against yeasts and molds, and not against bacteria as such (Gonzalez et al., 1998). In the animal food industry, organic acids were originally added to animal feeds, specifically as fungistats (Ricke, 2003). Organic acids are also the most commonly utilized antimicrobial agents in edible coatings (Durango, Soares, and Andrade, 2006). Preservation of bakery products commonly involves the use of preservatives such as propionates and sorbates, and sometimes benzoates (Marín et al., 2003). Sodium benzoate is used in soft drinks, baked goods, and lollipops (Poulter, 2007).

Organic acids are also beneficial for use in perfume production, pharmaceuticals, medicine, and the production of other chemicals. For example, cinnamic acid is synthesized on a commercial scale from styrene and carbon tetrachloride (Roller, 1995). Due to biodegradable properties, there is a growing demand for organic acids for the production of polymeric materials (Bailly, 2002).

2.5   Food products naturally containing organic acids

Organic acids occur naturally in many foods as a result of metabolic processes, hydrolysis, or bacterial growth (Table 2.2). Alternatively they can be added directly to food as acidulants or flavorants (Gomis, 1992).

2.5.1   Fruit

Various organic acids are commonly found to be natural constituents of fruit. Citric acid can be found in almost all fruits, especially citrus fruit and pineapple (Nikbakht, Sadrzadeh, and Mohammadi, 2007). Malic acid is also

Table 2.2 Organic Acids Naturally Found in Foodstuffs

Organic acid	Foodstuff
Acetic acid	Vinegar
Benzoic acid	Cranberries, prunes, greengage plums, cinnamon, ripe cloves, apples
Butyric acid	Butter
Citric acid	Citrus fruits (oranges, lemons, grapefruit), blackcurrents, strawberries
Formic acid	Citrus essential oils
Gallic acid	Witch hazel, gallnuts, tea leaves
Lactic acid	Sour milk products (yogurt, cottage cheese)
Malic acid	Apples, cherries, plums
Oxalic acid	Tomatoes
Tartaric acid	Grape juices

widely found in fruits and vegetables (Barbosa-Cánovas et al., 2003). Benzoic acid is a common phenolic in cranberries, especially in freshly squeezed cranberry juice. It is considered to be the main antifungal agent in cranberries (Penney et al., 2004). Benzoic acid is also the natural constituent of raspberries, plums, prunes, lingonberries, cloudberries, cinnamon, and cloves (Barbosa-Cánovas et al., 2003; Be-Onishi et al., 2004). The benzoic concentration is especially high in lingonberries (0.6–1.3 g/l free benzoic acid) and the pH low (pH 2.6–2.9). Concentration of benzoic acid is increased further if the fruit or fruit juice is stored in warm conditions (Be-Onishi et al., 2004). Tartaric acid is commonly found in pineapples, grapes, and grape-derived products. It is, therefore, also present in relatively large amounts in wine vinegar, including balsamic vinegar, and in low concentration in cider vinegars (Morales, Gonzalez, and Troncoso, 1998).

Organic acids have been described as functioning as strong antimicrobial agents against psychrophylic and mesophilic microorganisms in fresh-cut fruits as well as vegetables. Ascorbic acid and β-carotene are antioxidants also present in the greatest quantities in fruits and vegetables, but are better known for their oxidative characteristics (Rico et al., 2007).

2.5.2   Juices

Fruit juices prepared from the pulp contain relatively large amounts of citric acid (Hayakawa, Linko, and Linko, 1999). In most juices citrate is omnipresent, and malic acid is present in high concentrations, with minute concentrations of other organic acids such as lactic acid (Arellano et al., 1997). Lemon juice is often used as an alternative to acidification with citric acid (Valero et al., 2000).

2.5.3   Wine and vinegar

Acetic acid is the major volatile acid in wine. At levels above 1.2–1.4 g/l, it is known to be one of the main reasons for wine spoilage (Gonzalez et al., 2005). Acetic acid is also an active ingredient in household vinegar, and common white household vinegar consists of approximately 5% acetic acid (Kerth and Braden, 2007). It is speculated that acidity, alcohol content, and content of polyphenolic flavonoid compounds including tannins and reveratrol may be responsible for antimicrobial activities reported for wine against foodborne pathogens. This has recently evoked much interest and the question arises whether wine formulations can be devised and applied in improving food safety. These wine substances would then be regarded as nontoxic, food-compatible, and plant-derived antimicrobials (Friedman et al., 2006).

2.5.4 Dairy

Citric acid is an organic acid found in abundance in raw milk, whereas lactic acid is found to be more prevalent in yogurt and cheese (Tormo and Izco, 2004).

2.5.5 Coffee

Malic and citric acids are already present in the green coffee bean, but most of the acidity in the coffee bean is generated at the beginning of the roasting process (Rodrigues et al., 2007).

2.5.6 Bakery products

Sourdough has long been known to improve the shelf life of bread and bakery products, primarily because of the existence of lactic acid produced by lactic acid bacteria (Lavermicocca, Valerio, and Visconti, 2003).

2.5.7 Honey

The natural occurrence of varying amounts of organic acids in honeys from different geographical areas have been documented. Much research is being done to determine the use of honey as an antimicrobial agent in various food systems (Suarez-Luque et al., 2002).

References

AMCA (Agrobest Multi Chelaton Advantage). 2005. Agrobest, Australia (PTY), Ltd. http://www.agrobest.com.au/amca.pdf.

Arellano, M., Andrianary, J., Dedieu, F., Couderc, F., and Puig, P. 1997. Method development and validation for the simultaneous determination of organic and inorganic acids by capillary zone electrophoresis. *Journal of Chromatography A* 765:321–328.

Bailly, M. 2002. Production of organic acids by bipolar electrodialysis: Realizations and perspectives. *Desalination* 144:157–162.

Banerjee, M. and Sarkar, P.K. 2004. Antibiotic resistance and susceptibility to some food preservative measures of spoilage and pathogenic microorganisms from spices. *Food Microbiology* 21:335–342.

Barbosa-Cánovas, G.V., Fernández-Molina, J.J., Alzamora, S.M., Tapia, M.S., López-Malo, A., and Chanes, J.W. 2003. General considerations for preservation of fruits and vegetables. In: *Handling and Preservation of Fruits and Vegetables by Combined Methods for Rural Areas*. Rome: Food and Argriculture Organization of the United Nations.

Bauer, B.E., Rossington, D., Mollapour, M., Mamnun, Y., Kuchler, K., and Piper, P.W. 2003. Weak organic acid stress inhibits aromatic amino acid uptake by yeast, causing a strong influence of amino acid auxotrophies on the phenotypes of membrane transporter mutants. *European Journal of Biochemistry* 270:3189–3195.

Benthin, S. and Villadsen, J. 1995. Production of optically pure d-lactate by *Lactobacillus bulgaricus* and purification by crystallization and liquid-liquid extraction. *Applied Microbiology and Biotechnology* 42:826–829.

Be-Onishi, Y., Yomota, C., Sugimoto, N., Kubota, H., and Tanamoto, K. 2004. Determination of benzoyl peroxide and benzoic acid in wheat flour by high-performance liquid chromatography and its identification by high-performance liquid chromatography-mass spectrometry. *Journal of Chromatography A* 1040:209–214.

Bolton, D.J., Catarame, T., Byrne, C., Sheridan, J.J., McDowell, D.A., and Blair, I.S. 2002. The ineffectiveness of organic acids, freezing and pulsed electric fields to control *Escherichia coli* O157:H7 in beef burgers. *Letters in Applied Microbiology* 34:139–143.

Brul, S. and Coote, P. 1999. Preservative agents in foods. Mode of action and microbial resistance mechanisms. *International Journal of Food Microbiology* 50:1–17.

Brul, S., Klis, F.M., Oomes, S.J.C.M., et al. 2002. Detailed process design based on genomics of survivors of food preservation processes. *Trends in Food Science and Technology* 13:325–333.

Bunch, P.K., Mat-Jan, F., Lee, N., and Clark, D.P. 1997. The *ldhA* gene encoding the fermentative lactate dehydrogenase of *Escherichia coli*. *Microbiology* 143: 187–195.

Cai, Y., Benno, Y., Ogawa, M., and Kumai, S. 1999. Effect of applying lactic acid bacteria isolated from forage crops on fermentation characteristics and aerobic deterioration of silage. *Journal of Dairy Science* 82:520–526.

Caplice, E. and Fitzgerald, G.F. 1999. Food fermentations: Role of microorganisms in food production and preservation. *International Journal of Food Microbiology* 50:131–149.

Carrasco, E., Garcia-Gimeno, R., and Seselovsky, R. 2006. Predictive model of *listeria monocytogenes'* growth rate under different temperatures and acids. *Food Science and Technology International* 12:47–56.

Chaveerach, P., Keuzenkamp, D.A., Urlings, H.A., Lipman, L.J., and Van, K.F. 2002. In vitro study on the effect of organic acids on *Campylobacter jejuni/coli* populations in mixtures of water and feed. *Poultry Science* 81:621–628.

ChemicalLand21.com. 2008. Succinic Acid. http://www.chemicalland21.com. (Accessed September 2009)

Choi, S.H. and Chin, K.B. 2003. Evaluation of sodium lactate as a replacement for conventional chemical preservatives in comminuted sausages inoculated with *Listeria monocytogenes*. *Meat Science* 65:531–537.

Comes, J.E. and Beelman, R.B. 2002. Addition of fumaric acid and sodium benzoate as an alternative method to achieve a 5-log reduction of *Escherichia coli* O157:H7 populations in apple cider. *Journal of Food Protection* 65:476–483.

Couto, S.R. and Sanroman, M.A. 2006. Application of solid-state fermentation to food industry—A review. *Journal of Food Engineering* 76:291–302.

De la, R.P., Cordoba, G., Martin, A., Jordano, R., and Medina, L.M. 2005. Influence of a test preservative on sponge cakes under different storage conditions. *Journal of Food Protection* 68:2465–2469.

Doyle, E.M. 1999. *Use of organic acids to control Listeria in meat.* American Meat Institute Foundation.

Durango, A.M., Soares, N.F.F., and Andrade, N.J. 2006. Microbiological evaluation of an edible antimicrobial coating on minimally processed carrots. *Food Control* 17:336–341.

Dziezak, J.D. 1990. Acidulants: Ingredients that do more than meet the acid test. *Food Technology* 44:78–83.

El-Ziney, M.G., De, M.H., and Debevere, J.M. 1997. Growth and survival kinetics of *Yersinia enterocolitica* IP 383 0:9 as affected by equimolar concentrations of undissociated short-chain organic acids. *International Journal of Food Microbiology* 34:233–247.

Farkas, J. 1998. Irradiation as a method for decontaminating food. A review. *International Journal of Food Microbiology* 44:189–204.

Fite, A., Dykhuizen, R., Litterick, A., Golden, M., and Leifert, C. 2004. Effects of ascorbic acid, glutathione, thiocyanate, and iodide on antimicrobial activity of acidified nitrite. *Antimicrobial Agents and Chemotherapy* 48:655–658.

Fowlds, R.W.R. 2002. Production of a food acid mixture containing fumaric acid. http://www.freepatentsonline.com. (Accessed August 29, 2005)

Friedman, M., Henika, P., Levin, C., and Mandrell, R. 2006. Antimicrobial wine formulations active against the foodborne pathogens *Escherichia coli* O157: H7 and *Salmonella enterica*. *Journal of Food Science* 71:M245–M251.

Gomis, D.B. 1992. HPLC analysis of organic acids. In: L.M.L. Nollet (Ed.), *Food Analysis by HPLC*, New York: Marcel Dekker, pp. 371–385.

Gonzalez, A., Hierro, N., Poblet, M., Mas, A., and Guillamon, J.M. 2005. Application of molecular methods to demonstrate species and strain evolution of acetic acid bacteria population during wine production. *International Journal of Food Microbiology* 102:295–304.

Gonzalez, M., Gallego, M., and Valcarcel, M. 1998. Simultaneous gas chromatographic determination of food preservatives following solid-phase extraction. *Journal of Chromatography A* 823:321–329.

Gonzalez-Fandos, E. and Dominguez, J.L. 2007. Effect of potassium sorbate washing on the growth of *Listeria monocytogenes* on fresh poultry. *Food Control* 18:842–846.

Gould, G.W. and Jones, M.V. 1989. Combination and synergistic effects. In: *Mechanisms of Action of Food Preservation Procedures*, G.W. Gould (Ed.), London: Elsevier Science.

Graham, A.F. and Lund, B.M. 1986. The effect of citric acid on growth of proteolytic strains of *Clostridium botulinum*. *Journal of Applied Bacteriology* 61:39–49.

Gravesen, A., Diao, Z., Voss, J., Budde, B.B., and Knochel, S. 2004. Differential inactivation of *Listeria monocytogenes* by D- and L-lactic acid. *Letters in Applied Microbiology* 39:528–532.

Greenacre, E.J., Brocklehurst, T.F., Waspe, C.R., et al. 2003. *Salmonella enterica* Serovar Typhimurium and *Listeria monocytogenes* acid tolerance response induced by organic acids at 20°C; optimization and modeling. *Applied and Environmental Microbiology* 69:3945–3951.

Guerzoni, M.E., Lanciotti, R., and Marchetti, R. 1993. Survey of the physiological properties of the most frequent yeasts associated with commercial chilled foods. *International Journal of Food Microbiology* 17:329–341.

Hawley's Condensed Chemical Dictionary, 12th ed. 1988. New York: Elsevier Science, Van Nostrand, p. 1099.

Hayakawa, K., Linko, Y.Y., and Linko, P. 1999. Mechanism and control of food allergy. *Lebensmittel-Wissenschaft und-Technologie* 32:1–11.

Hazan, R., Levine, A., and Abeliovich, H. 2004. Benzoic acid, a weak organic acid food preservative, exerts specific effects on intracellular membrane trafficking pathways in *Saccharomyces cerevisiae*. *Applied and Environmental Microbiology* 70:4449–4457.

Hinton, A., Jr. 2006. Growth of *Campylobacter* in media supplemented with organic acids. *Journal of Food Protection* 69:34–38.

Hoffman, P. and Possin, I. 2000. Adding organic acids to high moisture corn. Focus on Forage 2, 1. http://www.uwex.edu/ces/crops/uwforage/HMC-OA.pdf. (Accessed August 28, 2005)

Hsiao, C.P. and Siebert, K.J. 1999. Modeling the inhibitory effects of organic acids on bacteria. *International Journal of Food Microbiology* 47:189–201.

Imai, K., Banno, I., and Iijima, T. 1970. Inhibition of bacterial growth by citrate. *Journal of General Applied Microbiology* 16:479–489.

Juneja, V.K. and Thippareddi, H. 2004. Inhibitory effects of organic acid salts on growth of *Clostridium perfringens* from spore inocula during chilling of marinated ground turkey breast. *International Journal of Food Microbiology* 93:155–163.

Kerth, C. and Braden, C. 2007. Using acetic acid rinse as a CCP for slaughter. *Virtual Library—Meat, Poultry & Egg Processors.*

Kouassi, Y. and Shelef, L.A. 1996. Metabolic activities of *Listeria monocytogenes* in the presence of sodium propionate, acetate, lactate and citrate. *Journal of Applied Bacteriology* 81:147–153.

Krishnakumar, V. 1994. Tartaric acid. Worldwide business review. *International Food Ingredients* 3:17–21.

Lavermicocca, P., Valerio, F., and Visconti, A. 2003. Antifungal activity of phenyllactic acid against molds isolated from bakery products. *Applied and Environmental Microbiology* 69:634–640.

Lemay, M.J., Choquette, J., Delaquis, P.J., Claude, G., Rodrigue, N., and Saucier, L. 2002. Antimicrobial effect of natural preservatives in a cooked and acidified chicken meat model. *International Journal of Food Microbiology* 78:217–226.

Marín, S., Abellana, M., Rubinat, M., Sanchis, V., and Ramos, A.J. 2003. Efficacy of sorbates on the control of the growth of *Eurotium* species in bakery products with near neutral pH. *International Journal of Food Microbiology* 87:251–258.

Marz, U. 2002. World markets for citric, ascorbic, isoascorbic acids: Highlighting antioxidants in food. *Food and Beverage.* McWilliam Leitch, E.C. and Stewart, C.S. 2003. *Escherichia coli* O157 and non-O157 isolates are more susceptible to L-lactate than to D-lactate. *Applied and Environmental Microbiology* 68:4676–4678.

McWilliam Leitch, E.C. and Stewart, C.S. 2003. *Escherichia coli* O157 and non-O157 isolates are more susceptible to l-lactate than to d-lactate. *Applied and Environmental Microbiology* 68:4676–4678.

Min, J.S., Lee, S.O., Jang, A., Jo, C., and Lee, M. 2007. Irradiation and organic acid treatment for microbial control and the production of biogenic amines in beef and pork. *Food Chemistry* 104:791–799.

Morales, M.L., Gonzalez, A.G., and Troncoso, A.M. 1998. Ion-exclusion chromatographic determination of organic acids in vinegars. *Journal of Chromatography A* 822:45–51.

Moresi, M. and Sappino, F. 1998. Economic feasibility study of citrate recovery by electrodialysis. *Journal of Food Engineering* 35:75–90.

Murphy, R.Y., Hanson, R.E., Johnson, N.R., Chappa, K., and Berrang, M.E. 2006. Combining organic acid treatment with steam pasteurization to eliminate *Listeria monocytogenes* on fully cooked frankfurters. *Journal of Food Protection* 69:47–52.

Nakai, S.A. and Siebert, K.J. 2003. Validation of bacterial growth inhibition models based on molecular properties of organic acids. *International Journal of Food Microbiology* 86:249–255.

Nielsen, M.K. and Arneborg, N. 2007. The effect of citric acid and pH on growth and metabolism of anaerobic *Saccharomyces cerevisiae* and *Zygosaccharomyces bailii* cultures. *Food Microbiology* 24:101–105.

Niho, N., Shibutani, M., Tamura, T., et al. 2001. Subchronic toxicity study of gallic acid by oral administration in F344 rats. *Food and Chemical Toxicology* 39:1063–1070.

Nikbakht, R., Sadrzadeh, M., and Mohammadi, T. 2007. Effect of operating parameters on concentration of citric acid using electrodialysis. *Journal of Food Engineering* 83:596–604.

Otero-Losada, M.E. 1999. A kinetic study on benzoic acid pungency and sensory attributes of benzoic acid. *Chemical Senses* 24:245–253.

Otero-Losada, M.E. 2003. Differential changes in taste perception induced by benzoic acid pickling. *Physiology & Behavior* 78:415–425.

Papadimitriou, M.N., Resende, C., Kuchler, K., and Brul, S. 2007. High Pdr12 levels in spoilage yeast (*Saccharomyces cerevisiae*) correlate directly with sorbic acid levels in the culture medium but are not sufficient to provide cells with acquired resistance to the food preservative. *International Journal of Food Microbiology* 113:173–179.

Penney, V., Henderson, G., Blum, C., and Johnson-Green, P. 2004. The potential of phytopreservatives and nisin to control microbial spoilage of minimally processed fruit yogurts. *Innovative Food Science and Emerging Technologies* 5:369–375.

Perez-Ruiz, T., Martinez-Lozano, C., and Garcia, M.D. 2007. High-performance liquid chromatography-post-column chemiluminescence determination of aminopolycarboxylic acids at low concentration levels using tris(2,2'-bipyridyl)ruthenium(III). *Journal of Chromatography A* 1169:151–157.

Permprasert, J. and Devahastin, S. 2005. Evaluation of the effects of some additives and pH on surface tension of aqueous solutions using a drop-weight method. *Journal of Food Engineering* 70:219–226.

Piper, P.W. 1999. Yeast superoxide dismutase mutants reveal a pro-oxidant action of weak organic acid food preservatives. *Free Radical Biology and Medicine* 27:11–12.

Plumridge, A., Hesse, S.J., Watson, A.J., Lowe, K.C., Stratford, M., and Archer, D.B. 2004. The weak acid preservative sorbic acid inhibits conidial germination and mycelial growth of *Aspergillus niger* through intracellular acidification. *Applied and Environmental Microbiology* 70:3506–3511.

Poulter, S. 2007. The proof food additives ARE as bad as we feared. *Daily Mail* MailOnline.

Ramanchandran, S., Fontanille, P., Pandey, A., and Larroche, C. 2006. Gluconic acid: Properties, applications and microbial production. *Food Technology and Biotechnology* 44:185–195.

Rammel, C.G. 1962. Inhibition by citrate of growth of coagulase-positive staphylococci. *Journal of Bacteriology* 84:1123–1124.

Ricke, S.C. 2003. Perspectives on the use of organic acids and short chain fatty acids as antimicrobials. *Poultry Science* 82:632–639.

Rico, D., Martin-Diana, A.B., Barat, J.M., and Barry-Ryan, C. 2007. Extending and measuring the quality of fresh-cut fruit and vegetables: A review. *Trends in Food Science & Technology* 18:373–386.

Rodrigues, C.I., Marta, L., Maia, R., Miranda, M., Ribeirinho, M., and Maguas, C. 2007. Application of solid-phase extraction to brewed coffee caffeine and organic acid determination by UV/HPLC. *Journal of Food Composition and Analysis* 20:440–448.

Roller, S. 1995. The quest for natural antimicrobials as novel means of food preservation: Status report on a European research project. *International Biodeterioration & Biodegradation* 36:333–345.

Schnürer, J. and Magnusson, J. 2005. Antifungal lactic acid bacteria as bio-preservatives. *Trends in Food Science and Technology* 16:70–78.

Serrano, A., Palacios, C., Roy, G., et al. 1998. Derivatives of gallic acid induce apoptosis in tumoral cell lines and inhibit lymphocyte proliferation. *Archives of Biochemistry and Biophysics* 350:49–54.

Shtenberg, A.J. and Ignat'ev, A.D. 1970. Toxicological evaluation of some combination of food preservatives. *Food and Cosmetics Toxicology* 8:369–380.

Şimşek, O., Çon, A.H., and Tulumoğlu, S. 2006. Isolating lactic starter cultures with antimicrobial activity for sourdough processes. *Food Control* 17:263–270.

Smulders, F.J. and Greer, G.G. 1998. Integrating microbial decontamination with organic acids in HACCP programmes for muscle foods: Prospects and controversies. *International Journal of Food Microbiology* 44:149–169.

Steiner, P. and Sauer, U. 2003. Overexpression of the ATP-dependent helicase RecG improves resistance to weak organic acids in *Escherichia coli*. *Applied Microbiology and Biotechnology* 63:293–299.

Stratford, M. 1999. Weak acid and "weak-acid preservative" inhibition of yeasts. In: A.C.J. Tuijtelaars, R.A. Samson, F.M. Rombouts, and S. Notermans (Eds.), *Food Microbiology and Food Safety in the Next Millenium*. Foundation Food Micro '99. Veldhoven, The Netherlands, pp. 315–319.

Stratford, M. and Anslow, P.A. 1998. Evidence that sorbic acid does not inhibit yeast as a classic 'weak acid preservative'. *Letters in Applied Microbiology* 27:203–206.

Suarez-Luque, S., Mato, I., Huidobro, J.F., Simal-Lozano, J., and Sancho, M.T. 2002. Rapid determination of minority organic acids in honey by high-performance liquid chromatography. *Journal of Chromatography A* 955:207–214.

Suhr, K.I. and Nielsen, P.V. 2004. Effect of weak acid preservatives on growth of bakery product spoilage fungi at different water activities and pH values. *International Journal of Food Microbiology* 95:67–78.

Tang, Y. and Wu, M. 2007. The simultaneous separation and determination of five organic acids in food by capillary electrophoresis. *Food Chemistry* 103:243–248.

Theron, M.M. and Lues, J.F.R. 2007. Organic acids and meat preservation: A review. *Food Reviews International* 23:141–158.

Tormo, M. and Izco, J.M. 2004. Alternative reversed-phase high-performance liquid chromatography method to analyse organic acids in dairy products. *Journal of Chromatography A* 1033:305–310.

Valero, M., Leontidis, S., Fernandez, P.S., Martinez, A., and Salmeron, M.C. 2000. Growth of Bacillus cereus in natural and acidified carrot substrates over the temperature range 5–30°C. *Food Microbiology* 17:605–612.

Valli, M., Sauer, M., Branduardi, P., Borth, N., Porro, D., and Mattanovich, D. 2006. Improvement of lactic acid production in *Saccharomyces cerevisiae* by cell sorting for high intracellular pH. *Applied and Environmental Microbiology* 72:5492–5499.

Van der Heijden, C.A., Janssen, P.J.C.M., and Strik, J.J.T.W.A. 1986. Toxicology of gallates: A review and evaluation. *Food and Chemical Toxicology* 24:1067–1070.

Visti, A., Viljakainen, S., and Laakso, S. 2003. Preparation of fermentable lingonberry juice through removal of benzoic acid by *Saccharomyces cerevisiae* yeast. *Food Research International* 36:597–602.

Wang, X., Sun, L., Wei, D., et al. 2005. Reducing by-product formation in l-lactic acid fermentation by *Rhizopus oryzae*. *Journal of Industrial Microbiology & Biotechnology* 32:38–40.

Wen, Y., Wang, Y., and Feng, Y.Q. 2007. A simple and rapid method for simultaneous determination of benzoic and sorbic acids in food using in-tube solid-phase microextraction coupled with high-performance liquid chromatography. *Analytical and Bioanalytical Chemistry* 388:1779–1787.

Yang, M.H. and Choong, Y.M. 2001. A rapid gas chromatographic method for direct determination of short-chain (C2–C12) volatile organic acids in foods. *Food Chemistry* 75:101–108.

Zhang, Z.Y., Jin, B., and Kelly, J.M. 2007. Production of lactic acid from renewable materials by *Rhizopus* fungi. *Biochemical Engineering Journal* 35:251–263.

chapter three

Application of organic acids in food preservation

3.1   Introduction

Much research has been done on the antimicrobial efficacy of organic acids, but there is relatively little data available on its use in commercial practice (Smulders and Greer, 1998). Studies have predominantly been conducted in research abattoirs and with artificial carcasses, or small pieces of meat (Smulders, 1987; Siragusa, 1995; Smulders and Greer, 1998; Theron and Lues, 2007). In evaluating the effectiveness of organic acids for specific applications, a better understanding of general and specific response capacities of foodborne pathogens is essential, as this may pave the way to the development of more targeted strategies in controlling foodborne pathogens (Ricke, 2003; Carroll et al., 2007). Food-processing treatments in preventing pathogen contamination in food products must also be effective against postprocessing contamination (Calicioglu, Sofos, and Kendall, 2003b).

Although organic acids are food ingredients that are often utilized at will (Bégin and Van Calsteren, 1999), thorough risk assessment should be the first action taken by all leading food industries in the preservation of food products (Brul and Coote, 1999; Van Gerwen et al., 2000; Hoornstra and Notermans, 2001). It is crucial to first analyze the production process as a whole, or the so-called from farm to fork or from raw material to consumer. Such an assessment requires four processes that must be systematically followed (1) hazard identification, (2) hazard categorization, (3) exposure evaluation, and (4) risk assessment (Brul and Coote, 1999).

3.2   Foodstuffs

3.2.1   Meat

In 1970 findings were reported that treating meat with organic acids can provide other means of extending the distribution and visual attraction of fresh meat (Bauernfeind and Pinkert, 1970; Barker and Park, 2001; Huang, Ho, and McMillin, 2005).

3.2.1.1   Cured meat

Brines, containing a combination of monolaurin and lactate, pumped into microwave-ready beef roasts, have been found to be more effective in inhibiting *Listeria monocytogenes* during cooking in bags than brines with only lactate. Cured meats (sausage, ham, and frankfurters) contain salt and other preservatives which all serve to enhance the listericidal effects of organic acids (Doyle, 1999).

3.2.1.2   Poultry

Unlike the breast muscle (pH 5.7–5.9) of the poultry carcass, poultry leg muscle has a pH of 6.4–6.7. The higher pH has an impact on the growth of certain spoilage bacteria and as a result, the leg tends to spoil more quickly than breast meat. High pH of the skin also plays a role in such contamination (Gonzalez-Fandos and Dominguez, 2007).

3.2.1.3   Seafood

Variations have been found in the antimicrobial effect of organic acids and salts on fish, which may be dependent on several factors, including (1) the concentration of organic acid, (2) dipping time, (3) fish species, (4) fish product, (5) degree of contamination, and also (6) the storage condition (Sallam, 2007a). Reports indicate a lack of effect on the psychrotrophic populations in fish products, such as shrimp (Zhuang, Huang, and Beuchat, 1996), and rainbow trout (Nykänen et al., 1998) during refrigeration. On the other hand, sodium acetate (2%) is reported to provide significant reduction in the growth of psychrotrophic bacteria in catfish fillets over 12 days at 4°C, but not in shrimp. Different organic acid salt treatments have all shown to cause significant reduction ($P < 0.05$) in *Pseudomonas* populations in sliced salmon (Sallam, 2007a).

Salmon slices treated with different organic salts also contained a lower count of H_2S-producing bacteria throughout storage. Complete inhibition of these bacteria has been reported in fresh cod fillets after sprayed application of 10% acetate buffer and stored under modified atmospheres for 12 days at 7°C. Complete inhibition of Enterobacteriaceae is achieved in fresh cod fillets, also after spraying with 10% acetate buffer during storage under the same conditions (Boskou and Debevere, 2000). Dipping of salmon slices in aqueous solutions (2.5%) of sodium salts of organic acids is generally effective and was efficient against proliferation of various spoilage microorganisms. This delays lipid oxidation and as such, extends the shelf life of the product during refrigerated storage and can be utilized as a safe organic preservative for fish under refrigerated storage (Sallam, 2007a). The quality of fish may rapidly degrade, due to a range of complex physical, chemical, and microbiological forms of deterioration (Sallam, 2007b).

3.2.2   Acidic foods

Acidified food products have been produced safely for many years without heat treatments, but it has become evident that the organic acid(s) present in such products have made a huge contribution to achieve this (Zagory and Garren, 1999). Acidic foods (apple cider, dry-fermented sausage, mayonnaise, and yogurt) have all been implicated in *E. coli* O157:H7 food poisoning outbreaks (Jordan, Oxford, and O'Byrne, 1999).

3.2.3   Confectionery

In bakery products with relatively high pH (7), the use of preservatives is often redundant, as they do not have any effect on the shelf life. Many of these products may contain unnecessary preservatives that could just as well be excluded from their recipes, as an appropriately low a_w serves as the main preserving factor (Marín et al., 2003).

Bakery products have a short shelf life, but can be maximized by good hygiene practices and preservatives (De la Rosa et al., 2005). In intermediate moisture bakery products, preservatives and moisture producers are often added to prevent spoilage by fungi. Potassium sorbate and calcium propionate are used at concentrations 3000 and 300 ppm (wt./flour wt.) (Arroyo, Aldred, and Magan, 2005).

3.2.4   Fruits and vegetables

Preservation of fresh-cut fruits and vegetables requires improved effective washing treatments. Alternative methods that have been proposed and investigated, include (1) organic acids, (2) antioxidants, (3) irradiation, (4) ozone, (5) modified atmosphere packaging, and (6) whey permeate, but none of these has thus far gained much widespread acceptance in the industry. Main objectives of the implementation of minimal processing techniques are to meet the challenges of maintaining nutritional and sensory quality, even when sufficiently preserved (Rico et al., 2007). However, the minimal processing to which fresh-cut fruits and vegetables are generally subjected, delivers products that are highly perishable, and need to be chilled during storage to ensure a reasonable extended shelf life (Garcìa and Barrett, 2002; Rico et al., 2007).

3.2.5   Fruit juices

Fruits have an inherent acidity and contain various organic acids that protect against bacterial invasion (Fielding, Cook, and Grandison, 1997). Fruit juices and other fruit products may be easily spoiled as a result of contamination due to high water activity, which allows microbial growth

(Gliemmo, Campos, and Gerschenson, 2006). These products are, there-fore, preserved by a combination of hurdles that include low pH, suppression of water activity by solute addition, added preservatives, and heat (Stiles et al., 2002; Gliemmo, Campos, and Gerschenson, 2006).

3.2.6 Salads

Salad bars are becoming increasingly popular and are also found in supermarkets and convenience stores. They have introduced new growth environments for microbial spoilage and also the prevalence of foodborne pathogens (Allende et al., 2007). *L. Monocytogenes* has been isolated from fresh or minimally processed vegetables (Francis and O'Beirne, 2006; Crépet et al., 2007), and has also been implicated in several outbreaks of foodborne infections, as a result of contaminated vegetables (Beuchat, Berrang, and Brackett, 1990; Beuchat, 1996; Sewell and Farber, 2001). It is, therefore, essential that washing and sanitizing treatments be adequate to prevent proliferation of *L. monocytogenes* during storage (Allende et al., 2007).

Organic acids are strong antimicrobial agents against both psychrophilic and mesophilic microorganisms in fresh-cut vegetables (Rico et al., 2007). For example, citric and ascorbic acids are often used on salad vegetables, and ascorbic acid (L-ascorbic acid) and its various salts and derivatives have for many years been popular GRAS (generally recognized as safe) antioxidants used on fruits and vegetables, and in fruit juices to prevent browning and other oxidative reactions (Rico et al., 2007). The use of weak acids is also, for example, permitted in table olives, although there is a lack of scientific information about their preserving effects. This is in part responsible for reduced efficacy obtained with their application (Arroyo Lopez, Duran Quintana, and Garrido, 2006).

3.2.7 Vegetables

Minimally processed vegetables can be inhabited by invasive microorganisms that may be the cause of deterioration (Martins et al., 2004).

3.2.8 Dairy

Spoilage bacteria and fungi (especially yeasts) on or in fruit hinders the production of minimally processed fruit yogurt. However, the use of synthetic preservatives is known to reduce the natural sensory appeal of these yogurts (Penney et al., 2004).

3.2.9   Soft drinks

Some organic acids are naturally present in soft drinks, but may also be intentionally added to provide sharpness, tang, and brightness (Chen and Wang, 2001).

3.2.10   Sport drinks

Citric acid, lactic acid, and ascorbic acid are commonly found in sports drinks. In most juices citrate is omnipresent, and vitamin C is present in beverages, juices, and medicines (Wu et al., 1995).

3.2.11   Animal feed

In animal feed organic acids were originally added to inhibit fungal contamination. However, in the past 40 years organic acids and various combinations have been investigated for potential bactericidal activity in feeds and feed ingredients (Ricke, 2003).

3.3   Industrial applications

The FDA and U.S. Department of Agriculture (USDA) both provide regulatory oversight in ensuring a safe food supply by conducting mandatory and rigorous safety (toxicological and pharmacokinetic) studies prior to approval of an antibiotic for use and also by monitoring the correct use of such antibiotic (Donoghue, 2003). After the USDA and Food Safety and Inspection Service (FSIS) established the requirement for *E. coli* biotype 1 enumeration as a means of verifying the control of slaughter processes, renewed interest developed in meat decontamination technologies and included the spraying of carcasses with either (1) organic acid solutions, (2) hot water, (3) steam, or (4) nonacid chemical solutions in passing the regulatory criteria (Samelis et al., 2002b).

3.3.1   Labeling

Sorbic acid and benzoic acid, as well as their salts (sodium sorbate, potassium sorbate, calcium sorbate, sodium benzoate, potassium benzoate, and calcium benzoate) are commonly allowed as food additives by the European Legislative Directive no. 98/72/CE, but their presence must be declared on the food label (Mota et al., 2003).

3.3.2   Vacuum packaging

In the production of boxed meat for domestic distribution as well as export markets, the safety of vacuum-packaged products receives critical consideration. It is envisaged that organic acids may play an important role in future processing treatments directly prior to vacuum packaging. Practical treatment conditions are essential and need to be established to facilitate improvement of safety and storage without sacrificing satisfactory sensory characteristics (Smulders and Greer, 1998).

3.3.3   Meat

In 1997 W. Dorsa commented on the vast difference between the large numbers of laboratory results and the actual application of organic acids as antimicrobials in industrial practices (Dorsa, Cutter, and Siragusa, 1997; Smulders and Greer, 1998). It would, therefore, not be appropriate to extrapolate these results. However, organic acids are not widely applied for meat decontamination. Meat hygiene regulations stipulated by the European Union do not allow any method of decontamination except for washing with potable water. Legislators are reluctant to grant permission for implementing other technologies, as it is alleged a means of compensating for inadequate hygienic practices in the abattoir. In 1996, a report was issued by the European Union's Advisory Scientific Veterinary Committee where the advantages of decontamination by irradiation, organic acids, alkaline compounds (e.g., trisodium phosphate), hyperchlorinated water, steam, or hot water were evaluated (European Union, 1996). Although this treatment is currently considered only for poultry carcasses, it is generally accepted that in other major meat processing, adherence to strict hygiene measures should be adequate (Smulders and Greer, 1998).

In Table 3.1 some factors are stipulated that should be considered in the decontamination of meat (European Union, 1996). Spraying with 1.5–2.5% organic acids such as acetic or lactic acid is effectively applied on red meat carcasses (Canadian Food Inspection Agency, 2004). Lactic and acetic acid solutions are commonly used by the red meat slaughtering industry as an antimicrobial spray wash on freshly slaughtered beef carcasses. These spray washes are used in the early steps of beef carcass processing, usually applied to carcasses after hide removal, before and after evisceration, but before chilling (Berry and Cutter, 2000).

Organic acid application is usually done by a spray cabinet. In some U.S. beef plants an organic acid wash is regarded as a good manufacturing practice (GMP) rather than a critical control point (CCP), because not all carcasses are treated. Organic acids are not applied to carcasses when they contain an open wound or a leaking abscess. The critical limits for organic

Table 3.1 Application of Decontaminants as Proposed by the Scientific
Veterinary Committee of the European Union (1996)

Factors that influence the application of decontaminants	Additional information
Chemical composition	pH, water, fat solubility
Safety aspects	Is it an approved additive? Is it a GRAS compound?
Adverse effects	Residues in product after decontamination Potential of toxic products forming
Effect on organoleptic quality of product	Taste, flavor, odor, appearance, etc.
Nutritional value	
Water retention	
Antimicrobial effectiveness	Effect on spoilage and pathogenic microorganisms
Pathogen load	Prospect of desired effect
Health aspects of workers	Hypersensitization
Environmental impact	
Microbial investigations	Susceptibility testing, determine efficacy
Application methods required	Cost effectiveness
Concentrations required	Within guidelines and consumer safety limits
Function	Food additive, or processing tool?

Source: Data from European Union, Report of the Scientific Veterinary Committee (Public Health Section) on the Decontamination of Poultry Carcasses. VI/7785/96 Final. 1996.

acid application are monitored once every hour, and the equipment used for monitoring is calibrated at least once per day. A combination of hot water and organic acid application consists of hot water pasteurization followed by organic acid decontamination, and has been developed for use in beef slaughter processes. A chilled water spray may be required after treatment to reduce carcass surface temperature prior to entering the chiller to prevent condensation in the chiller and also to rapidly cool the carcass (Salleh-Mack and Roberts, 2007).

It is necessary to use organic acids to disinfect the carcasses in an environmentally friendly way and still produce the maximum shelf life for fresh meat. These methods must, therefore, be simple, long-lasting, and with a minimal effect on the sensory, chemical, and physical characteristics of the meat, and be inexpensive (Ogden et al., 1995).

In Table 3.2 the application directives of organic acid sprays according to the Canadian Food Inspection Agency (2004) are summarized, and in Table 3.3 some of the factors in consideration of organic acid

Table 3.2 Application of Organic Acid Sprays

In decontamination of red meat carcasses the use of 1.5–2.5% of the following organic acids is permitted	Acetic acid Lactic acid Citric acid	Application/description
Conditions for application: May be applied where the application of water to a product is accepted practice	1. Final carcass rinse 2. Pre-evisceration rinse systems	1. Before, after, during 2. Consist of a potable water rinse (low pressure), and a second rinse with organic acid (mist, fog, or small droplets)
Treatment followed by appropriate measures:	To ensure the removal of organic acids to negligible levels	Final rinse with potable water
Documentation:	To describe the process and controls to monitor the concentration of the treatment solution	• Equipment used • Methods used to ensure that acid fumes are not hazardous • Type of acid • Acid concentration • Preparation of acid solution • Solution flow rate and pressure • Temperature of acid solution at contact point • Methods to remove residual acid • Actions taken if system is operating out of compliance

Source: Data from Canadian Food Inspection Agency. *Meat Hygiene Manual of Procedures.* 2004.

selection in meat decontamination are summarized (Smulders and Greer, 1998).

3.3.4   Processed meats

Not many studies have reported on the potential use of organic acids or the salts of organic acids as postprocessing antimicrobials, either individually or combined with other compounds prior to packaging. However,

Table 3.3 Factors That Determine the Selection of Organic Acids in
Meat Decontamination

Factors that influence antimicrobial effects	
1. Acid concentration (pH)	• pH dependent • Dissociation of the acid • Ability to penetrate the cell
2. Type of tissue	• Type of meat • Lean versus fat
3. Type of bacteria	• Sensitive to organic acid activity • Resistant to organic acids • Acid tolerance
4. Slaughter technology	• Extent of microbial contamination before slaughtering • Type of contaminating matter
5. Decontamination technique	• Application method • Time of acid exposure • Temperature • Pressure and angle of spray

Source: Data from Smulders, F.J. and Greer, G.G. *International Journal of Food Microbiology* 44:149–169, 1998.

dipping solutions containing concentrations of 2.5–5 g/100 ml lactic acid, acetic acid, sodium acetate, sodium diacetate, and potassium sorbate have been found to provide extensive protection against *L. monocytogenes* contamination in refrigerated processed meat products (Samelis et al., 2005).

3.3.5   Seafood

In frozen prawns or shrimp the concentration of L-ascorbic acid is used according to GMP, but is limited to such a level so as not to affect the product negatively, or to disturb its compliance with legal requirements (Van Heerden, 2003).

3.3.6   Poultry

In poultry carcass decontamination it seems a very attractive idea to inactivate potential pathogens in the end product, because preservatives are then applied just before retail distribution. Alternative sanitizing procedures to ionization radiation in poultry processing operations include: (1) spray application of lactic acid on carcasses, (2) surface heating, (3), in-plant chlorination, or (4) trisodium-phosphate dip, but efficiency of these methods is limited compared to ionizing radiation. Treating carcasses with organic acid spray has also been reported to have limited efficiency in controlling *Escherichia coli* O157:H7 (Farkas, 1998).

3.3.7   Dipping/spraying

Acetic, citric, lactic, and tartaric acids and their salts (potassium sorbate, potassium or sodium benzoate, sodium propionate, sodium acetate or diacetate, and sodium lactate) have been tested and approved as dipping or spraying treatments (Geornaras et al., 2005).

3.3.8   Acidified foods

Organic acids and their conjugate bases are widely applied in the prevention of spoilage in fermented and acid or acidified foods. However, U.S. FDA regulations do not allow the use of organic acids or preservatives as primary barriers to microbial pathogens in acidified foods. In the Code of Federal Regulations (21 CFR part 114) it is stated only that acid or acid ingredients be added to maintain the pH at or below 4.6, together with a heat treatment, if necessary. These regulations were designed to control the growth and toxin production by *Clostridium botulinum* and do not take into account the amount or type of organic acid already present in acidified food. Regulations (21 CFR part 114) were established in 1979 in the United States to ensure safe production of acidified foods. However, at the time, no pathogenic microorganisms were known to survive at or below pH 4.6 (Breidt, Jr., Hayes, and McFeeters, 2004).

3.4   Salts of organic acids

In contrast to numerous research reports on organic acids, comparatively few studies have reported on the antimicrobial activity of their salts. Moreover, available research results emerged mainly from laboratory investigations and infrequently from commercial trials. However, from these studies, the actions of lactic and sorbic acid salts have been shown to be primarily bacteriostatic, and it has been established that the inactivation of organic acids may be both bacteriostatic and bacteriocidic. To our knowledge, little research has been done on their application to red meat carcasses. It has, however, been indicated that lactates, for example, are not very active antimicrobials in raw meats stored in either air or vacuum. These salts may, therefore, play a more significant role when used as preservatives in cooked or cured products in combination with other preservation hurdles (Smulders and Greer, 1998). The closer the environment is to the pK_a of an acid, the greater is the lag phase of microbial growth. However, the salts of organic acids appear to inhibit microbial growth without changing the pH of a solution. This is advantageous in terms of reducing purge and cook losses and would ultimately reduce production costs (Mohd Adnan and Tan, 2007).

There have been reports of potassium sorbate showing promise in preventing fungal spoilage, but only at maximum concentrations at which it was tested (0.3%). Similar concentrations of other salts, such as calcium propionate and sodium benzoate have been found to be effective only at low a_w levels. Potassium sorbate activity was also slightly reduced at pH 5.5, as a concentration of 0.3% was only effective at 0.8 a_w. In other *in vitro* studies calcium propionate, sodium benzoate, and potassium sorbate have also been found to inhibit some isolates from bakery products at pH 4.5 when applied at a concentration of 0.3% (Guynot et al., 2005). Sodium benzoate has, however, been found to be more effective at inactivating *E. coli* O157:H7 than potassium sorbate (Comes and Beelman, 2002). Higher solubility of potassium sorbate renders it the preferred form of sorbic acid to use in a wide range of food products (Marín et al., 2003). Although little data exists as to their specific mode of inhibitory action, salts of organic acids have been widely approved for use as food ingredients. Organic acid salts are traditionally utilized to enhance the quality of cooked or cured meat. These salts have also been employed as emulsifiers, color and flavor enhancers, humectants, and in pH control (Smulders and Greer, 1998).

Organic acid salts are increasingly used by meat processors as flavor enhancers and also microbial inhibitors, specifically in controlling *Listeria monocytogenes*. Buffered sodium citrate alone or in combination with sodium diacetate or sodium salts of lactic or acetic acids are often used in processed meat products to provide additional protection against *C. perfringens* contamination during the chilling of meat products (Juneja and Thippareddi, 2004). Guidelines in the chilling of thermally processing meat are explained in Table 3.4. Lactates are often used in meat and meat products, but may play a more significant role as preservatives in cooked or cured products in combination with other preservative hurdles (Smulders and Greer, 1998). Sodium salts, specifically, of low molecular weight organic acids (such as acetic, lactic, or citric) are commonly used to control microbial growth, improve sensory attributes, and extend shelf life of, among others, various meat, poultry, and fish products. These salts are widely available, economical, and generally recognized as safe (GRAS). Sodium acetate is remarkably effective in extending the shelf life of various refrigerated fish. However, the antimicrobial effect of other sodium salts, such as sodium lactate on fish have shown some variations, which may be due to the various factors usually associated with the application of organic acids as preservatives (Sallam, 2007a).

3.4.1   Potassium sorbate

Sorbates are effective against many food spoilage organisms, and have many uses because of a milder taste, greater effectiveness, and broader pH range of inhibitory activity in comparison with either benzoate or

Table 3.4 Compliance Guidelines for Chilling of Thermally Processed Meat and Poultry Products

	Step 1		Step 2 (further chilling)	
Starting temperature (°C)	Stop temperature (°C)	Maximum time (h)	End temperature (°C)	Maximum time (h)
OPTION 1:				
54.4	26.7	1.5 h	4.4	5 h
OPTION 2:				
48	26.7	1 h	12.7	6 h
			4.4	cooling continued*
OPTION 3[#]:				
54.4	26.7	5 h	4.4	10 h

Sources: U.S. Department of Agriculture, Food Safety and Inspection Service. *Federal Register* 64:732–749, 1999 and U.S. Department of Agriculture, Food Safety and Inspection Service. *Federal Register* 66:12589–12636, 2001.

* Stipulated that product not be shipped until it reaches 4.4°C.

[#] For cooked, cured meat products with minimum of 100 ppm sodium nitrite (ingoing).

(These guidelines are independent of the addition of organic acids or salts.)

propionate (Barbosa-Cánovas et al., 2003). Potassium sorbate is most effective in preventing fungal spoilage at 0.3%, as opposed to the same concentration of other salts, such as calcium propionate and sodium benzoate that are effective only at low a_w levels. However, the activity of potassium sorbate is slightly reduced at pH 5.5 and 0.3% is only effective at 0.8 a_w. Potassium sorbate is suitable to inhibit deterioration of FBPA of slightly acidic pH (close to 4.5) by xerophilic fungi. *In vitro* studies have demonstrated the effectiveness of calcium propionate, sodium benzoate, and potassium sorbate to inhibit isolates from bakery products at pH 4.5, applied at a concentration of 0.3%. Potassium sorbate has been found to be effective, even at concentrations as low as 0.03% (Guynot et al., 2005).

Notably high concentrations of acetic acid (80–150 mM) are needed to totally inhibit the growth of *Saccharomyces cerevisiae* at pH 4.5, whereas only 1–3 mM sorbate may be needed, even when they both have the same pK_a (degree of dissociation) (Piper et al., 2001). Although the activity of sorbic acid as well as its salt against *L. monocytogenes* has been studied in laboratory media and in some foods such as cheese, meat products, or fish (Dorsa, Marshall, and Semien, 1993; Samelis et al., 2001), few studies have reported on their effect on *L. monocytogenes* in poultry (Gonzalez-Fandos and Dominguez, 2007). Poultry legs treated with 5% potassium sorbate can preserve a reasonable sensorial quality after storage at 4°C for seven days, whereas 10% potassium sorbate solutions were required to reduce mesophiles as well as psychrotroph counts in pork meat. No significant differences were found in pH values after the different treatments, and panel members could not distinguish between those legs dipped in 5% sorbate and those dipped in distilled water. Poultry leg muscle has a pH of 6.4–6.7, whereas other parts such as breast muscle have lower pH values (5.7–5.9) and, therefore, spoil more quickly than breast meat. The skin of a chicken carcass has a high pH. Shelf life of fresh broilers at 3°C can be extended by dipping a freshly chilled carcass in 5% (w/v) solution of potassium sorbate for 1 min (Gonzalez-Fandos and Dominguez, 2007).

L. monocytogenes is predominantly isolated from chicken legs and chicken wings, the parts that are still covered with skin, as this pathogen is mainly located on the skin surface and the higher pH of leg meat may provide favorable conditions for multiplication of *L. monocytogenes* (Barnes, 1976). Sorbate has a bacteriostatic effect on *L. monocytogenes* (Buncic et al., 1995) and inconsistent reports on activity against *L. monocytogenes* may be attributed to variations in media or food, pH, or concentration of the added sorbate (Gonzalez-Fandos and Dominguez, 2007).

Potassium sorbate has greater solubility in water than sorbic acid and is used accordingly in dipping or spraying fruit and vegetable products. The antimycotic action of potassium sorbate is similar to sorbic acid, but approximately 25% more potassium sorbate is usually needed to secure the same protection. Stock solutions of potassium sorbate can be made in water and concentrated up to 50%, to be mixed later with liquid food products, or diluted in dips and sprays. In foods with very low pH, low sorbate levels have excellent protective activity. Sorbates are applied in beverages, syrups, fruit juices, wines, jellies, jams, salads, pickles, etc. (Barbosa-Cánovas et al., 2003). Salts of propionic and sorbic acids are often added to intermediate moisture bakery products (Arroyo, Aldred, and Magan, 2005).

3.4.2   Sodium benzoate

Sodium benzoate as a food preservative is generally limited to products that are naturally acidic and is more effective in food systems where the pH ≤ 4. It is mainly used as an antimycotic agent and most yeasts and molds are inhibited by a concentration of 0.05–0.1% (Barbosa-Cánovas et al., 2003). Benzoates and parabenzoates are used primarily in fruit juices, chocolate syrup, candied fruit peel, pie fillings, pickled vegetables, relishes, horseradish, and cheese. Sodium benzoate produces benzoic acid once it is dissolved in water (Marsili, Sobrero, and Goicoechea, 2003).

3.4.3   Sodium lactate

Sodium lactate is used and approved as: (1) a flavor enhancer, (2) humectant, and (3) pH control agent (Duxbury, 1988; Lemay et al., 2002). The salts lactate and diacetate on their own are known to exhibit antilisterial activity in RTE (ready-to-eat) meat, but enhanced inhibition has also been reported when these salts are used in combination (Mbandi and Shelef, 2002).

Sodium lactate (3–4%) in cooked beef is effective in limiting proliferation of *Salmonella typhimurium*, *L. monocytogenes*, and *E. coli* O157:H7 (Miller and Acuff, 1994). Sodium lactate (3%) added to vacuum-packaged beef top rounds is effective as an oxidant and color stabilizer for cooked beef (Maca et al., 1999; Choi and Chin, 2003).

3.4.4   Other

Buffered sodium citrate and buffered sodium citrate supplemented with sodium diacetate can be effectively used to inhibit *C. perfringens* outgrowth during chilling of roast beef (Juneja and Thippareddi, 2004). There are also various reports on the antimicrobial effect of sodium salts of organic acids on *C. botulinum* outgrowth and toxin production (Maas, Glass, and Doyle, 1989; Meng, 1992; Miller, Call, and Whiting, 1993; Juneja and Thippareddi, 2004). There is generally no correlation found between the effect of sodium citrate on bacterial growth and the pH of the foodstuff (Sabah, Juneja, and Fung, 2004; Macpherson et al., 2005). Citrate, lactate, and ascorbate are also commonly used in sports drinks (Wu et al., 1995).

In the application of fumarate to ruminant feed, the sodium salt is generally used because fumaric acid decreases ruminal pH. Sodium fumarate is effectively used as a buffer, due to the properties of fumaric acid, and also the sodium part that serves to raise the pH similar to the role of sodium in sodium bicarbonate (Castillo et al., 2004).

General statements regarding the antimicrobial effect of combinations of spices and organic acid salts cannot readily be made and each organic acid salt and spice combination should be evaluated separately to determine the appropriate level of organic acid salt. Such combinations are useful and can help food processors meet FSIS requirements, while providing acceptable-tasting food products (Sabah, Juneja, and Fung, 2004).

3.5   Organic acid combinations

In extending the shelf life of minimally processed foods, the hurdle concept uses a combination of suboptimal growth factors, such as (1) reduced temperature, (2) reduced pH, (3) reduced a_w, (4) addition of organic acids, and (5) modified atmosphere packaging (Uyttendaele, Taverniers, and Debevere, 2001). Treatment protocols that are based only on organic acid application have been found ineffective in reducing *E. coli* O157:H7 numbers (Bolton et al., 2002). It is believed that in carcass decontamination it is better to implement more hurdles. Even a potent and commonly used organic acid, such as lactic acid alone, was found to have no antimicrobial effect against *Staphylococcus aureus* (Taniguchi et al., 1998).

Benzoic acid, mainly used in many types of acidic food products, is also used in combination with sorbic acid in other types of products, such as confectionery (Suhr and Nielsen, 2004). Sorbic and benzoic acids are often used in combination, due to acclaimed antimicrobial synergy between them (Cole and Keenan, 1986; Lambert and Bidlas, 2007).

3.5.1 Combinations in general

Increased effectiveness of organic acids may be achieved when used in lower concentrations, but in combination with additional inhibitors. For example, combining nisin with reduced organic acid concentrations may decrease potential flavor flaws in meat products (Samelis et al., 2005). Citric and ascorbic acids, in combination, are effective in inhibiting growth and toxin production of *C. botulinum* type B in vacuum-packed foodstuffs (Barbosa-Cánovas et al., 2003; Samelis et al., 2005). In other studies, a combination of equal amounts of ascorbic acid and citric acid did not have additional benefits, compared with using ascorbic acid alone (Mancini et al., 2007).

Mixtures of ethanoic and lactic acids exhibit synergistic antimicrobic action, which is mainly attributed to lactic acid (the stronger acid). However, the effect of ethanoic acid is potentiated by increasing the proportion present in the undissociated form (Adams and Nicolaides, 1997). Combination treatments are also known as hurdle technology. These treatments rely on the assumption that the susceptibility of organisms to stresses to which they may be exposed can be accumulative (Fielding, Cook, and Grandison, 1997). Organic acids and their salts greatly potentiate the antimicrobial activity of bacteriocins, whereas acidification enhances antibacterial activity of both organic acids and bacteriocins (Galvez et al., 2007).

Combinations of two antimicrobials in a formulation provide greater inhibition of *L. monocytogenes* growth, compared with antimicrobials used individually. When two antimicrobials in such a formulation were then also combined with a dipping in organic acid solutions, listericidal effects were actually observed after processing. However, combinations of antimicrobials did not decrease water activity, and sometimes even increased it. Combinations of sodium lactate (1.8%) and sodium diacetate (0.25%) have been found to completely prevent growth throughout storage, whereas lactate (1.8%) and diacetate (0.125%) have been shown to cause complete inhibition of *L. monocytogenes* during the first eight days of storage. Subsequent growth is then also known to be significantly reduced (Barmpalia et al., 2004).

A combination of fumaric acid (0.15%) and sodium benzoate (0.05%) was more effective against *E. coli* O157:H7 than a combination of fumaric acid (0.15%) and potassium sorbate (0.05%). This combination has been found to possess even more potent antimicrobial activity than sodium benzoate (0.05%) plus potassium benzoate, which is a common commercial practice (Comes and Beelman, 2002).

Chill-wash combined with citric acid and followed by low-temperature incubation with or without the presence of nisin, is also useful in the poultry industry for raw poultry products (Phillips and Duggan, 2002). Hot water washings, followed by an organic acid rinse have been widely evaluated on beef carcasses. The inclusion of organic acid rinses further reduces the bacterial populations. The greatest reductions in *E. coli* populations are found when combining hot water and acid treatment (Eggenberger-Solorzano et al., 2002).

3.5.2   Salt combinations

Although much information is available on antimicrobial activities of lactate and acetate salts, there is limited information on the effect achieved when using them in combination. In a study done in 2002, enhanced inhibition was found at various storage temperatures when used in combination (Mbandi and Shelef, 2002).

Buffered sodium citrate in combination with sodium diacetate or sodium salts of lactic or acetic acids can be used as ingredients in processed meat products to provide an additional measure of safety against *C. perfringens* contamination of chilled meat products (Juneja and Thippareddi, 2004). Addition of lactate does not affect meat pH, whereas diacetate reduces pH from 6.3 to 5.9, and a combination of the two salts reduces pH to 6.1 (Mbandi and Shelef, 2002).

3.5.2.1   Possible adverse effects

3.5.2.1.1 *Sodium chloride (NaCl)*   The "hurdle" effect is commonly applied to fermented and minimally processed foods in reducing the risk of foodborne illness. It entails a combination of two or more inhibitory agents being more inhibitory than alone. However, this is not always the effect obtained against foodborne pathogens. Adding salt (NaCl) and organic acid to foods is common practice. However, NaCl reduces the inhibitory effect of lactic acid on *E. coli* O157:H45 by raising its cytoplasmic pH. *E. coli* can use NaCl to cancel out the acidification of its cytoplasm by organic acids. This protective mechanism is also found in other *E. coli* as well as with other organic acids. A combination of NaCl and acid pH is, therefore, less effective than acid pH alone in reducing the numbers of *E. coli* O157:H45. Sodium chloride also reduces the bactericidal effect of lactic acid on *E. coli* O157:H45 cells in the exponential growth phase. However, the addition of NaCl some time after acid challenge will not protect the cells, and, therefore, NaCl cannot reverse acid damage (Casey and Condon, 2002).

   3.5.2.1.2 Other   Combination of sodium lactate with chili, or garlic and herbs, has also been found to exert possible antagonist actions (Sabah, Juneja, and Fung, 2004).

3.5.3   Aromatic compounds

When using individual aromatic compounds as preservatives, the high concentration needed to inhibit spoilage bacteria often exceeds the flavor threshold acceptable to consumers. Again, combining different preservative techniques is needed as an alternative to maintain high flavor quality. Synergistic inhibition is indicated when a combination produces an additional one $\log_{10}$ reduction compared to the sum of the lethal effects of each preservative on its own (Nazer et al., 2005). Although not much information is available on acidic and aromatic compound combinations, sodium citrate is known to increase the inhibitory action of eugenol and monolaurin (Blaszyk and Holley, 1998). Combinations of plant oils and derivatives of benzoic acid have also been found to exert satisfactory activity on *Salmonella* sv. Enteritidis (Nazer et al., 2005). However, combinations containing methylparaben demonstrated higher levels of inhibition (Fyfe, Armstrong, and Stewart, 1998).

3.5.4   Ethanol

Ethanol or lactate in combination have been found to be more bactericidal than on their own against *L. monocytogenes*. In fact, for a variety of organic acids, a combination with 5% ethanol resulted in a dramatic decline in viability, and this was always greater than observed with either agent alone. Although benzoate and formate used alone are effective at killing *L. monocytogenes* at pH 3, the addition of ethanol will result in shorter killing times. Ethanol has GRAS status and can enhance the inactivation rate of *L. monocytogenes* during exposure to low pH, organic acids, and also osmotic stress (Barker and Park, 2001).

3.5.5   Irradiation

Organic acid and ionizing radiation can both disrupt the membranes of bacterial cells. Acetic acid in combination with irradiation results in increased membrane damage. In addition, the effect of irradiation on *E. coli* (at all doses) is enhanced by the presence of acetic acid (0.02–1%). At least three possibilities have been proposed for the response of *E. coli* to the presence of acetic acid during irradiation: (1) acid may sensitize the organism to irradiation by disruption of metabolic functions such as energy production or enzyme activity, (2) the increased demand for energy that is placed on the cell by homeostasis may inhibit DNA repair, and (3) another theory is that the effect of irradiation on the acetic acid molecule produces a lethal free radical which subsequently acts on the cell. The radiation process may also sensitize a cell to the presence of an organic acid by injuring the cell membrane to allow uncontrolled influx of acid molecules. However, undissociated acetic acid has the ability anyway to freely permeate the cell membrane and the synergism is then an indirect effect, as neither irradiation nor acetic acid is producing the effect (Fielding, Cook, and Grandison, 1997).

A combination of ascorbic acid and gamma irradiation has also been applied to achieve antibacterial effects (Ouattara et al., 2002).

3.5.6   Emulsifiers

Many pharmaceutical emulsifiers are GRAS in food applications. For example, a combination of lactic acid and propylene glycol used on broiler carcasses has been found highly effective in eliminating any salmonellae present. The surfactant 12-butyryloxy-9-octadecenoic acid (BOA) is composed of citric acid, EDTA, and sodium lauryl sulfate, and 0.6% of this medium has been found to reduce levels of *S. typhimurium* cells by >5 $\log_{10}$. Sodium lauryl sulfate resulted in the greatest increase in activity. The addition of ethanol, DMSO, or Span 20 to acids had little or no effect on their antimicrobial activity. In some cases it actually resulted in decreased activity of the acid (Tamblyn and Conner, 1997).

3.5.7   Spices

It is not realistically possible to make general statements regarding the antimicrobial activity of organic acid salts and spice combinations. Such combinations should each be evaluated separately to determine the appropriate level of organic acid salt to effectively control the growth of *C. perfringens* (Sabah, Juneja, and Fung, 2004).

3.5.8   Liquid smoke

Liquid smoke, which contains approximately 10% acetic acid, has been used as an antimicrobial agent against the pathogens *E. coli* O157:H7, *Salmonella, Listeria,* and *Streptococcus.* It has been confirmed that combined use of liquid smoke and steam had a synergistic thermal lethality effect on

L. monocytogenes and also inhibited a potential effect on processed meat (Murphy et al., 2006).

3.6   Considerations in the selection of organic acids

3.6.1   Sensory properties

Organic acids are generally used to acidify various food products, but have different abilities to lower the pH that depend on (1) the number of acid functions, (2) the K_a of the acid, and (3) the pH of the medium. In addition, the effect of the type of acid on the taste of the product must always be considered (Derossi et al., 2004). Although organic acids are popular because of the lack of toxicological implications when applied at the prescribed concentrations, there is constant consumer pressure toward decreasing their concentrations or using as little preservative as possible or even none at all (Marín et al., 2003). Organic acids have been reported to produce adverse sensory changes, but diluted solutions of organic acids (1–3%) are generally without effect on the desirable sensory properties of meat (Smulders and Greer, 1998; Min et al., 2007).

3.6.2   Color stability

The organic acids, ascorbic acid and citric acid, are often applied to improve color stability of fresh pork (CFR 2005) by dipping the meat into a solution containing one of these agents. This dipping also helps to maintain the lightness of the meat (Huang, Ho, and McMillin, 2005). Ascorbic acid acts as both an antioxidant and a pro-oxidant and the appropriate levels for preventing muscle discoloration depend on a number of factors (Mancini et al., 2007). Inconsistency is the result of a presence of metals within a food. The pro-oxidant nature of ascorbic acid could be due to the production of ferrous heme proteins, which may be more reactive and more oxidative than ferric derivatives (Yamamoto, Takahashi, and Niki, 1987; Mancini et al., 2007).

3.6.3   Flavor

Sodium salts of low molecular weight organic acids are used to improve sensory attributes and extend shelf life of various food systems in addition to their preservative characteristics (Sallam, 2007a). However, the use of acetic, propionic, and lactic acids might be limited due to their flavor and taste (Taniguchi et al., 1998). Benzoic addition is promising in flavoring of reduced sucrose content products but with consideration of possible unpleasant side sensations and also potential toxicological aspects (Otero-Losada, 2003).

Preservative-treated ciders have been found to have an overall consumer rating between "like slightly" and "like moderately" (early season cider) and "neither like nor dislike" and "like slightly" (late season cider) (Comes and Beelman, 2002).

3.6.4   Carcass decontamination

In carcass decontamination dilute solutions of organic acids (1–3%) normally do not have any effect on the sensory properties of meat. However, it is known that lactic and acetic acid can produce unfavorable sensory changes if applied directly to meat cuts, which may be irreversible. Salts of organic acids are approved for use as food ingredients such as emulsifiers, color and flavor enhancers, and humectants. They are also used to enhance the quality of cooked or cured meat products and to control the pH (Smulders and Greer, 1998). Sodium lactate is approved for use as (1) a flavor enhancer, (2) humectant, and (3) pH control agent (Lemay et al., 2002).

3.6.5   Chemical stability

Dipping of fresh salmon slices in aqueous solutions of the sodium salts of organic acids (approximately 2.5%) have been shown to maintain chemical quality, and also extend the shelf life of the product with only minor sensory changes during refrigerated storage (Sallam, 2007b).

3.7   Organic acids in antimicrobial packaging

The immobilization of antimicrobials into structuring solutions has become an advantageous technology in food preservation (Ouattara et al., 2002). Antimicrobial films or coatings have been found to be more effective than addition of antimicrobial agents directly to food as these may gradually migrate from the package onto the surface of the food, providing concentrated protection when most needed (Durango, Soares, and Andrade, 2006).

3.7.1   Antimicrobial films

A major origin of food spoilage is microbial contamination of food surfaces (Weng, Chen, and Chen, 1999). The incorporation of organic acids or other preservatives into a film is, therefore, an economically effective method used in manufacturing antimicrobial packaging. For example, sorbic and benzoic acid are incorporated into methylcellulose, hydroxylpropylcellulose, or chitosan films (Bégin and Van Calsteren, 1999). The interaction between the organic acids and the film-forming material may, however, affect casting of the film and also the discharge acids (Chen, Yeh, and Chiang, 1996). Film thickness and film density have been found to correlate well with the van der Waals molecular volume of the anion. When the solutions are dried, the polymer concentration increases, and formation is achieved, followed by film formation. Meanwhile, ionic strength is increased, which results in a higher association between the polyelectrolyte and the counter ions. This also causes an increase in the shielding effect of the counter ions. In addition to this, the structure of the counter ions can influence intramolecular and intermolecular interactions (Bégin and Van Calsteren, 1999).

Lactic acid is a monocarboxylic acid and cannot act as a reticulating agent. When considering the elongation of films in regard to anion volume, the acids belong to two classes: those of small volume, for example, hydrochloric acid (HCl), formic and acetic acids, and those with larger volume (lactic and citric acids). During drying, citric acid and lactic acids will promote gel formation at lower polysaccharide concentrations, resulting in thicker films that are actually weaker, whereas films containing HCl and formic acid are stronger, brittle, and of similar strength. Citric acid is capable of forming multiple linkages, although this does not improve strength. When using antimicrobial agents with a molecular volume greater than that of acetic acid, soft films are produced which are only suitable for use in multilayered films or as coating. Films containing chloride or formate may be used as support for films containing antimicrobial agents or as biodegradable packaging. An example of antimicrobial film is PEMA (polyethylene-co-methacrylic acid), based on incorporation of sorbic acid and benzoic acid (Bégin and Van Calsteren, 1999).

3.7.2 Active packaging

"Active packaging" is defined as packaging that extends shelf life and improves safety or flavor characteristics while maintaining the quality of the food. Antimicrobial packaging is a popular form of active food packaging, especially in meat products. Active packaging is a challenging technology, specifically designed to have an impact on shelf life extension of various meat products. It is envisaged that an antimicrobial be released from the package during an extended period, in order for activity also to be extended, even entering the transport and storage phases of food distribution. Antimicrobial compounds that have been evaluated in films (polymers and edible) include organic acids, organic acid salts, enzymes, bacteriocins, and other compounds such as triclosan, silver zeolites, and fungicides (Quintavalla and Vicini, 2002).

Examples of active package/food systems are individually wrapped ready-to-eat meat products, or deli products, where antimicrobial agents may be initially incorporated into the packaging materials and migrate into the food through diffusion and partitioning. There are several approaches to incorporating antimicrobial compounds into the film (Quintavalla and Vicini 2002):

1. The antimicrobial is added to the film by adding it in the extruder when the film is produced. Disadvantages of this are that it is not cost-effective and the surface of the film is not actually involved in the activity.
2. The antimicrobial additive may be applied in a controlled way where material is not lost, such as incorporation into the food-contact layer of a multilayer packaging material.

In the design of antimicrobial films or packages various factors may be of importance. These include the chemical nature of films and lasting antimicrobial activity, storage temperature, mass transfer coefficients, and physical properties of packaging materials (Quintavalla and Vicini, 2002).

3.7.3   Edible films

Films and coatings are useful in maintaining desirable characteristics of food, such as color, flavor, spiciness, acidity, sweetness, and saltiness, and only very small amounts of additives are needed. In addition, edible coatings are conducive to the use of antimicrobials as well as antioxidants, whereas these compounds are concentrated at the product surface, where protection is needed. Edible films made from pectinate, pectate, or zein, and containing citric acid, have been shown to be useful in preventing rancidity and maintaining the attractive texture of, for example, nuts (Guilbert, Gontard, and Gorris, 1996).

Edible antimicrobial coatings and films are also effectively used in controlling food contamination (Debeaufort, Quezada-Gallo, and Volley, 1998). The most commonly used organic acids in edible coatings are sorbic, propionic, potassium sorbate, benzoic acid, sodium benzoate, and citric acid (Quintavalla and Vicini, 2002; Durango, Soares, and Andrade, 2006). Others include bacteriocins (nisin and pediocin) (Sebti and Coma, 2002), enzymes (peroxidase and lysozyme) (Padgett, Han, and Dawson, 1998), and polysaccharides displaying natural antimicrobial properties (chitosan) (Debeaufort, Quezada-Gallo, and Volley, 1998; Durango, Soares, and Andrade, 2006).

The organic acids, sorbic, benzoic, and propionic acids, and anhydrates have been proposed and tested for antimicrobial activity in edible films. Organic acids are also being applied as an edible film, together with essential oils, into a chitosan matrix (Ouattara et al., 2000). From here they are initially released quickly (with the high gradient of ion concentration between the inside of a polymer matrix and the environment) but more slowly as the release of acids progresses. Most effective inhibition is found on surfaces with lower water activity, onto which the acid is more slowly released (Quintavalla and Vicini, 2002). Incorporation of ascorbic acid into ground beef, and spice powder incorporated into edible coating film, have been found to stabilize lipid oxidation and production of –SH radicals during postirradiation storage (Ouattara et al., 2002).

Edible films can have very low permeability for O_2, creating anaerobic conditions on the food surface, which in turn, may lead to a risk of contamination by anaerobic pathogens such as *C. botulinum*. The inclusion of an antimicrobial such as sorbic acid is mostly advised to control this pathogen. It is also important to know the ability of films to modify gas transport in order to tailor films to specific applications. For example, coatings or films applied to fresh fruits or vegetables that are characterized by an active metabolism, even during refrigerated storage, should provide correct modification of the gaseous environment inside the package. This may entail allowing O_2 to penetrate into the package and unnecessary CO_2 to escape from it (Guilbert, Gontard, and Gorris, 1996).

3.7.4   Modified atmosphere packaging (MAP)

The meat industry recently moved toward meat cuts ready to be packaged in a modified atmosphere. Bone discolorization has, as a result, increased, particularly those cuts packaged in high-oxygen MAP. Hemoglobin is responsible for bone-marrow color in bone-in beef and it has been found that untreated vertebrae will significantly discolor within 6 h of packaging. However, this rapid discoloration was inhibited by organic acids such as ascorbic acid. With ascorbic acid, the concentration was not critical. On the contrary, an increase of citric acid concentration from 1–3% improved the color early in display, whereas a 10% citric acid concentration had a significant discoloration effect. Ascorbic acid can, however, reverse the oxidizing effects of 10% citric acid, but treating vertebrae with combinations of ascorbic acid and 10% citric acid is not advised (Mancini et al., 2007).

3.8   Organic acids in animal feed preservation

3.8.1   The essence of preserving feed

Large quantities of feed are produced daily, transported, and stored, and even a minor contamination with pathogens, for example, *Salmonella*, has the potential to affect many herds (Sauli et al., 2005). To enhance the nutritional value of feed, large-scale producers treat their feed with organic acids or heat, or both. Dietary inclusion of organic acids has been found to positively affect the growth rate and efficiency of feed utilization (Giesting and Easter, 1991; Mroz et al., 2000), whereas heat treatment of the feed increases digestibility of nitrogen and amino acids (Sauli et al., 2005). Mainly only the smaller producers do not treat their feed and although having a controlled market, they too can have a large impact on the spread of infection. Organic acids are useful because of their ability to protect against outgrowth after recontamination with pathogens such as *Salmonella*, which is not achieved after heat treatment (Sauli et al., 2005).

3.8.2   The postantibiotic era

In a "farm to fork" food safety concept, safe feed is the first step for ensuring safe food (Sauli et al., 2005). Organic acids were originally added to animal feeds to inhibit fungal growth, but in the past 30–40 years, various organic acids and combinations have been examined for their effects on bacterial contamination in feeds and feed ingredients, more specifically in targeting foodborne pathogens such as *Salmonella* spp. (Ricke, 2003). Organic acids are increasingly accepted as an alternative to antibiotics and are introduced as feed additives for ruminants, although application is less extensive in ruminants than in other farm animals. Organic acids are, however, known to buffer ruminal pH. Organic acids proposed and approved for use include aspartic, citric, succinic, and pyruvic acids. Malic and fumaric acids have not yet been intensively studied. Organic acids fall in the preservative group on the list of feed additives authorized by E.U. legislation. They are considered safe substances and produce no detectable abnormal residues in meat and their use is allowed in all species of livestock. The high cost does, however, pose a major problem, especially for malic acid. Organic acids also reduce methanogenesis by reducing energy losses associated with CH_4 production in rumen (Castillo et al., 2004).

Not much literature is available on the ideal dosage, management, or application of organic acids in feed or on the effects of organic acids on beef cattle performance. The optimal amount of organic acid needed will depend on the final form of the feed, and whether forage is included. However, the proper dosage is 1–5% by body fat. Less than 0.1% will not be sufficient, and higher than 10% will affect the palatability of the diet. When the diet is mixed and contains roughage, the concentration is preferably about 2–10% by body weight (Castillo et al., 2004).

Organic acids have been investigated in the control of postweaning edema disease (PWOD), which is caused by enterotoxigenic *E. coli* (ETEC) and occurs after weaning. Acidic conditions in the GIT normally have bactericidal effects on potentially harmful bacteria and produce a more favorable environment for lactobacilli, which may inhibit the colonization and proliferation of *E. coli* and also secrete metabolites that act against Gram-negative bacteria. Organic acids incorporated into the feed or drinking water have been found to prevent PWOD. PWOD is an important disease in the modern pig industry and control has become imperative. Supplementation of a diet with pure lactic acid (1.6%) or 1.5% pure citric acid (1.5%), started on weaning day, has been found to give satisfactory results (Tsiloyiannis et al., 2001b).

The effect of dietary additions of organic acids on the control of postweaning "colibacillosis" (diarrhea) in piglets also include a reduction in the incidence as well as severity of diarrhea in weaned pigs. This could be as a result of a decrease of the pH in the gut after ingestion of acids. However, the reduction of intestinal pH is not a primary effect of feeding organic acids, and they have little effect on the intestinal microflora. Unfortunately, organic acids do have odors that are not always pleasant and as such reduce the palatability of the diet (Tsiloyiannis et al., 2001a).

3.8.3   Chicken feed

Feed with formic and propionic acids has been found to reduce *Salmonella* colonization in broilers (Thomson and Hinton, 1997), whereas decontamination of chicken carcasses with acetic or lactic acid reduced *Campylobacter* on carcasses or meat (Van Netten et al., 1994; Chaveerach et al., 2002). Organic acids, when added to feeds, should be protected to avoid dissociation in the crop and in the intestine that are known to possess higher pH and to reach far into the GIT, where the targeted bacterial population is situated (Gauthier, 2005).

The addition of organic acids to the feed or drinking water has also been reported to decrease contamination of chicken carcasses as well as egg contamination. Poultry feed is a major source for *Salmonella* introduction to the farm, but feed, treated with organic acids, when eaten by chickens, is both warmed and moistened and the organic acid activity increases. These added acids are capable of exerting an antimicrobial effect in the crop and gizzard rather than the intestine. Moreover, drinking water is also a major source of infection and it is essential to keep drinking water *Salmonella* free. Organic acids are also used as sanitizers in drinking water (Van Immerseel, Russell, and Flythe, 2006).

3.8.4   In combination with heat treatment

A heat treatment, combined with organic acid treatment, is a method described as a solution for the control of *Salmonella* in pig feed. Pathways of production may follow different procedures (Sauli et al., 2005). One such production system in Switzerland has been described and consists of eight pathways followed in feed manufacturing (Table 3.5).

Table 3.5 Example of a Pathway Implemented in the Production of Feed

Pathway	Steps involved	Heat treatment	Organic acid
Directly stored	Mixed ingredients directly stored as feed	None	None
Acid treatment	Organic acids added to mixed ingredients	None	1.2% propionic acid
Heat treatment (1)	Mixed ingredients submitted to short heat treatment	71.1°C, 5–10 s	None
Heat + Acid (1)	Organic acid added to mixed ingredients and submitted to short heat treatment	71.1°C, 5–10 s	0.2% propionic acid
Heat treatment (2)	Mixed ingredients submitted to intermediate heat treatment	71.1°C, 25–35 s	None
Heat + Acid (2)	Organic acid added to mixed ingredients and submitted to intermediate heat treatment	71.1°C, 25–35 s	0.2% propionic acid
Heat treatment (3)	Mixed ingredients submitted to long heat treatment	71.1°C, 115–125 s	None
Heat + Acid (3)	Organic acid added to mixed ingredients and submitted to long heat treatment	71.1°C, 115–125 s	0.2% propionic acid

Source: Data from Sauli, Danuser, Geeraerd, et al. *International Journal of Food Microbiology* 100:289–310, 2005.

3.8.5 Propionic acid in feed

Propionic acid displays inconsistent antimicrobial activities against fungal populations in feed. This is attributed to possible buffering by protein ingredients such as soybean meal, and conversion to its less active form or batch differences in primary feed components, such as cornmeal. Feeds treated with organic acids, once ingested by the animal, also limit pathogen infestation. Propionic and formic acids have been widely reported to reduce *Salmonella*, coliforms, and *E. coli* in the small intestinal and fecal substances of chickens (Kerth and Braden, 2007).

3.8.6   Organic acids in animal nutrition

Studies done on the organic acids and their role in animal nutrition have mainly been done on pigs. Citric acid seems to have no effect on various parameters measured in the GI tract of pigs. Benzoic acid has not yet been approved as an additive or preservative for pig or chicken feed. Although benzoic acid has been found superior in bactericidal effect on coliform and lactic acid bacteria in the stomach and intestinal tract, the killing effect of organic acids in the GI tract of pigs has been investigated and the effects found to be, from low to high: propionic < formic < butyric < lactic < fumaric < benzoic (Øverland et al., 2007).

3.9   Concentrations

Concentrations of organic acids used as food preservatives range between approximately 100–300 mM (Kwon and Ricke, 1998). It is, however, often necessary to use organic acids at high levels (mM rather than μM levels) to prevent spoilage of low pH foods and beverages by yeasts. For example, a relatively high concentration of acetic acid (80–150 mM) is needed to totally inhibit the growth of *S. cerevisiae* at pH 4.5, whereas only 1–3 mM sorbate is needed, despite their having the same pK_a (Piper et al., 2001). Also the inhibition of *Z. bailii* by organic acids often requires concentrations nearer or greater than the legal limits (Hazan et al., 2004).

3.9.1   Pressure toward decreased concentrations

Organic acids are popular preservatives and have no toxicological implications when applied at prescribed concentrations. However, consumers are constantly pressuring for decreased concentrations. Concentrations of propionates, sorbates, and benzoates added to bakery products commonly do not exceed 0.3%. The use of sorbic acid in baked goods does not create any problems as it has no residual taste if baking powder is used to raise the dough and the acid is added at 0.1–0.2%. It has been found that 0.025%, 0.05%, and 0.1% concentrations almost without exception led to an increase in bacterial colonies, indicating counterproductivity and a negative effect on some foods (Marín et al., 2003). Reduction of preservatives to subinhibitory concentrations has been found to stimulate growth of fungi (Magan and Lacey, 1986; Marín et al., 1999) or mycotoxin production (Yousef and Marth, 1981; Bullerman, 1985; Suhr and Nielsen, 2004). In answer to this, propionic and sorbic acid may be added to bakery wares in concentrations of up to 3000 and 2000 ppm, respectively, and benzoic acid in concentrations of up to 1500 ppm is allowed (European Union, 1995; Suhr and Nielsen, 2004).

Potassium sorbate has been found to be effective in preventing fungal spoilage at a concentration of 0.3%, whereas the same concentration of calcium propionate and sodium benzoate is effective only at low a_w levels. However, potassium sorbate activity is slightly reduced at pH 5.5, and 0.3% is then only effective at a water activity of 0.8 a_w. In some *in vitro* studies calcium propionate, sodium benzoate, and potassium sorbate were effective in inhibiting isolates from bakery products at pH 4.5 when applied at a concentration of 0.3%. At a pH of 4.5 potassium sorbate is also effective at a concentration of 0.03% (Guynot et al., 2005). Long-term antilisterial effects after dipping treatments in organic acids are often dependent on the type and concentration of the acid or salt. A decrease in organic acid concentration (to 1 g/100 ml), causes decreased antilisterial effects after 20–35 days of storage, irrespective of the presence of nisin (Samelis et al., 2005).

3.9.2   Concentrations effective against common pathogens

In raw beef 2% fumaric acid was found to be more effective against *L. monocytogenes* than 1% acetic or 1% lactic acid. The most effective treatment for artificially contaminated raw chicken legs was demonstrated to be a wash with 10% lactic acid/sodium lactate buffer (pH 3.0). In this instance it was followed by packaging in 90% carbon dioxide and 10% oxygen. Dipping frankfurters in 5% acetic or 5% lactic acid was demonstrated to kill *L. monocytogenes* and even to prevent regrowth during 90 days storage. The addition of 1.8 or 2% lactic acid to raw or cooked ground beef does not have any appreciable effect on the survival or growth of *L. monocytogenes*. Sodium lactate (4%) suppresses growth of *L. monocytogenes* in cooked beef and beef roasts (Doyle, 1999). Vacuum-packed treatment of, for example, frankfurters, is being prescribed by the use of 2% acetic, 1% lactic, 0.1% propionic, and 0.1% benzoic acids, combined with steam surface pasteurization. This has been proven as an effective inhibition of *L. monocytogenes*. This combination of organic acid and steam pasteurization further inhibited the growth of surviving *L. monocytogenes* cells for 14 to 19 weeks after storage at 7°C and 4°C (Murphy et al., 2006).

3.9.3   Daily consumption of organic acids

Concentrations of 1–10 mM sorbic acid is routinely used in soft drinks and this acid is, therefore, consumed daily in potentially physiologically significant amounts (Piper, 1999). Acetic acid is an active ingredient of household vinegar. Normal white household vinegar consists of a concentration of approximately 5% acetic acid. When this is diluted to at least 2% it is actually recommended as a preservative (Kerth and Braden, 2007).

3.9.4   Legislation

The maximum permitted concentration of benzoic and sorbic acids in different foods are regulated by legislation (Wen, Wang, and Feng, 2007). Sodium lactate has been approved for meat products as an antimicrobial when applied at no more than 4.8% of the product weight. Sodium citrate is a GRAS food additive and is applied according to current GMP (Sabah, Juneja, and Fung, 2004).

3.10   A review of current methodologies

In the United States most processors of ready-to-eat meat and poultry products include levels of up to 2% sodium or potassium lactate in their products (Geornaras et al., 2005). Often these are combined with 0.05 to 0.015% sodium diacetate. Gluconic acid is used over a wide concentration range (0.2–200 mM) as a noninhibitory buffer (Bjornsdottir, Breidt, Jr., and McFeeters, 2006). Liquid smoke contains approximately 10% acetic acid and has been effective in inhibition of pathogens such as *E. coli* O157:H7, *Salmonella*, *Listeria*, and *Streptococcus* (Murphy et al., 2006). Sorbic acid, when added as a surface spray, must be carefully considered, in view of the spectrum of its action. A concentration of 0.62% had no marked effect on the microbiological quality and shelf life of the finished products (Hozova et al., 2002).

Approximately 3 g of benzoyl peroxide are added to 20 kg of wheat flour. This is then easily decomposed to benzoic acid and other substances such as biphenyl and phenylbenzoate (Be-Onishi et al., 2004). Aqueous solutions of 2.5% of the sodium salts of organic acids are adequate to maintain the chemical quality and extend the shelf life of fresh salmon with only minor sensory changes during refrigerated storage (Sallam, 2007b). In carcass surface applications 2% lactic, 1.5% acetic plus 1.5% propionic acids are usually applied (Dubal et al., 2004).

Application of phenyllactic acid (PLA) in the reduction of fungal mass in food is more desirable compared to other preservatives commonly used in bakery products, such as propionic acid and propionic salts. PLA concentrations effective against fungi from bakery products are usually lower than those required for inhibitory activity. Required concentrations have been reported as *L. monocytogenes*, 13 mg/ml; *S. aureus*, *E. coli*, *Aeromonas hydrophila*, 20 mg/ml, whereas at pH 4 a concentration of ≤ 7.5 mg/ml is enough to inhibit > 50% of bacterial growth (Lavermicocca, Valerio, and Visconti, 2003).

3.11   Recommended applications

There is little or no data available on organic acid application to red meat carcasses and these compounds may play a more significant role in cooked or cured products in combination with other hurdles (Smulders and Greer, 1998). However, inhibition of *L. monocytogenes* by sorbic acid and sorbic salts has been studied in laboratory media and in foods such as cheese, meat products, or fish (Dorsa, Marshall, and Semien, 1993; Moir and Eyles, 1992; Samelis et al., 2001), and only a few studies have been done on the effect of sorbic acid or its salt on *L. monocytogenes* growth on poultry (Gonzalez-Fandos and Dominguez, 2007).

There is a growing interest and also increased consumer selection of fresh processed products (or minimally processed), yet not much information is available on the nutrient stability or effectiveness of postharvest treatments on nutritive value retention during storage (Ahvenainen, 1996; Cocci et al., 2006). At this stage of the development of minimally processed foods, the maintenance of nutritional value has not been the primary concern of producers and transporters, although particular attention has been paid to safety and organoleptic aspects (Cocci et al., 2006).

Conflicting reports on the efficacy of sorbate against *L. monocytogenes* have been found and may be due to variations in media, food, pH, or sorbate concentration. It has, however, been observed that sorbate has a bacteriostatic effect on *L. monocytogenes*. No listericidal effect has yet been observed (Buncic et al., 1995; Gonzalez-Fandos and Dominguez, 2007).

3.11.1   Carcasses

Treated carcasses are known to possess pH values ranging from 3.3–5.8 (Kanellos and Burriel, 2005; Mehyar et al., 2005; Del Río et al., 2007). The pH may be influenced by different factors, such as (1) the type of organic acid, (2) treatment time, (3) organic acid concentration, and (4) combinations with other decontamination techniques (Del Río et al., 2007; Álvarez-Ordóñez et al., 2009).

More forceful use of water-based decontamination technologies, alone or in combination with acidic interventions, may greatly contribute to enhanced meat safety, by modulation of resistance of *E. coli* O157:H7 and other pathogens to acid. Water-based decontamination may also be rotated with acidic interventions (Samelis et al., 2002a).

3.11.2   Processed meats

Sodium lactate has been evaluated as a potential replacement for potassium sorbate or sodium benzoate. Sausages containing 3.3% sodium lactate were compared with a control and sausages containing 0.05 or 0.1%

sorbate or benzoate with regard to (1) changes of chemical composition, (2) physicochemical and textural properties, and (3) growth of inoculated *L. monocytogenes* when stored at 4°C for up to 8 weeks. Results indicated inhibition of the proliferation of *S. typhimurium, L. monocytogenes,* and *E. coli* O157:H7 by 3–4% sodium lactate (Miller and Acuff, 1994; Choi and Chin, 2003). The addition of 3% sodium lactate to vacuum-packaged beef top rounds served as an oxidant and color stabilizer for cooked beef (Maca et al., 1999; Choi and Chin, 2003).

Frankfurters containing 3.3% sodium lactate were also found to have lower microbial counts for lactic acid bacteria than those of a control (Murano and Rust, 1995). Frankfurters manufactured with potassium sorbate and sodium benzoate, at 0.05% and 1.0%, were compared with those containing 3.3% sodium lactate. After addition of 3.3% sodium lactate to the sausage formulations, decreased TBARS (thiobarbituric acid reacting substances) values were found. Storage time was found to affect the product quality and resulted in decreases in textural properties (Choi and Chin, 2003).

A combination of organic acids, freezing, and pulsed electric fields has been found to be insignificant in reducing *E. coli* O157:H7 on filter paper, burger components, or burger patties. However, one exception, lactic acid plus freezing provided a 6 $\log_{10}$ cfu/ml decrease in *E. coli* O157:H7 numbers, which may be attributed to synergistic interactions between low pH and low temperature (Bolton et al., 2002).

3.12 Control of common pathogens

Studies have reported on the strong antimicrobial activity of sodium citrate and sodium lactate in controlling *C. perfringens* during cooling procedures. In a study on the antimicrobial activity of sodium citrate, sodium lactate, and spices, alone or combined, in controlling *C. perfringens* in ground beef products, when following procedures in violation of FSIS safe cooling guidelines, no correlation was found between the effect of sodium citrate on *C. perfringens* growth and the pH of the sample. In some studies a lack of interaction was found between the influence of pH and the effect of organic acids or spices on bacterial growth during cooling (Sabah, Juneja, and Fung, 2004).

3.12.1 Chickens

Organic acids have been investigated in drinking water of young broiler chickens and no damaged epithelial cells were observed in the chicken gut after consumption of acidified drinking water. Acidified drinking water could therefore play a crucial role in preventing pathogens, such as the spread of *Campylobacter* via drinking water in broiler flocks, as drinking water is the most prominent risk factor implicated in the spread of *Campylobacter* infection in such flocks (Gibbens et al., 2001; Chaveerach et al., 2004). It was also demonstrated that acidified drinking water containing organic acids could have a potential effect on *Campylobacter* infection in young chickens (Chaveerach et al., 2004).

3.12.2   Fruit

From industry it has been reported that the shelf life of fruit produce such as Kensington mango slices may be increased to as much as six months by combining low oxygen and increased concentrations of carbon dioxide and organic acids (De Souza et al., 2006). The application of citric or ascorbic acid had no effect on the visual appearance, or firmness in the Kensington mango or on the shelf life of the Espado or Tommy Atkins mango. Contrary to these findings, significant responses have been reported with kiwi fruit (De Souza et al., 2006). Kensington mango slices treated with 2.0% citric acid indicated significantly better general appearance, especially toward the end of storage. Low-oxygen treatment controls darkening well in Kensington slices, and an additional antibrowning agent such as citric acid is only necessary once slices are removed from a low oxygen atmosphere and allowed to stand (De Souza et al., 2006).

3.12.3   Vegetables

There are few literature reports concerning the direct addition of organic acids to vegetable creams. This method of acidification is important in the reduction of variables that influence the pH in vegetable foods. Lactic and citric acids are effective in lowering the pH, but if the target is to bring down and keep the pH between 4.0 and 4.3, more than 0.12% lactic or citric acid is needed. When adding 0.1% (minimum value) lactic acid, a concentration of 0.26% citric acid is necessary to obtain a pH value of 4.3. The same pH value is obtained by adding 0.1% citric acid as when 0.23% of lactic acid is added. The use of citric acid has a strong organoleptic impact on acidified products, and can be reduced. However, lactic acid is more effective than citric acid at the same concentration (Derossi et al., 2004).

3.13   Organic acids as additives in chilled foods

Limited literature is available on the inhibitory effects of chemical antimicrobials on germination and outgrowth of pathogens during continuous chilling. However, buffered sodium citrate and buffered sodium citrate supplemented with sodium diacetate have been effectively used to inhibit *C. perfringens* outgrowth during chilling of roast beef and injected pork

(Thippareddi, 2003), and can be used as ingredients in processed meat products as an additional safety measure in combating *C. perfringens* contamination during chilling of meat products. According to USDA-FSIS compliance guidelines (USDA-FSIS, 1999, 2001) various options are stated for chilling of thermally processed meat and poultry products (Juneja and Thippareddi, 2004).

It is also stipulated that for germination and outgrowth of spore-forming bacteria germination and outgrowth should be <1.0 $\log_{10}$ CFU/g. *C. perfringens* can be used to demonstrate that the cooling performance standard is met for both *C. perfringens* and *C. botulinum*. Salts of organic acids are extensively used in meat and poultry products to enhance microbiological safety and to control *L. monocytogenes* (Juneja and Thippareddi, 2004).

Increasing storage temperature to 15 and 25°C significantly increased effective inhibition of *E. coli* O157:H7 populations by the fumaric acid/sodium benzoate preservative mixture in apple cider (Chikthimmah, LaBorde, and Beelman, 2003). *Yersinia enterocolitica* can grow aerobically and anaerobically at temperatures as low as –2°C (Mollaret and Thal, 1974) and at pH values as low as 4.4. Formic acid is the most effective organic acid for the inhibition of *Y. enterocolitica* (El-Ziney, De Meyer, and Debevere, 1997).

3.14   Marinating

The effect of marinating on meat is the result of the influence of the acid on the collagen connective tissue, which plays a very important part in the tenderness of beef (Bailey and Light, 1989; Aktas and Kaya, 2001b). The acid causes the acid labile cross-linkages in the collagen molecule to be released, which then results in a loosened structure (Aktas and Kaya, 2001b). Organic acids in marinades can negatively affect cooked meat quality. These influences include water-holding capacity, marinade retention, percentage cook loss, and percentage moisture and bind ability (Carroll et al., 2007).

Marinating is particularly effective in tenderizing muscles containing large amounts of connective tissue. However, a long soaking period and full penetration of acid solutions are necessary to achieve this. But care should be taken, as meat treated with a marinade of high acid concentration (> 1.0%) can be sour. The overall acceptability of meat treated with NaCl and $CaCl_2$ is higher than meat treated with high concentrations of lactic and citric acid (> 1.0) (Aktas and Kaya, 2001b).

Sensory characteristics (texture, flavor, aroma, and color) of a food product are the most important attributes for the consumer (Aktas and Kaya, 2001b). The texture of food is mostly determined by moisture and fat content, as well as the types and amounts of structural carbohydrates and proteins. There are several physical and chemical methods of tenderizing meat. However, the mechanism of muscle tissue tenderization in solutions of organic acids is not clearly described, as limited work has been done on muscle tissue and further research is needed to evaluate the effects of specific ions on collagen shrinkage (Judge and Aberle, 1982). The treatment of intramuscular connective tissue with lactic and citric acid at 24, 48, and 72 h causes significant decreases in both denaturation onset temperature (T_o) and denaturation peak temperature (T_p), suggesting that the aldimine cross-links are very sensitive to a decrease in pH. For example, it has been reported that acid treatment of intramuscular connective tissue decreased T_o by approximately 25°C and T_p by about 23°C, whereas the collapse of the triple helix structure of collagen has been recorded to occur around 39°C. Thermal transitions of collagens treated with organic acids were also around 39°C, suggesting that marinating with weak acids promotes disruption of noncovalent intermolecular bonds that reinforce the collagen fibril structure. This improves the swelling effect and decreases the temperature of denaturation (Aktas and Kaya, 2001a).

Modified marinades may be applied in jerky processing, as this provides antimicrobial effects against possible postprocessing contamination with *L. monocytogenes* at low water activity. Presence of *L. monocytogenes* in jerky (Leviné et al., 2001) may be attributed to survival of the pathogen during the drying process or may be the result of postdrying contamination (Calicioglu, Sofos, and Kendall, 2003a). Acidic marinades may have a negative effect on some cooked product attributes, but they decrease *L. monocytogenes* growth (Carroll et al., 2007). Many poultry processors add organic acids to marinades to control the growth of *L. monocytogenes* in the final cooked deli loaf product. Color, water-holding capacity, and binding ability in cooked turkey deli loaves can be affected by pH. There is a decrease in pH as lactic acid increases in the muscle because of the glycolytic breakdown of glucose during rigor mortis (Lawrie, 1991). A decrease in pH can cause problems with protein functionality and postmarinating pH should be monitored (Carroll et al., 2007).

References

Adams, M.R. and Nicolaides, L. 1997. Review of the sensitivity of different foodborne pathogens to fermentation. *Food Control* 8:227–239.

Ahvenainen, R. 1996. New approaches in improving the shelf life of minimally processed fruits and vegetables. *Trends in Food Science and Technology* 7:179–186.

Aktas, N. and Kaya, M. 2001a. Influence of weak organic acids and salts on the denaturation characteristics of intramuscular connective tissue. A differential scanning calorimetry study. *Meat Science* 58:413–419.

Aktas, N. and Kaya, M. 2001b. The influence of marination with weak organic acids and salts on the intramuscular connective tissue and sensory properties of beef. *European Food Research and Technology* 213:88–94.

Allende, A., Martinez, B., Selma, V., Gil, M.I., Suarez, J.E., and Rodriguez, A. 2007. Growth and bacteriocin production by lactic acid bacteria in vegetable broth and their effectiveness at reducing *Listeria monocytogenes* in vitro and in fresh-cut lettuce. *Food Microbiology* 24:759–766.

Álvarez-Ordóñez, A., Fernandez, A., Bernardo, A., and Lopez, M. 2009. Comparison of acids on the induction of an acid tolerance response in *Salmonella typhimurium*, consequences for food safety. *Meat Science* 81:65–70.

Arroyo, M., Aldred, D., and Magan, N. 2005. Environmental factors and weak organic acid interactions have differential effects on control of growth and ochratoxin A production by *Penicillium verrucosum* isolates in bread. *International Journal of Food Microbiology* 98:223–231.

Arroyo Lopez, F.N., Duran Quintana, M.C., and Garrido, F.A. 2006. Microbial evolution during storage of seasoned olives prepared with organic acids with potassium sorbate, sodium benzoate, and ozone used as preservatives. *Journal of Food Protection* 69:1354–1364.

Bailey, A.J. and Light, N.D. 1989. *Connective tissue in meat and meat products*, p. 95. New York: Elsevier Applied Science.

Barbosa-Cánovas, G.V., Fernández-Molina, J.J., Alzamora, S.M., Tapia, M.S., López-Malo, A., and Chanes, J.W. 2003. General considerations for preservation of fruits and vegetables. In: *Handling and Preservation of Fruits and Vegetables by Combined Methods for Rural Areas*. Rome: Food and Argriculture Organization of the United Nations.

Barker, C. and Park, S.F. 2001. Sensitization of *Listeria monocytogenes* to low pH, organic acids, and osmotic stress by ethanol. *Applied and Environmental Microbiology* 67:1594–1600.

Barmpalia, I.M., Geornaras, I., Belk, K.E., et al. 2004. Control of *Listeria monocytogenes* on frankfurters with antimicrobials in the formulation and by dipping in organic acid solutions. *Journal of Food Protection* 67:2456–2464.

Barnes, E.M. 1976. Microbiological problems of poultry at refrigerator temperatures—A review. *Journal of the Science of Food and Agriculture* 27:777–782.

Bauernfeind, J.C. and Pinkert, D.M. 1970. Food processing with added ascorbic acid. *Advances in Food Research* 18:219–315.

Be-Onishi, Y., Yomota, C., Sugimoto, N., Kubota, H., and Tanamoto, K. 2004. Determination of benzoyl peroxide and benzoic acid in wheat flour by high-performance liquid chromatography and its identification by high-performance liquid chromatography-mass spectrometry. *Journal of Chromatography A* 1040:209–214.

Bégin, A. and Van Calsteren, M.R. 1999. Antimicrobial films produced from chitosan. *International Journal of Food Microbiology* 26:63–67.

Berry, E.D. and Cutter, C.N. 2000. Effects of acid adaptation of *Escherichia coli* O157:H7 on efficacy of acetic acid spray washes to decontaminate beef carcass tissue. *Applied and Environmental Microbiology* 66:1493–1498.

Beuchat, L.R. 1996. *Listeria monocytogenes*: Incidence on vegetables. *Food Control* 7:223–228.

Beuchat, L.R., Berrang, M.E., and Brackett, R.E. 1990. Presence and public health implications of *Listeria monocytogenes* on vegetables. In: A.L. Miller, J.L. Smith, and G.A. Somkuti (Eds.), *Foodborne Listeriosis*. New York: Society for Industrial Microbiology, Elsevier, pp. 175–181.

Bjornsdottir, K., Breidt, F. Jr., and McFeeters, R.F. 2006. Protective effects of organic acids on survival of *Escherichia coli* O157:H7 in acidic environments. *Applied and Environmental Microbiology* 72:660–664.

Blaszyk, M. and Holley, R.A. 1998. Interaction of monolaurin, eugenol, and sodium citrate on growth of common meat spoilage and pathogenic organisms. *International Journal of Food Microbiology* 39:175–183.

Bolton, D.J., Catarame, T., Byrne, C., Sheridan, J.J., McDowell, D.A., and Blair, I.S. 2002. The ineffectiveness of organic acids, freezing and pulsed electric fields to control *Escherichia coli* O157:H7 in beef burgers. *Letters in Applied Microbiology* 34:139–143.

Boskou, G. and Debevere, J. 2000. Shelf life extension of cod fillets with an accurate buffer spray prior to packaging under modified atmosphere. *Food Additives and Contaminants* 17:17–25.

Breidt, F., Jr., Hayes, J.S., and McFeeters, R.F. 2004. Independent effects of acetic acid and pH on survival of *Escherichia coli* in simulated acidified pickle products. *Journal of Food Protection* 67:12–18.

Brul, S. and Coote, P. 1999. Preservative agents in foods. Mode of action and microbial resistance mechanisms. *International Journal of Food Microbiology* 50:1–17.

Bullerman, L.B. 1985. Effects of potassium sorbate on growth and ochratoxin production by *Aspergillus ochraceaus* and *Penicillium* spp. *Journal of Food Protection* 48:162–165.

Buncic, S., Fitzgerald, C.M., Bell, R.G., and Hudson, J.A. 1995. Individual and combined listericidal effects of sodium lactate, potassium sorbate, nisin and curing salts at refrigeration temperature. *Journal of Food Safety* 15:247–264.

Calicioglu, M., Sofos, J.N., and Kendall, P.A. 2003a. Influence of marinades on survival during storage of acid-adapted and nonadapted *Listeria monocytogenes* inoculated post-drying on beef jerky. *International Journal of Food Microbiology* 86:283–292.

Calicioglu, M., Sofos, J.N., and Kendall, P.A. 2003b. Fate of acid-adapted and nonadapted *Escherichia coli* O157:H7 inoculated post-drying on beef jerky treated with marinades before drying. *Food Microbiology* 20:169–177.

Canadian Food Inspection Agency. Meat Hygiene Manual of Procedures. 2004. http://www.inspection.gc.ca/english/anima/meavia/mmopmmhv/direct/2004/direct44e.shtml. (Accessed September 6, 2005).

Carroll, C.D., Alvarado, C.Z., Brashears, M.M., Thompson, L.D., and Boyce, J. 2007. Marination of turkey breast fillets to control the growth of *Listeria monocytogenes* and improve meat quality in deli loaves. *Poultry Science* 86:150–155.

Casey, P.G. and Condon, S. 2002. Sodium chloride decreases the bacteriocidal effect of acid pH on *Escherichia coli* O157:H45. *International Journal of Food Microbiology* 76:199–206.

Castillo, C., Benedito, J.L., Mendez, J., et al. 2004. Organic acids as a substitute for monensin in diets for beef cattle. *Animal Feed Science and Technology* 115:101–116.

Chaveerach, P., Keuzenkamp, D.A., Lipman, L.J., and Van, K.F. 2004. Effect of organic acids in drinking water for young broilers on *Campylobacter infection*, volatile fatty acid production, gut microflora and histological cell changes. *Poultry Science* 83:330–334.

Chaveerach, P., Keuzenkamp, D.A., Urlings, H.A., Lipman, L.J., and Van, K.F. 2002. In vitro study on the effect of organic acids on *Campylobacter jejuni/coli* populations in mixtures of water and feed. *Poultry Science* 81:621–628.

Chen, M.C., Yeh, G.H.C., and Chiang, B.H. 1996. Antimicrobial and physicochemical properties of methylcellulose and chitosan films containing a preservative. *Journal of Food Processing and Preservation* 20:379–390.

Chen, Q-C. and Wang, J. 2001. Simultaneous determination of artificial sweeteners, preservatives, caffeine, theobromine and theophylline in food and parmaceutical preparations by ion chromatography. *Journal of Chromatography A* 937:57–64.

Chikthimmah, N., LaBorde, L.F., and Beelman, R.B. 2003. Critical factors affecting the destruction of *Escherichia coli* O157:H7 in apple cider treated with fumaric acid and sodium benzoate. *Journal of Food Science* 68:1438–1442.

Choi, S.H. and Chin, K.B. 2003. Evaluation of sodium lactate as a replacement for conventional chemical preservatives in comminuted sausages inoculated with *Listeria monocytogenes*. *Meat Science* 65:531–537.

Cocci, E., Rocculi, P., Romani, S., and de la Rosa, M. 2006. Changes in nutritional properties of minimally processed apples during storage. *Postharvest Biology and Technology* 39:265–271.

Cole, M.B. and Keenan, M.H.J. 1986. Synergistic effects of weak-acid preservatives and pH on the growth of *Zygosaccharomyces bailii*. *Yeast* 2:93–100.

Comes, J.E. and Beelman, R.B. 2002. Addition of fumaric acid and sodium benzoate as an alternative method to achieve a 5-log reduction of *Escherichia coli* O157:H7 populations in apple cider. *Journal of Food Protection* 65:476–483.

Crépet, A., Albert, I., Dervin, C., and Carlin, F. 2007. Estimation of microbial contamination of food from prevalence and contamination data: Application to *Listeria monocytogenes* in fresh vegetables. *Applied and Environmental Microbiology* 73:250–258.

Debeaufort, F., Quezada-Gallo, J.A., and Volley, A. 1998. Edible films and coatings: Tomorrow packaging: A review. *Critical Reviews in Food Science* 38:299–313.

De la Rosa, P., Cordoba, G., Martin, A., Jordano, R., and Medina, L.M. 2005. Influence of a test preservative on sponge cakes under different storage conditions. *Journal of Food Protection* 68:2465–2469.

Del Río, E., Panizo-Morán, M., Prieto, M., Alonso-Calleja, C., and Capita, R. 2007. Effect of various chemical decontamination treatments on natural microflora and sensory characteristics of poultry. *International Journal of Food Microbiology* 115:268–280

Derossi, A., Palmieri, L., De Giorgi, A., Loiudice, R., and Severini, C. 2004. Study on pH lowering in preserved asparagus cream. *Italian Journal of Food Science* 16:457–464.

De Souza, B.S., O'Hare, T.J., Durigan, J.F., and de Souza, P.S. 2006. Impact of atmosphere, organic acids, and calcium on quality of fresh-cut 'Kensington' mango. *Postharvest Biology and Technology* 42:161–167.

Donoghue, D.J. 2003. Antibiotic residues in poultry tissues and eggs: Human health concerns? *Poultry Science* 82:618–621.

Dorsa, W.J., Marshall, D.L., and Semien, M. 1993. Effect of potassium sorbate and citric acid sprays on growth of *Listeria monocytogenes* on cooked crawfish tail meat at 4°C. *Lebensmittel-Wissenschaft und-Technologie* 26:480–483.

Dorsa, W., Cutter, C., and Siragusa, G. 1997. Effects of acetic acid, lactic acid and trisodium phosphate on the microflora of refrigerated beef carcass surface tissue inoculated with *Escherichia coli* O157:H7, *Listeria innocua* and *Clostridium sporogenes. Journal of Food Protection* 60:619–624.

Doyle, E.M. 1999. *Use of Organic Acids to Control Listeria in Meat.* American Meat Institute Foundation. Washington, DC.

Dubal, Z.B., Paturkar, A.M., Waskar, V.S., et al. 2004. Effect of food grade organic acids on inoculated *S. aureus, L. monocytogenes, E. coli* and *S. Typhimurium* in sheep/goat meat stored at refrigeration temperature. *Meat Science* 66:817–821.

Durango, A.M., Soares, N.F.F., and Andrade, N.J. 2006. Microbiological evaluation of an edible antimicrobial coating on minimally processed carrots. *Food Control* 17:336–341.

Duxbury, D.D. 1988. Natural sodium lactate extends shelf-life of whole and ground meats. *Food Processing* 49:91–92.

Eggenberger-Solorzano, L., Niebuhr, S.E., Acuff, G.R., and Dickson, J.S. 2002. Hot water and organic acid interventions to control microbiological contamination on hog carcasses during processing. *Journal of Food Protection* 65:1248–1252.

El-Ziney, M.G., De Meyer, M.H., and Debevere, J.M. 1997. Growth and survival kinetics of *Yersinia enterocolitica* IP 383 0:9 as affected by equimolar concentrations of undissociated short-chain organic acids. *International Journal of Food Microbiology* 34:233–247.

European Union. 1995. European Parliament and Council Directive No. 95/2/EC of 20 February 1995 on food additives other than colours and sweeteners. http://www.europa.eu.int/eur-lex/en/consleg/pdf/1995L0002_do_001.pdf. (Accessed January 26, 2008)

European Union. 1996. Report of the Scientific Veterinary Committee (Public Health Section) on the Decontamination of Poultry Carcasses. VI/7785/96 Final.

Farkas, J. 1998. Irradiation as a method for decontaminating food. A review. *International Journal of Food Microbiology* 44:189–204.

Fielding, L.M., Cook, P.E., and Grandison, A.S. 1997. The effect of electron beam irradiation, combined with acetic acid, on the survival and recovery of *Escherichia coli* and *Lactobacillus curvatus. International Journal of Food Microbiology* 35:259–265.

Francis, G.A. and O'Beirne, D. 2006. Isolation and pulsed-field gel electrophoresis typing of *Listeria monocytogenes* from modified atmosphere package fresh-cut vegetables collected in Ireland. *Journal of Food Protection* 69:2524–2528.

Fyfe, L., Armstrong, F., and Stewart, J. 1998. Inhibition of *Listeria monocytogenes* and *Salmonella enteritidis* by combinations of plant oils and derivatives of benzoic acid: The development of synergistic antimicrobial combinations. *International Journal of Antimicrobial Agents* 9:195–199.

Galvez, A., Abriouel, H., Lopez, R.L., and Ben, O.N. 2007. Bacteriocin-based strategies for food biopreservation. *International Journal of Food Microbiology* 120:51–70.

Garcìa, E. and Barrett, D.M. 2002. Preservative treatments for fresh-cut fruits and vegetables. In: O. Lamikanra (Ed.), *Fresh-Cut Fruits and Vegetables. Science, Technology and Market.* Boca Raton, FL: CRC Press.

Gauthier, R. 2005. Organic acids and essential oils, a realistic alternative to antibiotic growth promoters in poultry. I Forum Internacional de avicultura 17–19 August 2005, pp. 148–157.

Geornaras, I., Belk, K.E., Scanga, J.A., Kendall, P.A., Smith, G.C., and Sofos, J.N. 2005. Postprocessing antimicrobial treatments to control *Listeria monocytogenes* in commercial vacuum-packaged bologna and ham stored at 10 degrees C. *Journal of Food Protection* 68:991–998.

Gibbens, J.C., Pascoe, S.J., Evans, S.J., Davies, R.H., and Sayers, A.R. 2001. A trial of biosecurity as a means to control *Campylobacter* infection of broiler chickens. *Preventive Veterinary Medicine* 48:85–99.

Giesting, D.W. and Easter, R.A. 1991. Effect of protein source and fumaric acid supplementation on apparent ileal digestability of nutrients by young pigs. *Journal of Animal Science* 69:2497–2503.

Gliemmo, M.F., Campos, C.A., and Gerschenson, L.N. 2006. Effect of sweet solutes and potassium sorbate on the thermal inactivation of *Z. bailii* in model aqueous systems. *Food Research International* 39:480–485.

Gonzalez-Fandos, E. and Dominguez, J.L. 2007. Effect of potassium sorbate washing on the growth of *Listeria monocytogenes* on fresh poultry. *Food Control* 18:842–846.

Guilbert, S., Gontard, N., and Gorris, L.G.M. 1996. Prolongation of the shelf-life of perishable food products using biodegradable films and coatings. *Lebensmittel-Wissenschaft und-Technologie* 29:10–17.

Guynot, M.E., Ramos, A.J., Sanchis, V., and Marín, S. 2005. Study of benzoate, propionate, and sorbate salts as mould spoilage inhibitors on intermediate moisture bakery products of low pH (4.5–5.5). *International Journal of Food Microbiology* 101:161–168.

Hazan, R., Levine, A., and Abeliovich, H. 2004. Benzoic acid, a weak organic acid food preservative, exerts specific effects on intracellular membrane trafficking pathways in *Saccharomyces cerevisiae*. *Applied and Environmental Microbiology* 70:4449–4457.

Hoornstra, E. and Notermans, S. 2001. Quantitative microbiological risk assessment. *International Journal of Food Microbiology* 66:21–29.

Hozova, B., Kukurova, I., Turicova, R., and Dodok, L. 2002. Sensory quality of stored croissant-type bakery products. *Czech Journal of Food Science* 20:105–112.

Huang, N., Ho, C., and McMillin, K.W. 2005. Retail shelf-life of pork dipped in organic acid before modified atmosphere or vacuum packaging. *Journal of Food Science* 70:M382–M387.

Jordan, K.N., Oxford, L., and O'Byrne, C.P. 1999. Survival of low-pH stress by *Escherichia coli* O157:H7: Correlation between alterations in the cell envelope and increased acid tolerance. *Applied and Environmental Microbiology* 65:3048–3055.

Judge, M.D. and Aberle, E.D., 1982. Effect of chronological age and postmortem aging on thermal shrinkage temperature of bovine intramuscular collagen. *Journal of Animal Science* 54:68–71.

Juneja, V.K. and Thippareddi, H. 2004. Inhibitory effects of organic acid salts on growth of *Clostridium perfringens* from spore inocula during chilling of marinated ground turkey breast. *International Journal of Food Microbiology* 93:155–163.

Kanellós, T.S. and Burriel, A.R. 2005. The bactericidal effect of lactic acid and trisodium phosphate on *Salmonella enteritidis* serotype pt4, total viable counts and counts of Enterobacteriaceae. *Food Protection Trends* 25:346–350.

Kerth, C. and Braden, C. 2007. Using Acetic Acid Rinse as a CCP for Slaughter. *Virtual Library—Meat, Poultry & Egg Processors.*

Kwon, Y.M. and Ricke, S.C. 1998. Induction of acid resistance of *Salmonella typhimurium* by exposure to short-chain fatty acids. *Applied and Environmental Microbiology* 64:3458–3463.

Lambert, R.J.W. and Bidlas, E. 2007. An investigation of the Gamma hypothesis: A predictive modelling study of the effect of combined inhibitors (salt, pH and weak acids) on the growth of *Aeromonas hydrophila. International Journal of Food Microbiology* 115:12–28.

Lavermicocca, P., Valerio, F., and Visconti, A. 2003. Antifungal activity of phenyllactic acid against molds isolated from bakery products. *Applied and Environmental Microbiology* 69:634–640.

Lawrie, R.A. 1991. *Meat Science*, 6th ed. Cambridge, UK: Woodhead, pp. 55–60.

Lemay, M.J., Choquette, J., Delaquis, P.J., Claude, G., Rodrigue, N., and Saucier, L. 2002. Antimicrobial effect of natural preservatives in a cooked and acidified chicken meat model. *International Journal of Food Microbiology* 78:217–226.

Leviné, P., Rose, B., Green, S., Ransom, G., and Hill, W. 2001. Pathogen testing of ready-to-eat meat and poultry products collected at federally inspected establishments in the United States, 1990 to 1999. *Journal of Food Protection* 64:1188–1193.

Maas, M.R., Glass, K.A., and Doyle, M.P. 1989. Sodium lactate delays toxin production by *Clostridium botulinum* in cook-in-bag turkey products. *Applied and Environmental Microbiology* 55:2226–2229.

Maca, J.M., Miller, R.K., Bigner, M.E., Lucia, L.M., and Acuff, G.R. 1999. Sodium lactate and storage temperature effects on shelf-life of vacuum packed beef top round. *Meat Science* 53:23–29.

Macpherson, N., Shabala, L., Rooney, H., Jarman, M.G., and Davies, J.M. 2005. Plasma membrane H+ and K+ transporters are involved in the weak-acid preservative response of disparate food spoilage yeasts. *Microbiology* 151:1995–2003.

Magan, N. and Lacey, J. 1986. The effects of two ammonium propionate formulations on growth in vitro of *Aspergillus* species isolated from hay. *Journal of Applied Bacteriology* 60:221–225.

Mancini, R.A., Hunt, M.C., Seyfert, M., et al. 2007. Effects of ascorbic and citric acid on beef lumbar vertebrae marrow colour. *Meat Science* 76:568–573.

Marín, S., Abellana, M., Rubinat, M., Sanchis, V., and Ramos, A.J. 2003. Efficacy of sorbates on the control of the growth of *Eurotium* species in bakery products with near neutral pH. *International Journal of Food Microbiology* 87:251–258.

Marín, S., Sanchiz, V., Sanz, D., et al. 1999. Control of growth and fumonisin B1 production by *Fusarium verticillioides* and *Fusarium proliferatum* isolates in moist maize with propinic preservatives. *Food Additives and Contamination* 16:555–563.

Marsili, N.R., Sobrero, M.S., and Goicoechea, H.C. 2003. Spectrophotometric determination of sorbic and benzoic acids in fruit juices by a net analyte signal-based method with selection of the wavelength range to avoid non-modelled interferences. *Analytical and Bioanalytical Chemistry* 376:126–133.

Martins, C.G., Behrens, J.H., Destro, M.T., et al. 2004. Gamma radiation in the reduction of *Salmonella* spp. inoculated on minimally processed watercress (*Nasturtium officinalis*). *Radiation Physics and Chemistry* 71:89–93.

Mbandi, E. and Shelef, L.A. 2002. Enhanced antimicrobial effects of combination of lactate and diacetate on *Listeria monocytogenes* and *Salmonella* spp. in beef bologna. *International Journal of Food Microbiology* 76:191–198.

Mehyar, G., Blank, G., Han, J.H., Hydamaka, A., and Holley, R.A. 2005. Effectiveness of trisodium phosphate, lactic acid and commercial antimicrobials against pathogenic bacteria on chicken skin. *Food Protection Trends* 25:351–362.

Meng, J. 1992. Effect of sodium lactate on probability of *Clostridium botulinum* growth in BHI broth after heat shock and on toxigenesis in cooked poultry meat products. M.S. Thesis. University of California, Berkeley.

Miller, A.J., Call, J.E., and Whiting, R.C. 1993. Comparison of organic acid salts for *Clostridium botulinum* control in an uncured turkey product. *Journal of Food Protection* 56:958–962.

Miller, R.K. and Acuff, G.R. 1994. Sodium lactate effects pathogens in cooked beef. *Journal of Food Science* 59:15–19.

Min, J.S., Lee, S.O., Jang, A., Jo, C., and Lee, M. 2007. Irradiation and organic acid treatment for microbial control and the production of biogenic amines in beef and pork. *Food Chemistry* 104:791–799.

Mohd Adnan, A.F. and Tan, I.K.P. 2007. Isolation of lactic acid bacteria from Malaysian foods and assessment of the isolates for industrial potential. *Bioresource Technology* 98:1380–1385.

Moir, C.J. and Eyles, M.J. 1992. Inhibition, injury and inactivation of four psychrotrophic foodborne bacteria by the preservatives methyl *p*-hydroxybenzoate and potassium sorbate. *Journal of Food Protection* 55:360–366.

Mollaret, H.E. and Thal, E. 1974. "Yersinia". In *Bergey's Manual of Determinative Bacteriology*, ed. R.E. Buchanan and N.E. Gibbons, pp. 330-332. Baltimore, MD: Williams and Wilkins.

Mota, F.J.M., Ferreira, I.M.P.L., Cunha, S.C., Beatriz, M., and Oliveira, P.P. 2003. Optimisation of extraction procedures for analysis of benzoic and sorbic acids in foodstuffs. *Food Chemistry* 82:469–473.

Mroz, Z., Jongbloed, A., Partanen, K.H., Vreman, K., Kemme, P.A., and Kogut, J. 2000. The effects of calcium benzoate in diets with or without organic acids on dietary buffering capacity, apparent digestibility, retention of nutrients, and manure characterization in swine. *Journal of Animal Science* 78:2622–2632.

Murano, E.A. and Rust R.E. 1995. General microbial profile of low-fat frankfurters formulated with sodium lactate and a texture modifier. *Journal of Food Quality* 18:313–323.

Murphy, R.Y., Hanson, R.E., Johnson, N.R., Chappa, K., and Berrang, M.E. 2006. Combining organic acid treatment with steam pasteurization to eliminate *Listeria monocytogenes* on fully cooked frankfurters. *Journal of Food Protection* 69:47–52.

Nazer, A.I., Kobilinsky, A., Tholozan, J.L., and Dubois-Brissonnet, F. 2005. Combinations of food antimicrobials at low levels to inhibit the growth of *Salmonella* sv. *Typhimurium*: A synergistic effect? *Food Microbiology* 22:391–398.

Nykänen, A., Lapveteläinen, A., Kallio, H., and Salminen, S. 1998. Effects of whey, whey-derived lactic acid and sodium lactate on the surface microbial counts of rainbow trout packed in vacuum pouches. *Lebensmittel-Wissenschaft und-Technologie* 31:361–365.

Ogden, S.K., Guerrero, I., Taylor, A.J., Escalona Buendia, H., and Gallardo, F. 1995. Changes in odour, colour and texture during the storage of acid preserved meat. *Lebensmittel-Wissenschaft und-Technologie* 28:521–527.

Otero-Losada, M.E. 2003. Differential changes in taste perception induced by benzoic acid prickling. *Physiology & Behavior* 78:415–425.

Ouattara, B., Giroux, M., Yefsah, R., et al. 2002. Microbiological and biochemical characteristics of ground beef as affected by gamma irradiation, food additives and edible coating film. *Radiation Physics and Chemistry* 63:299–304.

Outtara, B., Simard, R.E., Piette, G., Bégin, A., and Holley, R.A. 2000. Diffusion of acetic and propionic acids from chitosan-based antimicrobial packaging films. *Journal of Food Science* 65:768–773.

Øverland, M., Kjos, N.P., Borg, M., and Sørum, H. 2007. Organic acids in diets for entire male pigs. *Livestock Science* 109:170–173.

Padgett, T., Han, L.H., and Dawson, P.L. 1998. Incorporation of food-grade antimicrobial compounds into biodegradable packaging films. *Journal of Food Protection* 61:1330–1335.

Penney, V., Henderson, G., Blum, C., and Johnson-Green, P. 2004. The potential of phytopreservatives and nisin to control microbial spoilage of minimally processed fruit yogurts. *Innovative Food Science and Emerging Technologies* 5:369–375.

Phillips, C.A. and Duggan, J. 2002. The effect of temperature and citric acid, alone, and in combination with nisin, on the growth of *Arcobacter butzleri* in culture. *Food Control* 13:463–468.

Piper, P., Calderon, C.O., Hatzixanthis, K., and Mollapour, M. 2001. Weak acid adaptation: The stress response that confers yeasts with resistance to organic acid food preservatives. *Microbiology* 147:2635–2642.

Piper, P.W. 1999. Yeast superoxide dismutase mutants reveal a pro-oxidant action of weak organic acid food preservatives. *Free Radical Biology and Medicine* 27:11–12.

Quintavalla, S. and Vicini, L. 2002. Antimicrobial food packaging in meat industry. *Meat Science* 62:373–380.

Ricke, S.C. 2003. Perspectives on the use of organic acids and short chain fatty acids as antimicrobials. *Poultry Science* 82:632–639.

Rico, D., Martin-Diana, A.B., Barat, J.M., and Barry-Ryan, C. 2007. Extending and measuring the quality of fresh-cut fruit and vegetables: A review. *Trends in Food Science and Technology* 18:373–386.

Sabah, J.R., Juneja, V.K., and Fung, D.Y. 2004. Effect of spices and organic acids on the growth of *Clostridium perfringens* during cooling of cooked ground beef. *Journal of Food Protection* 67:1840–1847.

Sallam, K.I. 2007a. Antimicrobial and antioxidant effects of sodium acetate, sodium lactate, and sodium citrate in refrigerated sliced salmon. *Food Control* 18:566–575.

Sallam, K.I. 2007b. Chemical, sensory and shelf life evaluation of sliced salmon treated with salts of organic acids. *Food Chemistry* 101:592–600.

Salleh-Mack, S.Z. and Roberts, J.S. 2007. Ultrasound pasteurization: The effects of temperature, soluble solids, organic acids and pH on the inactivation of *Escherichia coli* ATCC 25922. *Ultrasonics Sonochemistry* 14:323–329.

Samelis, J., Bedie, G.K., Sofos, J.N., Belk, K.E., Scanga, J.A., and Smith, G.C. 2005. Combinations of nisin with organic acids or salts to control *Listeria monocytogenes* on sliced pork bologna stored at 4°C in vacuum packages. *Lebensmittel-Wissenschaft und-Technologie* 38:21–28.

Samelis, J., Sofos, J.N., Ikeda, J.S., Kendall, P.A., and Smith, G.C. 2002a. Exposure to non-acid fresh meat decontamination washing fluids sensitizes *Escherichia coli* O157:H7 to organic acids. *Letters in Applied Microbiology* 34:7–12.

Samelis, J., Sofos, J.N., Kain, M.L., Scanga, J.A., Belk, K.E., and Smith, G.C. 2001. Organic acid and their salts as dipping solutions to control *Listeria monocytogenes* inoculated following processing of sliced pork bologna stored at 4°C in vacuum packages. *Journal of Food Protection* 64:1722–1729.

Samelis, J., Sofos, J.N., Kendall, P.A., and Smith, G.C. 2002b. Effect of acid adaptation on survival of *Escherichia coli* O157:H7 in meat decontamination washing fluids and potential effects of organic acid interventions on the microbial ecology of the meat plant environment. *Journal of Food Protection* 65:33–40.

Sauli, I., Danuser, J., Geeraerd, A.H., et al. 2005. Estimating the probability and level of contamination with *Salmonella* of feed for finishing pigs produced in Switzerland—The impact of the production pathway. *International Journal of Food Microbiology* 100:289–310.

Sebti, I. and Coma, V. 2002. Active edible polysaccharide coating and interactions between solution coating compounds. *Carbohydrate Polymers* 49:139–144.

Sewell, A.M. and Farber, J.M. 2001. Foodborne outbreaks in Canada linked to produce. *Journal of Food Protection* 64:1863–1877.

Siragusa, G. 1995. The effectiveness of carcass decontamination systems for controlling the presence of pathogens on the surfaces of meat animal carcasses. *Journal of Food Safety* 15:229–238.

Smulders, F. 1987. Prospectives for microbial decontamination of meat and poultry by organic acids with special reference to lactic acid. In: F.J.M. Smulders (Ed.), *Elimination of Pathogenic Organisms from Meat and Poultry*, pp. 319–344, London: Elsevier Science, Biomedical Division.

Smulders, F.J. and Greer, G.G. 1998. Integrating microbial decontamination with organic acids in HACCP programmes for muscle foods: Prospects and controversies. *International Journal of Food Microbiology* 44:149–169.

Stiles, B.A., Duffy, S., and Schaffner, D. 2002. Modelling yeast spoilage on cold filled ready to drink beverages with *Saccharomyces cerevisiae*, *Zygosaccharomyces bailii* and *Candida lipolytica*. *Applied and Environmental Microbiology* 68:1901–1906.

Suhr, K.I. and Nielsen, P.V. 2004. Effect of weak acid preservatives on growth of bakery product spoilage fungi at different water activities and pH values. *International Journal of Food Microbiology* 95:67–78.

Tamblyn, K.C. and Conner, D.E. 1997. Bactericidal activity of organic acids in combination with transdermal compounds against *Salmonella typhimurium* attached to broiler skin. *Food Microbiology* 14:477–484.

Taniguchi, M., Nakazawa, H., Takeda, O., Kaneko, T., Hoshino, K., and Tanaka, T. 1998. Production of a mixture of antimicrobial organic acids from lactose by co-culture of *Bifidobacterium longum* and *Propionibacterium freudenreichii*. *Bioscience, Biotechnology and Biochemistry* 62:1522–1527.

Theron, M.M. and Lues, J.F.R. 2007. Organic acids and meat preservation: A review. *Food Reviews International* 23:141–158.

Thippareddi, H., Juneja, V.K., Phebus, R.K., Marsden, J.L., and Kastner, C.L. 2003. Control of *Clostridium perfringens* germination and outgrowth by buffered sodium citrate during chilling of roast beef and injected pork. *Journal of Food Protection* 66:376–381.

Thomson, J.L. and Hinton, M. 1997. Antibacterial activity of formic and propionic acids in the diet of hens on *Salmonella* in the crop. *British Poultry Science* 38:59–65.

Tsiloyiannis, V.K., Kyriakis, S.C., Vlemmas, J., and Sarris, K. 2001a. The effect of organic acids on the control of porcine post-weaning diarrhoea. *Research in Veterinary Sciences* 70:287–293.

Tsiloyiannis, V.K., Kyriakis, S.C., Vlemmas, J., and Sarris, K. 2001b. The effect of organic acids on the control of post-weaning oedema disease of piglets. *Research in Veterinary Sciences* 70:281–285.

U.S. Department of Agriculture, Food Safety and Inspection Service. 1999. Performance standards for the production of certain meat and poultry products. *Federal Register* 64:732–749.

U.S. Department of Agriculture, Food Safety and Inspection Service. 2000. Food additives for use in meat and poultry products: Sodium diacetate, sodium acetate, sodium lactate and potassium lactate: Direct final rule. *Federal Register* 65:3121–3123.

U.S. Department of Agriculture, Food Safety and Inspection Service. 2001. Performance standards for the production of meat and poultry products: Proposed rule. *Federal Register* 66:12589–12636.

Uyttendaele, M., Taverniers, I., and Debevere, J. 2001. Effect of stress induced by suboptimal growth factors on survival of *Escherichia coli* O157:H7. *International Journal of Food Microbiology* 66:31–37.

Van Gerwen, S.J., te Giffel, M.C., van't Riet, K., Beumer, R.R., and Zwietering, M.H. 2000. Stepwise quantitative risk assessment as a tool for characterization of microbiological food safety. *Journal of Applied Microbiology* 88:938–951.

Van Heerden, J. 2003. Circular with regards to frozen prawns (shrimps). South African Bureau of Standards, *Foods and Associated Industries*.

Van Immerseel, F., Russell, J.B., and Flythe, M.D. 2006. The use of organic acids to combat *Salmonella* in poultry: A mechanistic explanation of the efficacy. *Avian Pathology* 35:182–188.

Van Netten, P., Huis in't veld, J.H.J., and Mossel, D.A.A. 1994. The immediate bactericidal effect of lactic acid on meat-borne pathogens. *Applied Bacteriology* 77:490–496.

Wen, Y., Wang, Y., and Feng, Y.Q. 2007. A simple and rapid method for simultaneous determination of benzoic and sorbic acids in food using in-tube solid-phase microextraction coupled with high-performance liquid chromatography. *Analytical and Bioanalytical Chemistry* 388:1779–1787.

Weng, Y.M., Chen, M.J., and Chen, W. 1999. Antimicrobial food packaging materials from poly(ethylene-co-methacrylic acid). *Lebensmittel-Wissenschaft und-Technologie* 32:191–195.

Wu, C.H., Lo, Y.S., Lee, Y.-H., and Lin, T.-I. 1995. Capillary electrophoretic deter-
     mination of organic acids with indirect detection. *Journal of Chromatography
     A* 716:291–301.
Yamamoto, K., Takahashi, M., and Niki, E. 1987. Role of iron and ascorbic acid in
     the oxidation of methyl linoleate micelles. *Chemical Letter* 6:1149–1152.
Yousef, A.E. and Marth, E.H. 1981. Growth and synthesis of aflatoxin by
     Aspergillus paraciticus in the presence of sorbic acid. *Journal of Food Protection*
     44:736–741.
Zagory, D. and Garren, D. 1999. HACCP: What it is and what it isn't. *The Packer.*
     August 16. Online journal. http://www.thepacker.com (Accessed August
     29, 2005).
Zhuang, R-Y., Huang, Y.-W., and Beuchat, L.R. 1996. Quality changes during
     refrigerated storage of packed shrimp and catfish fillets treated with sodium
     acetate, sodium lactate or propyl gallate. *Journal of Food Science* 61:244–261.

chapter four

Microbial organic acid producers

4.1 Introduction

Foods that include the incorporation of microbial metabolites as part of their production are an intricate component of the world's food supply and for ethical and sensory-nutritional reasons it is essential for all the world's population to have access to this form of food. It is a process that has been in use since the early history of mankind. An Egyptian pot dating from 2300 BC (McGee, 1984) was found to contain residues of cheese and in passages in the Bible the use of some kind of fermentative starter culture is indicated. It is, therefore, possible that the use of bacteria such as the lactic acid bacteria (LAB) dates back at least four to five thousand years, although the exact principle behind the process may not have been known to the civilizations of those times (Davidson et al., 1995). Production of fermented foods, where organisms such as the LAB are involved, is a technological process that has been used for centuries at least (Herreros et al., 2005).

As a classical example, lactic acid bacteria occur naturally in various foodstuffs, their growth is enhanced, or they are added deliberately to produce a range of fermented foods. These include fish, meat, various dairy products, cereals, fruits, vegetables, and legumes. Certain treatments and storage conditions lead to the growth and predominance of LAB in a raw material and it has long been found that this presents desirable sensory characteristics as well as improved qualities and safety (Adams and Nicolaides, 1997). Since then, the LAB have been recognized, identified, and genetically described, and are continuously under investigation. They are a very important group of starter cultures, applied in the production of a wide range of fermented foods, they contribute to the enhancement of the organoleptic characteristics of food, and they have been recognized as contributing to the microbial safety of fermented food (Messens and De Vuyst, 2002; O'Sullivan et al., 2002). The LAB have an important antimicrobial function, due to their production of organic acids, CO_2, ethanol, diacetyl, acetoin, peroxide, hydrogen peroxide, and bacteriocins (Messens and De Vuyst, 2002).

The primary contributions of the LAB to a food product are to preserve the nutritive qualities of the raw material, by extending the shelf life and also by controlling the growth of spoilage and pathogenic bacteria. This is achieved by production of the inhibitory substances and by competing for nutrients (O'Sullivan et al., 2002). They occur naturally in foods and as such, have traditionally been used as natural biopreservatives of food and feed. The preservation activity is also recognized to be the result of acidification that results from their ability to synthesize and excrete lactic acid (Davidson et al., 1995; Herreros et al., 2005). The various compounds produced by the starter bacteria may all in some way be implicated in preventing the growth of unwanted bacteria, but the production of adequate quantities of lactic acid is the fundamental rule to successful fermentation (Davidson et al., 1995).

LAB are able to inhibit growth and survival of normal spoilage microflora as well as various pathogens that may be prevalent in raw materials and foodstuffs. Many different LAB play a role in the production of safe and stable fermented food, which in essence means that this principal inhibitory effect must be some metabolic feature that is present in all of them. This common feature has been determined to be the fermentative pathways that they use to generate cellular energy. In so doing, organic acids are produced (primarily lactic acid), no matter where they may grow, which then leads to a decrease in the pH of any surrounding medium (Adams and Nicolaides, 1997). Some LAB are also known to produce an array of antifungal compounds (Nes and Johnsborg, 2004).

LAB may also be added as pure cultures to various food products. LAB have GRAS (generally recognized as safe) status and are considered to be harmless. They are even being considered as probiotics, toward the improvement of human and animal health (Schnürer and Magnusson, 2005). LAB are also found in nutrient-rich food, and play an additional role in the food industry by significantly contributing to flavor, texture, and also the nutritional value of the food products, owing to metabolic characteristics (McKay and Baldwin, 1990; Topisirovic et al., 2006). Low pH, in particular, adds to the organoleptic properties of a fermented product (Davidson et al., 1995).

The LAB consist of a group of Gram-positive, nonspore-forming bacilli and cocci, that are taxonomically diverse, with the majority known for their applications in fermentation and processing of food (Gasser, 1994; Vankerckhoven et al., 2004). Some of the genera included in the LAB group are *Lactobacillus, Lactococcus, Streptococcus, Pediococcus, Leuconostoc, Enterococcus, Carnobacterium,* and *Propionibacterium* (Table 4.1). A wide variety of strains is routinely used as starter cultures in the manufacture of dairy, meat, and vegetable products (O'Sullivan et al., 2002). LAB are further classified as homofermentative (producing only lactic acid from glucose) or heterofermentative (producing lactic acid, acetic acid, ethanol, and CO_2 from glucose). Both homo- and heterofermentative LAB are found in sourdough and play an important role in improving the sensory and technological properties of bread (Gianotti et al., 1997; Meignen et al., 2001; Şimşek, Çon, and Tulumoğlu, 2006).

LAB can also spoil foods by fermentation of sugars to form lactic acid, slime, and CO_2. This often causes significantly low pH and also

Table 4.1 Some Commonly Known Lactic Acid Bacteria Important in Food
Processing, Preservation, and Spoilage

Genera	Species
Lactobacillus	
Group I: Homofermentative	*L. acidopilus, L. helveticus*
Group II: Facultative Heterofermentative	*L. casei, L. plantarum*
Group III: Obligate Heterofermentative	*L. brevis, L. fermentum*
Leuconostoc	*L. oenos, L. mesenteroides, L. lactis*
Pediococcus	*P. acidilactici, P. cellicola*
Lactococcus	*L. garvieae, L. lactis, L. piscium*
Streptococcus	*S. macedonicus*
Aerococcus	*A. viridans*
Carnobacterium	*C. divergens, C. maltaromaticum*
Enterococcus	*E. faecium, E. faecalis*
Oenococcus	*E. oeni*
Propionibacterium	*P. freudenreichii, P. acidipropionici*
Sporolactobacillus	*S. inulinus, S. dextrus*
Teragenococcus	*T. halophilus*
Vagococcus	*V. carniphilus, V. fluvialis,* *V. salmonarium*
Weisella	*W. paramesenteroides, W. confusa*

adverse flavors in the food (Huis in't Veld, 1996). LAB are considered
to be one of the major causes of bacterial spoilage in vacuum-packed
cooked cured meat products (Tompkin, 1986; Radin, Niebuhr, and
Dickson, 2006). There are increasing reports on the LAB as opportunistic pathogens (Sims, 1964; Adams, 1999). Lactobacilli that have been
implicated in human disease include *Lactobacillus casei, Lactobacillus plantarum, Lactobacillus rhamnosus,* and *Lactobacillus acidophilus* group, which
are associated with septicemia, meningitis, lung abscesses, infective
endocarditis, rheumatic diseases, vascular diseases, peritonitis, and urinary tract infections (Aguirre and Collins, 1993; Harty et al., 1994; Sidhu,
Langsrud, and Holck, 2001).

4.2   Predominant antimicrobial substances produced by LAB

4.2.1   Lactic acid

A principal characteristic of LAB is their ability to produce organic
acids and to thereby decrease the pH in food (Rossland et al., 2005). The
organic acids (mainly lactic acid and acetic acid) are therefore the major

antimicrobial and biopreservative substance produced by the LAB (Nes and Johnsborg, 2004). Lactic acid is the major LAB metabolite (Schnürer and Magnusson, 2005), and it exists in equilibrium with its undissociated forms (Ammor et al., 2006). The degree of dissociation normally depends on the pH of the environment or the foodstuff (Lindgren and Dobrogosz, 1990). Heterofermentative LAB may produce large amounts of acetic acid in the presence of external acceptors. Propionic acid is also produced, although in trace amounts (Schnürer and Magnusson, 2005). Heterofermentative LAB are, however, able to ferment various organic acids, such as citrate, malate, and pyruvate (Zaunmuller et al., 2006). See Table 4.2.

Lactic acid is not naturally present in foods, but is formed by the LAB during the fermentation of various foodstuffs (Barbosa-Cánovas et al., 2003). The lactic acid that results from fermentation can reach relatively high levels in some fermented foods (levels can exceed 100 mM). Depending on the buffering capacity of the environment or foodstuff, the resulting pH may be as low as 3.5–4.5. In spite of a range of antimicrobial substances that are known to be produced by LAB, it is evident that the production of organic acids (and their presence in an undissociated form) in combination with the resulting decrease in pH, is the primary cause of any inhibitory effect observed (Adams and Nicolaides, 1997).

Not all LAB produce the same lactic acid isomer. For example, *Lactococcus* and *Carnobacterium* are known to produce L-lactic acid, whereas *Leuconostoc* produces D-lactic acid (Liu, 2003; Gravesen et al., 2004). The levels and also the type of organic acids that are produced during any fermentation process are, therefore, dependent on LAB species or strains,

Table 4.2 Main Antifungal Compounds
Produced by Lactic Acid Bacteria

Active compounds of antifungal LAB
Lactic acid
Acetic acid
Caproic acid
Phenyllactic acid
Carbon dioxide
3-hydroxy fatty acids
Proteinaceous compounds (fungicins)
Cyclic dipeptides
Reuterin
Hydrogen peroxide
Diacetyl

growth conditions, and food composition (Lindgren and Dobrogosz, 1990; Ammor et al., 2006).

4.2.2 Bacteriocins

Antimicrobial peptides (also referred to as bacteriocins), are more important substances produced by lactic acid bacteria. Bacteriocins are currently commercially applied in various industrial food systems for the control of pathogens (McEntire, Montville, and Chikindas, 2003). These substances are also known for their attractive characteristics that make them suitable for food preservation (Galvez et al., 2007). They are nontoxic for eukaryotic cells, they have little influence on the gut microflora, have a broad antimicrobial spectrum, have a bactericidal mode of action, and are pH and heat tolerant. Bacteriocins are often used in combination with other antimicrobial agents such as organic acids. Nisin is an important commercially available bacteriocin and is produced by strains of *Lactococcus lactis* subsp. *Lactis*. Nisin is GRAS and the only bacteriocin that is permitted in food. It has been successfully applied in various foods, such as dairy products and salad dressings (Samelis et al., 2005). Other bacteriocins have been identified and are also produced by *L. lactis*. These include Lacticin 481, Lactococcin A, Lactococcin B, and Lactococcin M (O'Sullivan et al., 2002; Tome, Teixeira, and Gibbs, 2006).

4.3 Principles of lactic acid fermentation

In a fermentation process, the production of antimicrobial metabolites by starter cultures is crucial to successfully produce a desired food product (Nes and Johnsborg, 2004). LAB are widely used for the fermentation of various foodstuffs, in particular milk, meat, and vegetable foods. The main characteristic of LAB, which renders this group of organisms ideal as a starter culture in the fermentation of food, is their ability to produce organic acids and thereby also to decrease pH in food (Rossland et al., 2005). For example, during the fermentation of dairy products, these cultures metabolize lactose to lactic acid. This acid then decreases the pH to create an environment that would inhibit growth of spoilage organisms as well as foodborne pathogens (Aslim et al., 2005). In order to improve shelf life, slow down bread staling, and improve flavor, homo- and heterofermentative LAB have been evaluated for their acidification potential, production of volatile substances, proteolytic and amylolytic activity, and their ability to prevent microbial spoilage (Şimşek, Çon, and Tulumoğlu, 2006).

Lactic acid fermentation has received much attention for many years and is widely applied in the food, pharmaceutical, leather, and textile industries, as well as a chemical feed stock (Table 4.3). Two enantiomers of lactic acid are produced (L[+] and [D–]) of which L(+) can be used by the

Table 4.3 Importance of the Various Lactic Acid Bacteria and Their Products in Food Production and Preservation

Organism	Function
Lactobacillus	Production of yogurt, cheese, sauerkraut, pickles, beer, wine, cider, chocolate, and animal feed
Leuconostoc	Starter culture in dairy fermentation
	Antimicrobial activity by producing various bacteriocins
Pediococcus	Fermentation of cabbage to sauerkraut
	Lactic acid provides sour taste and extends shelf life
	Gives butterscotch aroma to some wines and beers
	Production of cheese and yogurt
Lactococcus	Common fermenters of dairy products such as cheese
	Major function is rapid
	Role in flavor of final product
Streptococcus	Dairy manufacturing
	Bacteriocin production
Aerococcus	Minor importance in food
	Have been associated with the greening of meat
Carnobacterium	Protective cultures in meat and seafood
	Bacteriocin production
Enterococcus	Ripening of traditional cheeses and contributing to typical taste and flavor
	Bacteriocin production
Oenococcus	Malolactic fermentation in wine
Teragenococcus	Fermentation of soy sauce
Weisella	Improving organoleptic properties of cheese, butter, and butter cheese

human metabolism because of L-lactate dehydrogenase and is, therefore, preferred for use in food (Couto and Sanroman, 2006). During homolactic fermentation, the fermentation of one mole of glucose yields two moles of lactic acid (Axelsson, 2004):

$$C_2H_{12}O_6 \rightarrow 2CH_3CHOHCOOH,$$

This fermentation reaction also produces >85% lactic acid from glucose (Figure 4.1).

During heterolactic fermentation, the fermentation of one mole of glucose yields one mole each of lactic acid, ethanol, and carbon dioxide:

$$C_2H_{12}O_6 \rightarrow CH_3CHOHCOOH + C_2H_5OH + CO_2$$

However, heterolactic fermentation produces only 50% lactic acid and considerable amounts of ethanol, acetic acid, and carbon dioxide (Figure 4.2).

The fermentation of 1 mole of glucose yields two moles of lactic acid:

$$C_2H_{12}O_6 \longrightarrow 2\ CH_3CHOHCOOH$$

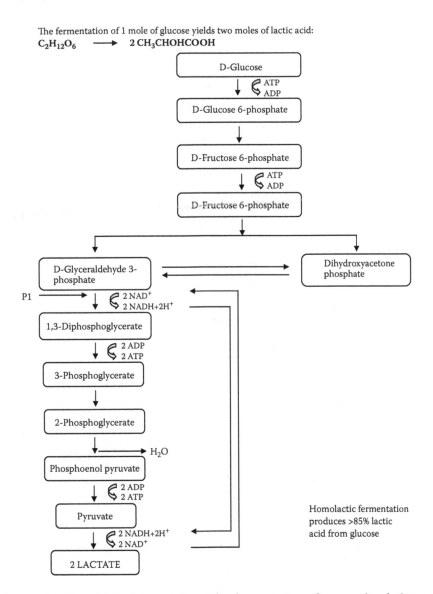

Figure 4.1 Homolactic fermentation. The fermentation of one mole of glucose yields two moles of lactic acid via the Embden–Meyerhof pathway.

Besides the production of L-lactic acid, the fermentation process can simultaneously produce various other metabolites such as acetic acid, fumaric acid, ethanol, malic acid, etc. However, the amount of these metabolites can have a significant influence on the downstream process and the quality of the L(+)-lactic acid produced (Wang et al., 2005). Fumaric acid is the main by-product and its accumulation is affected by many factors,

The fermentation of 1 mole of glucose yields 1 mole each of lactic acid, ethanol and carbon dioxide:

$$C_2H_{12}O_6 \longrightarrow CH_3CHOHCOOH + C_2H_5OH + CO_2$$

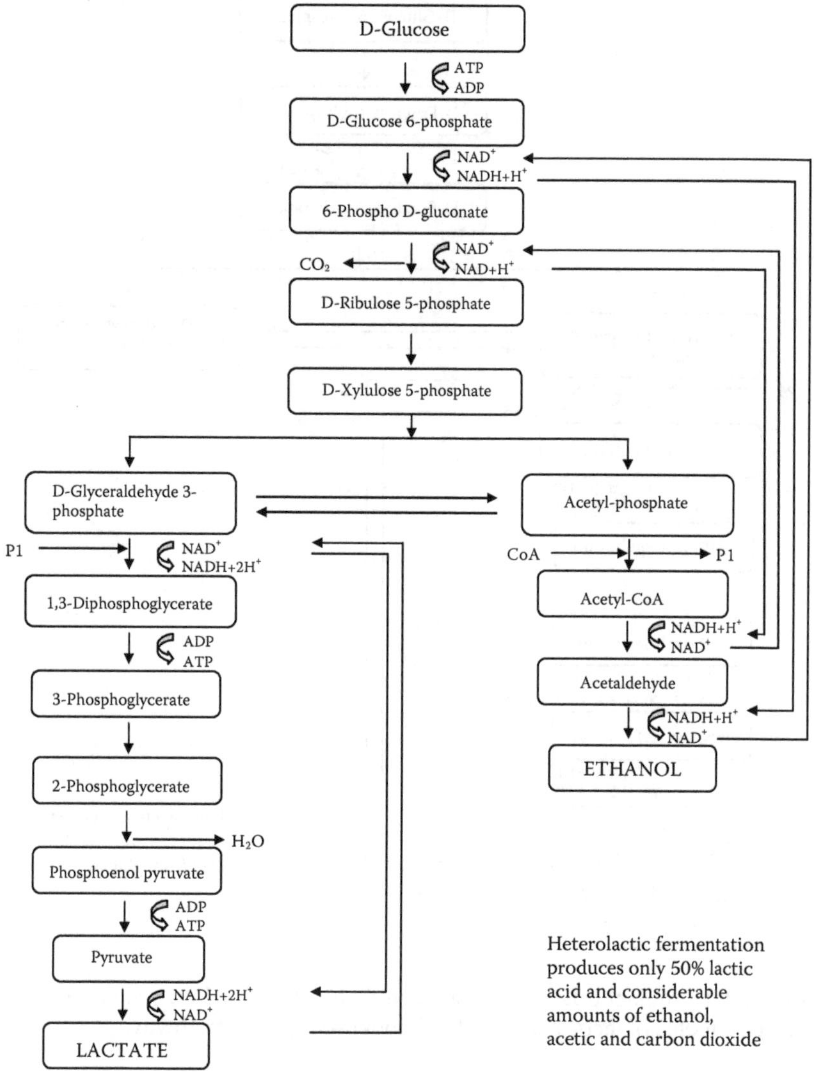

Figure 4.2 Heterolactic fermentation. The fermentation of one mole of glucose yields one mole each of lactic acid, ethanol, and carbon dioxide, via the 6-phosphoglucanate/phosphoketolase pathway.

such as different neutralizing agents (Zhou, Du, and Tsao, 2002). Reducing the amount of fumaric acid in a broth can benefit the downstream process and subsequently the quality of the l-lactic acid produced (Wang et al., 2005).

4.4 Other applications of LAB

LAB have, because of their numerous qualities, been evaluated for several other uses in addition to bioprotection. One such investigation involved the fermentation of raw fish to change its sensory qualities in attempting to obtain fish products that organoleptically resemble meat products and would meet the increasing concern of consumers for safety, quality, and nutritive value in seafood. *Leuconostoc mesenteroides* was indicated to be a promising candidate for use as such a starter culture for fish fermentation and for the development of new fish products (Gelman, Drabkin, and Glatman, 2000).

LAB are increasingly accepted as exhibiting probiotic properties. Probiotics are generally defined as viable microorganisms that, when ingested in specified quantities, exert health benefits in addition to inherent basic nutrition (Gotcheva et al., 2002). Probiotics aid in stimulating immune responses and as a result prevent infection by enteropathogenic bacteria and prevent diarrhea. Probiotics have also been recognized to enhance treatment of diarrhea (Reid, 1999; Mohd Adnan and Tan, 2007). Recently a *PROSAFE* collection of probiotic and human lactic acid bacteria was established. The collection was well documented and consisted of 907 LAB strains of nutritional or human origin. With this collection a project was launched to investigate the biosafety of LAB used for human consumption. The majority of these strains belonged to the genera *Lactobacillus*, *Bifidobacterium*, and *Enterococcus* (Vankerckhoven et al., 2004).

Antibiotic treatment often disrupts the healthy balance of the intestinal microflora of the human host, which may cause intestinal disorders. The administration of antibiotic-resistant LAB strains have also been proposed to help restore the normal bacterial ratio in the intestinal tract after or during intense antibiotic treatment (Gotcheva et al., 2002).

4.5 Genetic and bioinformatic characterization of LAB

Genetic analysis of lactic acid synthesis in LAB provides the necessary information and the potential to modify the rate of lactic acid production by genetic techniques (Davidson et al., 1995). It is, for example, possible, by selecting cells with higher pH_i, to identify strains exhibiting enhanced lactic acid production (Valli et al., 2006). Recently 29 ongoing genome-sequencing projects were described: these comprised LAB (24), bifidobacteria (3), propionic acid bacteria (1), and brevibacterium (1) (Nes and Johnsborg, 2004).

Identification of the unique metabolic properties of LAB are crucial in the synthesis of antimicrobial molecules and provide a starting point for improvement of the antimicrobial potential of a specific strain (Nes and Johnsborg, 2004). Prolonged batch fermentation with the cells of *L. plantarum* when using an immobilized biocatalyst indicate a shift in the metabolic pathway from homofermentative to heterofermentative. This also results in morphological changes in immobilized cells from the normal rod shape to coccoid. These changes are proposed to be related to a shift in the bacterial cell metabolism, which then ultimately results in a decrease in lactic acid yield (Krishnan et al., 2001). The *L. lactis* species is so far known to be the best characterized species among the LAB (Kim, Jeong, and Lee, 2007).

4.6   Acetic acid bacteria (AAB)

4.6.1   Acetic acid (vinegar) production

Vinegar is as old as the most ancient civilizations. It has been used since the earliest of time as a seasoning or preservative agent. Its domestic use may be as ancient as the use of wine (Tesfaye et al., 2002). Vinegar has been defined as "a liquid fit for human consumption, produced from a suitable raw material of agricultural origin, containing starch, sugars, or starch and sugars by the process of double fermentation, alcoholic and acetous, and contains a specified amount of acetic acid" (FAO/WHO, 1987). Two genera of acetic acid bacteria are known in the production of vinegar: *Acetobacter* and *Gluconobacter*. Acetic acid fermentation is a thermodynamically favorable strictly aerobic biochemical process of biological oxidation by which *Acetobacter* spp. oxidize ethanol into acetic acid (Tesfaye et al., 2002).

4.6.2   Microorganisms involved in the production of vinegar

4.6.2.1   Acetobacter and yeasts

Acetic acid bacteria (AAB) are known for their ability to oxidize different substrates containing ethanol into various types of vinegar (Trček, 2005). Substrates used as energy sources include glucose, ethanol, lactate, or glycerol. Most of these compounds are not completely oxidized into CO_2 and H_2O and many metabolites, such as acetic acid, may be accumulated in the growth medium. AAB are commonly found in nature and acetic acid is a key product of their metabolism. As a result of their presence and activity, acetic acid is found in many foods (Gonzalez et al., 2005).

Acetic acid is also secreted in high levels by certain yeasts, such as *Brettanomyces* and *Dekkera* (Piper et al., 2001). However, these have been implicated mostly in the spoilage of wine fermentations. Acetic acid is known to be the major volatile acid in wine and one of the main reasons for wine spoilage, when exceeding levels of 1.2–1.4 g/l. AAB are the main oxidative microorganisms that are able to survive in high acidic and high ethanol conditions, such as in wine (Gonzalez et al., 2005), with *Acetobacter* and *Gluconobacter* two of the AAB genera commonly associated with wine spoilage (Bartowsky et al., 2003).

The production of vinegar involves both yeasts and bacteria in a mixed fermentation process. This process is usually initiated by the yeasts, which convert glucose into ethanol and carbon dioxide

Yeast reaction in the production of acetic acid (FAO, 1998).

$$C_2H_{12}O_6 \rightarrow 2C_2H_5OH + 2CO_2$$
$$Glucose \quad ethanol\ carbon \quad dioxide$$

The bacterial part of the fermentation then follows and the *Acetobacter* oxidizes the ethanol to acetic acid and water:

Bacterial reaction in the production of acetic acid (FAO 1998).

$$2\ C_2H_5OH + O_2 \quad CH_3COOH + H_2O$$
$$Ethanol \quad acetic\ acid \quad water$$

The *Acetobacter* is, therefore, dependent on the yeasts for production of an oxidizable substance such as ethanol. Other organic acids may also be produced during this fermentation process. These are then converted to esters and may contribute to the odor, flavor, and color of the vinegar (FAO, 1998).

Decomposing plant materials are known to often contain high organic acid concentrations, which have been attributed to the growth and production by saprophytic fungi (Piper et al., 2001). AAB are also used for production of compounds such as sorbose and gluconic acids which are essential agents for biotechnology (Trček, 2005). But, again, their presence is often not welcome and regarded as spoilage of wine, juice, beer, and fruit. AAB are highly resistant to acidity and can use a wide variety of substrates, which places them under some of the main food spoilage microorganisms. However, the presence of AAB in food is mainly associated with the alteration of food or human activities related to food preservation (Gonzalez et al., 2005).

4.6.3   Industrial importance—essential versus undesirable

Apart from the presence of organic acids in foodstuffs such as vinegar, being important as a preservative, their level and nature in vinegars may also provide essential information with regard to the origin of the raw material and the status of microbiological growth present, as well as the processing techniques (Morales, Gonzalez, and Troncoso, 1998). AAB spoilage is more common during bulk storage and handling in the wineries and less often after bottling (Drysdale and Fleet, 1988), although unusual outbreaks of bottled wine spoilage by acetic acid bacteria have been reported. This may be caused by a combination of circumstances, including bottling without sterile filtration, lower levels of sulphur dioxide in the wine prior to bottling, or bottles stored in an upright position. The upright storage of wine bottles is known to encourage the entrance of air if cork closure is not adequate, and can also provide a diverse environment from which bacterial growth focus may develop and migrate to the surface of the wine (Bartowsky et al., 2003).

4.6.4   Glucose, acid, and ethanol tolerance

One of the greatest hurdles when studying AAB is culturing and subsequent maintenance in pure culture, especially when strains are isolated from sources containing high levels of acid (Entani et al., 1985; Sievers, Sellmer, and Teuber, 1992; Gullo et al., 2006). High levels of intracellular acetic acid are characteristic of AAB (Nakano, Fukaya, and Horinouchi, 2006). Resistance to an acid environment (pH 2.5–3.5) and the requirement by AAB of acetic acid for growth are characteristically variable among different species and genera (Trček et al., 2000; Gullo et al., 2006). Acetic acid resistance in species such as *Acetobacter aceti* may be caused by at least two mechanisms (1) the destruction of acetic acid by enzymes produced by these organisms, such as citrate synthase and aconitase, or (2) the export of acetic acid by a transporter, known as the ABC transporter. These mechanisms reduce the concentration of intracellular acetic acid. It is also known that the overexpression of *aatA* improves growth in the presence of high acetic acid concentration and increases the final yield of acetic acid. This may be the result of low levels of intracellular acetic acid. The gene *aatA* is widely distributed in the genera *Acetobacter* and *Gluconobacter* and it is well known that *aatA* plays an important role in acetic acid resistance in these bacteria (Nakano, Fukaya, and Horinouchi, 2006).

Various strains of the AAB are ethanol-tolerant, and the majority of strains have been found to be able to grow at 5% v/v ethanol, and some even at 10%. Glucose tolerance has often been described in various AAB, and is frequently found in strains isolated from traditional balsamic vinegar. One of the inhibiting substances for acetic acid bacteria is high sugar concentrations, but the majority are only inhibited by 25% glucose. Although sugar tolerance is not an important technological feature of AAB involved in industrial vinegar production, it may be required for the production of traditional balsamic vinegar (Gullo et al., 2006).

4.7   Susceptibility of and resistance to organic acids

Lactic acid bacteria are much more resistant to effects of weak organic acids (Fielding, Cook, and Grandison, 1997). Lactobacilli are much more resistant to acetic, benzoic, butyric, and lactic acids than the other organisms tested (Hsiao and Siebert, 1999). LAB are more resistant to acidic conditions because they can tolerate a lower intracellular pH than many other bacteria (Adams and Nicolaides, 1997). Acidic conditions in the human gastrointestinal tract are known to have a bactericidal effect on pathogenic harmful bacteria as they cause a more favorable environment for lactobacilli. The lactobacilli may then secrete metabolites that act against Gram-negative bacteria and also inhibit colonization and proliferation of *Escherichia coli* (Tsiloyiannis et al., 2001).

Although AAB are naturally resistant to acetic acid, notable differences in tolerance are seen between species (Trček et al., 2006). The highest resistance is described for *Gluconobacter europaues, Gluconobacter intermedius, Gluconobacter oboediens,* and *Gluconobacter entanii* (Sievers and Teuber, 1995). Differences in resistance are seen, for example, between *G. europaeus* and *G. intermedius* where the resistance appears to have a stable character, whereas resistance in *Acetobacter pasteurianus* is more temporary (Trček et al., 2006). A gene cluster responsible for acetic acid resistance in *Acetobacter acetii* has been identified. This cluster comprises *aarA* (encoding a citrate synthase), *aarB* (encoding a protein of which the function is not known), and *aarC* (encoding a protein involved in acetic acid absorption). Mutation and overexpression of the *aatA* gene have been reported to affect resistance to other organic acids such as formic and propionic acids, concurrently with resistance to acetic acid. The *aatA* functions as an efflux pump of acetic acid (Nakano, Fukaya, and Horinouchi, 2006).

Various protein production changes are also known to be implicated in the development of resistance to acetic acid. One important protein that has been identified is now known as "aconitase." It may also be possible that there exists some other mechanism, located in the bacterial cell membrane by which acetic acid resistance is conferred, as it is known that acetic acid causes toxicity by acting as an uncoupling agent, which would disturb the proton motive force. The presence of such a proton motive efflux system for acetic acid is present in *A. acetii* (Matsushita et al., 2005).

It is, however, unclear if it contributes to acetic acid resistance in acetic fermentation (Nakano, Fukaya, and Horinouchi, 2006).

In *A. acetii* a defect in the membrane-bound alcohol dehydrogenase has also been associated with a reduction in acetic acid resistance, implicating this enzyme in the development of resistance to this acid. It has also been found that resistance to acetic acid does not always result from resistance to low pH, as strains previously able to grow at a low pH cannot grow when the pH is decreased as a result of acetic acid (Nakano, Fukaya, and Horinouchi, 2006).

4.8   Other organisms

4.8.1   Fungi

Various fungi are also known to produce organic acids, some of which are important in industrial applications. *Aspergillus niger*, a filamentous fungus, secretes large quantities of organic acids, such as citric acid and gluconic acid, and may acidify the surrounding environment to pH values below pH 2.0. However, the intracellular pH of the fungus is not significantly altered (Plumridge et al., 2004). Although citric acid, one of the most commonly used organic acids in the food and pharmaceutical industries, can be obtained by chemical synthesis, the cost is much higher than using fermentation. As such it is mainly produced by submerged fermentation (SmF), by *A. niger*. However, solid-state fermentation (SSF) has recently been proposed as a potential alternative to SmF, to increase the efficiency of citric acid production (Couto and Sanroman, 2006).

A. niger holds GRAS status from the FDA for production of organic acids such as citric acid, despite reports of its ability to produce ochratoxin A. Of biotechnological interest the primary uses of *A. niger* have been found to be the production of enzymes and organic acids by fermentation, which have long been used by the food industry, without any evident adverse effects on human health. From the beginning of the previous century, *A. niger* has been used for biotechnological production of organic acids, such as citric acid and gluconic acid. These remain the only organic acids produced by mycological processes and are used in considerable quantities as food additives (Abarca et al., 2004). Microbial production of gluconic acid is the preferred method and dates back several decades (Ramanchandran et al., 2006).

Rhizopus oryzae is another important fungus involved in the production of organic acids in industrial fermentation. It is widely used to produce L-lactic acid as well as other organic acids. *R. oryzae*, produces only one stereospecific product (L-lactic acid), and not a racemic mixture (Wang et al., 2005). Much research has been done on the mechanism of lactic acid production by *Rhizopus* species, such as the confirmation of three lactate dehydrogenase enzymes present in *R. oryzae*, including one NAD-independent LDH and two NAD-dependent LDH isozymes (Pritchard, 1973; Skory, 2000). Strains of *Rhizopus* species have been divided into two groups, lactic acid and fumaric acid (malic acid) on the basis of the yield of the main organic acids and proportion of metabolites produced (Zhang, Jin, and Kelly, 2007).

Fumaric acid is another organic acid produced by *Rhizopus*, in particular *Rhizopus stolonifer*. The presence of fumaric acid has also been found to arise from the addition of synthetic malic acid and is sometimes considered as an index of adulteration when confirmed by analysis of D-malic acid (not present in malic acid from natural sources) (Trifiro et al., 1997).

In 2006 *Saccharomyces cerevisiae* was found for the first time to be able to produce large amounts of lactic acid, which is related to its ability to maintain a higher pH (Valli et al., 2006). Contrary to this finding, various yeasts are known to produce low amounts of acetic acid (Trifiro et al., 1997).

4.8.2   Other bacteria

Other bacterial genera that are popular organic acid producers include the propionibacteria, and more specifically *Propionibacterium freudenreichii*. This bacterium is known to produce propionic acid, lactic acid, and acetic acid (Tyree, Clausen, and Gaddy, 1991). *P. freudenreichii* produces a mixture of lactic and acetic acids from sugar, which is considered to be a potentially safe food preservative (Taniguchi et al., 1998). Propionibacteria are able to utilize lactate as a substrate much more quickly than glucose (Tyree, Clausen, and Gaddy, 1991). *P. freudenreichii* is also able to convert lactic acid to acetic and propionic acids (Taniguchi et al., 1998).

Listeria monocytogenes, which is considered more a pathogenic bacterium, is able to produce L-lactic acid (Gravesen et al., 2004). This is an additional mechanism of self-protection against the intrinsically produced isomer and could involve an enzymatic reaction (e.g., metabolization of the intracellular lactate) or the presence of a stereospecific efflux system for the produced isomer. *L. monocytogenes* can use lactate as a carbohydrate source and as a result, stereospecific antimicrobial activity of lactic acid does not have considerable impact as an antimicrobial agent against *L. monocytogenes*. There may, however, be species–species variation in sensitivity to the two lactic acid isomers, which may consequently influence the development of either beneficial or unwanted bacteria. A better understanding of the mechanisms involved could assist in developing optimal application guidelines for lactic acid or lactic acid producers in food production and preservation (Gravesen et al., 2004).

References

Abarca, M.L., Accensi, F., Cano, J., and Cabanes, F.J. 2004. Taxonomy and significance of black aspergilli. *Antonie Van Leeuwenhoek* 86:33–49.

Adams, M.R. 1999. Safety of industrial lactic acid bacteria. *Journal of Biotechnology* 68:171–178.

Adams, M.R. and Nicolaides, L. 1997. Review of the sensitivity of different foodborne pathogens to fermentation. *Food Control* 8:227–239.

Aguirre, M. and Collins, M.D. 1993. Lactic acid bacteria and human clinical infection. *Journal of Applied Microbiology* 75:95–107.

Ammor, S., Tauveron, G., Dufour, E., and Chevallier, I. 2006. Antibacterial activity of lactic acid bacteria against spoilage and pathogenic bacteria isolated from the same meat small-scale facility: 1-Screening and characterization of the antibacterial compounds. *Food Control* 17:454–461.

Aslim, B., Yuksekdag, Z.N., Sarikaya, E., and Beyatli, Y. 2005. Determination of the bacteriocin-like substances produced by some lactic acid bacteria isolated from Turkish dairy products. *Swiss Society of Food Science and Technology* 38: 691–694.

Axelsson, L. 2004. Lactic acid bacteria: Classification and physiology. In: S. Salminen, A. Von Wright, and A. Ouwehand (Eds.), *Lactic Acid Bacteria: Microbiological and Functional Aspects*, pp. 1–66. New York: Marcel Dekker.

Barbosa-Cánovas, G.V., Fernández-Molina, J.J., Alzamora, S.M., Tapia, M.S., López-Malo, A., and Chanes, J.W. 2003. General considerations for preservation of fruits and vegetables. In: *Handling and Preservation of Fruits and Vegetables by Combined Methods for Rural Areas*. Rome: Food and Argriculture Organization of the United Nations.

Bartowsky, E.J., Xia, D., Gibson, R.L., Fleet, G.H., and Henschke, P.A. 2003. Spoilage of bottled red wine by acetic acid bacteria. *Letters in Applied Microbiology* 36:307–314.

Couto, S.R. and Sanroman, M.A. 2006. Application of solid-state fermentation to food industry—A review. *Journal of Food Engineering* 76:291–302.

Davidson, B.E., Llanos, R.M., Cancilla, M.R., Redman, N.C., and Hillier, A.J. 1995. Current research on the genetics of lactic acid production in lactic acid bacteria. *International Dairy Journal* 5:763–784.

Drysdale, G.S. and Fleet, G.H. 1988. Acetic acid bacteria in winemaking: A review. *American Journal of Enology and Viticulture* 39:143–154.

Entani, E., Ohmori, S., Masai, H., and Suzuki, K.I. 1985. *Acetobacter polyoxogenes* sp. nov., a new species of an acetic acid bacterium useful for producing vinegar with high acidity. *Journal of General and Applied Microbiology* 31:475–490.

FAO Agricultural Services Bulletin. 1998. *Fermented Fruits and Vegetables: A Global Perspective*. Chapter 5.

FAO/WHO Food Standards Programme. 1987. Codex standards for sugars, cocoa products and chocolate and miscellaneous. Codex standard for vinegar. In Codex Alimentarius. Regional European Standard, Codex Stan 162. Ginebra.

Fielding, L.M., Cook, P.E., and Grandison, A.S. 1997. The effect of electron beam irradiation, combined with acetic acid, on the survival and recovery of *Escherichia coli* and *Lactobacillus curvatus*. *International Journal of Food Microbiology* 35:259–265.

Galvez, A., Abriouel, H., Lopez, R.L., and Ben, O.N. 2007. Bacteriocin-based strategies for food biopreservation. *International Journal of Food Microbiology* 120:51–70.

Gasser, F. 1994. Safety of lactic acid bacteria and their occurrence in human clinical infections. *Bulletin d'Institut Pasteur* 92:45–67.

Gelman, A., Drabkin, V., and Glatman, L. 2000. Evaluation of lactic acid bacteria, isolated from lightly preserved fish products, as starter cultures for new fish-based food products. *Innovative Food Science and Emerging Technologies* 1:219–226.

Gianotti, A., Vannini, L., Gobbetti, M., Corsetti, A., Gardini, F., and Guerzoni, M. 1997. Modeling of the activity of selected starters during sourdough fermentation. *Food Microbiology* 14:327–337.

Gonzalez, A., Hierro, N., Poblet, M., Mas, A., and Guillamon, J.M. 2005. Application of molecular methods to demonstrate species and strain evolution of acetic acid bacteria population during wine production. *International Journal of Food Microbiology* 102:295–304.

Gotcheva, V., Hristozova, E., Hristozova, T., Guo, M., Roshkova, Z., and Angelov, A. 2002. Assessment of potential probiotic properties of lactic acid bacteria and yeast strains. *Food Biotechnology* 16:211–225.

Gravesen, A., Diao, Z., Voss, J., Budde, B.B., and Knochel, S. 2004. Differential inactivation of *Listeria monocytogenes* by d- and l-lactic acid. *Letters in Applied Microbiology* 39:528–532.

Gullo, M., Caggia, C., De, V.L., and Giudici, P. 2006. Characterization of acetic acid bacteria in "traditional balsamic vinegar." *International Journal of Food Microbiology* 106:209–212.

Harty, D.W.S., Oakey, H.J., Patrikakis, M., Hume, E.B.H., and Knox, K.W. 1994. Pathogenic potential of lactobacilli. *International Journal of Food Microbiology* 24:179–189.

Herreros, M.A., Sandoval, H., González, L., Castro, J.M., Fresno, J.M., and Tornadijo, M.E. 2005. Antimicrobial activity and antibiotic resistance of lactic acid bacteria isolated from Armada cheese (a Spanish goat's milk cheese). *Food Microbiology* 22:455–459.

Hsiao, C.P. and Siebert, K.J. 1999. Modeling the inhibitory effects of organic acids on bacteria. *International Journal of Food Microbiology* 47:189–201.

Huis in't Veld, J.H.J. 1996. Microbial and biochemical spoilage of foods: An overview. *International Journal of Food Microbiology* 33:1–18.

Kim, J.E., Jeong, D.W., and Lee, H.J. 2007. Expression, purification, and characterization of arginine deiminase from *Lactococcus lactis* ssp. lactis ATCC 7962 in *Escherichia coli* BL21. *Protein Expression and Purification* 53:9–15.

Krishnan, S., Gowda, L.R., Misra, M.C., and Karanth, N.G. 2001. Physiological and morphological changes in immobilized *L. plantarum* NCIM 2084 cells during repeated batch fermentation for production of lactic acid. *Food Biotechnology* 15:193–202.

Lindgren, S.E. and Dobrogosz, W.J. 1990. Antagonistic activities of lactic acid bacteria in food and feed fermentation. *FEMS Microbiology Reviews* 7:149–163.

Liu, S.Q. 2003. Practical implications of lactate and pyruvate metabolism by lactic acid bacteria in food and beverage fermentations. *International Journal of Food Microbiology* 83:115–131.

Matsushita, K., Inoue, T., Adachi, O., and Toyama, H. 2005. *Acetobacter aceti* possesses a proton motive force-dependent efflux system for acetic acid. *Journal of Bacteriology* 187:4346–4352.

McEntire, J.C., Montville, T.J., and Chikindas, M.L. 2003. Synergy between nisin and select lactates against *Listeria monocytogenes* is due to the metal cations. *Journal of Food Protection* 66:1631–1636.

McGee, H. 1984. *On Food and Cooking*, p. 36. London: Unwin Hyman.

McKay, L.L. and Baldwin, K.A. 1990. Applications for biotechnology: Present and future improvements in lactic acid bacteria. *FEMS Microbiology Reviews* 87:3–14.

Meignen, B., Onno, B., Gelinas, P., Infontes, M., Guilois, S., and Cahagnier, B. 2001. Optimization of sourdough fermentation with *Lactobacillus brevis* and baker's yeast. *Food Microbiology* 18:239–245.

Messens, W. and De Vuyst, L. 2002. Inhibitory substances produced by *Lactobacilli* isolated from sourdoughs—A review. *International Journal of Food Microbiology* 72:31–43.

Mohd Adnan, A.F. and Tan, I.K.P. 2007. Isolation of lactic acid bacteria from Malaysian foods and assessment of the isolates for industrial potential. *Bioresource Technology* 98:1380–1385.

Morales, M.L., Gonzalez, A.G., and Troncoso, A.M. 1998. Ion-exclusion chromatographic determination of organic acids in vinegars. *Journal of Chromatography A* 822:45–51.

Nakano, S., Fukaya, M., and Horinouchi, S. 2006. Putative ABC transporter responsible for acetic acid resistance in *Acetobacter aceti*. *Applied and Environmental Microbiology* 72:497–505.

Nes, I.F. and Johnsborg, O. 2004. Exploration of antimicrobial potential in LAB by genomics. *Current Opinion in Biotechnology* 15:100–104.

O'Sullivan, L., Ross, R.P., and Hill, C. 2002. Potential of bacteriocin-producing lactic acid bacteria for improvements in food safety and quality. *Biochimi* 84:593–604.

Piper, P., Calderon, C.O., Hatzixanthis, K., and Mollapour, M. 2001. Weak acid adaptation: The stress response that confers yeasts with resistance to organic acid food preservatives. *Microbiology* 147:2635–2642.

Plumridge, A., Hesse, S.J., Watson, A.J., Lowe, K.C., Stratford, M., and Archer, D.B. 2004. The weak acid preservative sorbic acid inhibits conidial germination and mycelial growth of *Aspergillus niger* through intracellular acidification. *Applied and Environmental Microbiology* 70:3506–3511.

Pritchard, G.G. 1973. Factors affecting the activity and synthesis of NAD-dependent lactate dehydrogenase in *Rhyzopus oryzae*. *Journal of General Microbiology* 78:125–137.

Radin, D., Niebuhr, S.E., and Dickson, J.S. 2006. Impact of the population of spoilage microflora on the growth of *Listeria monocytogenes* on frankfurters. *Journal of Food Protection* 69:679–681.

Ramanchandran, S., Fontanille, P., Pandey, A., and Larroche, C. 2006. Gluconic acid: Properties, applications and microbial production. *Food Technology and Biotechnology* 44:185–195.

Reid, G. 1999. The scientific basis for probiotic strains of *Lactobacillus*. *Applied and Environmental Microbiology* 65:3763–3766.

Rossland, E., Langsrud, T., Granum, P.E., and Sorhaug, T. 2005. Production of anti-microbial metabolites by strains of *Lactobacillus* or *Lactococcus* co-cultured with *Bacillus cereus* in milk. *International Journal of Food Microbiology* 98:193–200.

Samelis, J., Bedie, G.K., Sofos, J.N., Belk, K.E., Scanga, J.A., and Smith, G.C. 2005. Combinations of nisin with organic acids or salts to control *Listeria monocytogenes* on sliced pork bologna stored at 4°C in vacuum packages. *Lebensmittel-Wissenschaft und-Technologie* 38:21–28.

Schnürer, J. and Magnusson, J. 2005. Antifungal lactic acid bacteria as bio-preservatives. *Trends in Food Science and Technology* 16:70–78.

Schüller, G., Hertel, C., and Hammes, W.P. 2000. *Gluconoacetobacter entanii* sp. nov., isolated from submerged high-acid industrial vinegar fermentations. *International Journal of Systematic and Evolutionary Microbiology* 50:2013–2020.

Sidhu, M.S., Langsrud, S., and Holck, A. 2001. Disinfectant and antibiotic resistance of lactic acid bacteria isolated from the food industry. *Microbial Drug Resistance* 7:73–83.

Sievers, M. and Teuber, M. 1995. The microbiology and taxonomy of *Acetobacter europaeus* in commercial vinegar production. *Journal of Applied Bacteriology* 79:84–95.

Sievers, M., Sellmer, S., and Teuber, M. 1992. *Acetobacter europaeus* sp. nov., a main component of industrial vinegar fermenters in central Europe. *Systematic and Applied Microbiology* 15:386–392.

Sims, W. 1964. A pathogenic *Lactobacillus*. *Journal of Pathology and Bacteriology* 87:99–105.

Şimşek, Ö., Çon, A.H., and Tulumoğlu, Ş. 2006. Isolating starter cultures with anti-microbial activity for sourdough processes. *Food Control* 17:263–270.

Skory, C.D. 2000. Isolation and expression of lactate dehydrogenase genes from *Rhizopus oryzae*. *Applied and Environmental Microbiology* 66:2343–2348.

Taniguchi, M., Nakazawa, H., Takeda, O., Kaneko, T., Hoshino, K., and Tanaka T. 1998. Production of a mixture of antimicrobial organic acids from lactose by co-culture of *Bifidobacterium longum* and *Propionibacterium freudenreichii*. *Bioscience, Biotechnology and Biochemistry* 62:1522–1527.

Tesfaye, W., Morales, M.L., Garcia-Parrilla, M.C., and Troncoso, A.M. 2002. Wine vinegar: Technology, authenticity and quality evaluation. *Trends in Food Science and Technology* 13:12–21.

Tome, E., Teixeira, P., and Gibbs, P.A. 2006. Anti-listerial inhibitory lactic acid bacteria isolated from commercial cold smoked salmon. *Food Microbiology* 23:399–405.

Tompkin, R.B. 1986. Microbiology of ready-to-eat meat and poultry products. In: A.M. Pearson and T.R. Dutson (Eds.), *Advances in Meat Research, Vol. 2: Meat and Poultry Microbiology*, pp. 89–122. Westport, CT: AVI.

Topisirovic, L., Kojic, M., Fira, D., Golic, N., Strahinic, I., and Lozo, J. 2006. Potential of lactic acid bacteria isolated from specific natural niches in food production and preservation. *International Journal of Food Microbiology* 112:230–235.

Trček, J. 2005. Quick identification of acetic acid bacteria based on nucleotide sequences of the 16S-23S rDNA internal transcribed spacer region and of the PQQ-dependent alcohol dehydrogenase gene. *Systematic & Applied Microbiology* 28:735–745.

Trček, J., Raspor, P., Teuber, M. 2000. Molecular identification of *Acetobacter* isolates from submerged vinegar production, sequence analysis of plasmid $_p$JK2-1 and application in the development of a cloning vector. *Applied Microbiology and Biotechnology* 53:289–295.

Trček, J., Toyama, H., Czuba, J., Misiewicz, A., and Matsushita, K. 2006. Correlation between acetic acid resistance and characteristics of PQQ-dependent ADH in acetic acid bacteria. *Applied Microbiology and Biotechnology* 70:366–373.

Trifiro, A., Saccani, G., Gherardi, S., et al. 1997. Use of ion chromatography for monitoring microbial spoilage in the fruit juice industry. *Journal of Chromatography A* 770:243–252.

Tsiloyiannis, V.K., Kyriakis, S.C., Vlemmas, J., and Sarris, K. 2001. The effect of organic acids on the control of post-weaning oedema disease of piglets. *Research in Veterinary Sciences* 70:281–285.

Tyree, R.W., Clausen, E.C., and Gaddy, J.L. 1991. The production of propionic acid from sugars by fermentation through lactic acid as an intermediate. *Journal of Chemical Technology and Biotechnology* 50:157–166.

Valli, M., Sauer, M., Branduardi, P., Borth, N., Porro, D., and Mattanovich, D. 2006. Improvement of lactic acid production in *Saccharomyces cerevisiae* by cell sorting for high intracellular pH. *Applied and Environmental Microbiology* 72:5492–5499.

Vankerckhoven, V.V., Van Autgaerden, T., Huys, G., Vancanneyt, M., Swings, J., and Goossens, H. 2004. Establishment of the PROSAFE collection of probiotic and human lactic acid bacteria. *Microbial Ecology in Health and Disease* 16:131–136.

Wang, X., Sun, L., Wei, D., and Wang, R. 2005. Reducing by-product formation in l-lactic acid fermentation by *Rhizopus oryzae*. *Journal of Industrial Microbiology and Biotechnology* 32:38–40.

Zaunmuller, T., Eichert, M., Richter, H., and Unden, G. 2006. Variations in the energy metabolism of biotechnologically relevant heterofermentative lactic acid bacteria during growth on sugars and organic acids. *Applied Microbiology and Biotechnology* 72:421–429.

Zhang, Z.Y., Jin, B., and Kelly, J.M. 2007. Production of lactic acid from renewable materials by *Rhizopus* fungi. *Biochemical Engineering Journal* 35:251–263.

Zhou, Y., Du, J., and Tsao, G.T. 2002. Comparison of fumaric acid production by *Rhizopus oryzae* using different neutralizing agents. *Bioprocess and Biosystems Engineering* 25:179–181.

chapter five

Mechanisms of microbial inhibition

5.1   Introduction

A number of inhibition mechanisms have been proposed for weak-acid preservatives, including action on nutrient uptake or energy metabolism, action as uncouplers, or through accumulation (Stratford and Anslow, 1998). It is becoming increasingly important to understand the mechanisms involved in the action of preservatives (Papadimitriou et al., 2007). It has been demonstrated, however, that weak-acid preservation only inhibits growth by causing extended lag phases and does not actually kill the microorganisms (Fernandes et al., 2005). Organic acids are, however, effective food preservatives by primarily acting as acidulants and reducing bacterial growth by lowering the pH of food products to levels that will inhibit bacterial growth (Dziezak, 1990; Hinton, Jr., 2006).

Although the weak organic acids have been used as food preservatives for centuries, only recently have the mechanisms involved in bacterial inhibition been investigated (Hirschfield, Terzulli, and O'Byrne, 2003). It has become imperative to achieve a better understanding of how microbial cells respond to preservatives in order to design improved preservation strategies (Papadimitriou et al., 2007). It has, however, become clear that the mechanism of microbial growth inhibition executed by the organic acids is complicated and has not yet been elucidated (Chaveerach et al., 2002; Taniguchi et al., 1998).

The key basic principle of the mode of action is that the acid in its undissociated state is able to penetrate the microbial cell wall to disrupt the normal physiology of the cell as it is not able to tolerate a major change in its internal pH (Gauthier, 2005). The acids can accumulate within the cell and in this concentrated state reduce the internal pH. This pH drop then destabilizes cell proteins (Price-Carter et al., 2005). Under acidic conditions the undissociated organic acids are believed to freely cross the microbial membrane to the cytoplasm. Inside the cell the more neutral pH results in a release of protons, acidification of the cytoplasm, and also dissipation of the membrane pH gradient (Gravesen et al., 2004). However, this mechanism does not account entirely for activity. The accumulation of anions has also been shown to exert a toxic effect (Russell, 1992). Scientific data are lacking to describe the exact role of an organic acid in determining conditions, independent of pH, that will cause growth inhibition of acid-resistant pathogens in acidified foods, without other

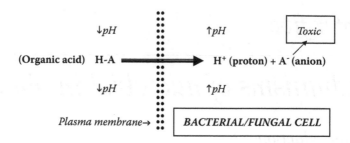

Figure 5.1 A schematic illustration of the transport system and consequent proposed action of organic acids on microbial cells.

preservative treatments. Organic acid type and concentration as well as pH can independently affect the growth and death of bacterial cells, or they can interact to accomplish more effective inhibition (Breidt, Jr., Hayes, and McFeeters, 2004).

In Figure 5.1 the transport and proposed antimicrobial activity of organic acids as antimicrobial agents is schematically illustrated.

5.2   Activity of organic acids

The various antimicrobial substances available target a range of cellular loci, ranging from the cytoplasmic membrane to respiratory functions, enzyme production, and genetic components (Cloete, 2003). In any organism it is essential that a stable intracellular pH be maintained for normal cellular functions such as gene expression, protein synthesis, and enzyme activity (Plumridge et al., 2004). The mode of action of organic acids is related to several inhibitory reactions. The following reactions have thus far been proposed (Barbosa-Cánovas et al., 2003; Russell, 2003; Gauthier, 2005):

1. Undissociated organic acids entering the bacterial cell. A schematic representation of this action is illustrated in Figure 5.1.
2. Acidification of internal components of cell membranes by the undissociated acid molecule.
3. Bacterial membrane disrupting (leakage, transport mechanisms). Therefore, loss of active transport of nutrients through the membrane.
4. Disruption of acid transport by alteration of cell membrane permeability.
5. Lowering the water activity (a_w).
6. Inhibition of essential metabolic reactions (e.g., glycolysis).
7. Accumulation of toxic anions.
8. Energy stress response to restore homeostasis.

9. Chelation as permeabilizing agent of outer membrane and zinc binding.
10. Act as inhibitors of other stress responses, for example, heat-shock response (Piper et al., 2001).

In explaining the acidity of organic acids, an acid in solution sets up the equilibrium: $AH + H_2O \leftrightarrow A^- + H_3O^+$, where a hydroxonium ion is formed together with the negative ion (anion) from the acid. During chemical reactions a hydrogen ion is always attached to something. Organic acids are weak in the sense that this ionization is very incomplete, and at any one time, most of the acid will be present as un-ionized molecules (Price-Carter et al., 2005). The amount of undissociated acid decreases as the pH value increases. Furthermore, the proportion of an undissociated molecule can be calculated from the dissociation constant by the following formula:

$$a = \frac{[H^+]}{[H^+] + D}$$

where

a	= amount of undissociated acid
$[H^+]$	= Hydrogen concentration
D	= dissociation constant

5.3   *Physiological actions of organic acids*

5.3.1   *Introduction*

There is still little understanding of the effectiveness of the use of the classical and natural preservatives, such as the organic acids, in conjunction with other common components of food preservation systems. It is imperative to understand whether microorganisms die, survive, adapt, or grow and also to understand the physiological and molecular mechanisms within microbial cells that result in these phenotypes: (1) the signal transduction systems involved, (2) which stress proteins are induced, (3) how these systems are induced, (4) how much cellular energy (ATP) is involved, and (5) the amount of available energy that will determine the extent to which a given microbial cell can have various stress response pathways activated (Brul and Coote, 1999).

The antimicrobial effect and toxicity of organic acids have been attributed to (1) a combination of the hydrogen ion concentration (lowering the pH), (2) consequent decreases in proton motive force, and (3) the action of undissociated molecules that may freely permeate the cell membrane

(Carrasco et al., 2006; Fielding, Cook, and Grandison, 1997). These undissociated organic acids are lipophilic and can easily diffuse across the cell membrane (Skrivanova and Marounek, 2007). Once inside the bacterial cell they dissociate because of the pH of the cytoplasm (>7), and cause metabolic uncoupling (Kashket, 1987). The higher pH environment of the cell cytosol favors rapid dissociation of the acid molecules into charged protons and anions. These cannot diffuse back across the plasma membrane (Plumridge et al., 2004). Organic acids or the subsequent low pH may denature proteins or cause oxidative damage (carbonylation). The acids, similarly to high temperature, partially unfold proteins, which makes them more vulnerable to oxidative damage (Price-Carter et al., 2005).

Most foodborne pathogenic bacteria are neutrophiles with their optimum growth pH ranging from pH 6–pH 7. These bacteria will only encounter acid-related stress in food systems as a result of the presence of lipid-permeable weak acids (Hill, O'Driscoll, and Booth, 1995).

Although the cell membrane is intrinsically impermeable to protons, undissociated weak acids are able to pass through the membrane with relative ease and may dissociate to then liberate protons in the cytoplasm (Hill, O'Driscoll, and Booth, 1995). In a low pH solution weak acids are mainly in their undissociated form; they can then freely diffuse across the cell membrane to enter the cytosol (Papadimitriou et al., 2007). In fact, at pH below 4 most of the molecules of these organic acids are undissociated (Kasemets et al., 2006). If the pH gradient across the membrane is two units more (e.g., pH_i 7 and external pH 5), the internal concentration of the weak acid can be more than 100× greater than the external concentration, resulting in a very high concentration of protons released in the cytoplasm. For example, if the external acid concentration is 1 mM, theoretically the internal concentration would be approximately 100 mM should the cell be able to maintain a pH_i of 7.0, and would result in the release of approximately 100 mM protons into the cytoplasm (Hill, O'Driscoll, and Booth, 1995). In the cytosol the pH is neutral, and the acids dissociate to produce protons and anions and acidify the cytoplasm (Papadimitriou et al., 2007). The intracellular pH is decreased below the normal physiological range that can be tolerated by the cell, and cell growth is consequently inhibited (Kasemets et al., 2006). The release of protons then outstrips the buffering capacity of the cell, the cytoplasmic pH drops, growth is inhibited, and cell death ultimately occurs (Hill, O'Driscoll, and Booth, 1995).

Transporters for monocarboxylic acids include (1) the bacterial lactate permease LctP family, (2) eukaryotic proton-linked monocarboxylate transporter MCT family, and (3) monocarboxylate permease. These transporters contain no ATP binding cassette (ABC) motifs and are considered to transport monocarboxylic acids via a proton-coupled reaction (Nakano, Fukaya, and Horinouchi, 2006).

5.3.2 Bacterial membrane disruption

Gram-negative bacteria possess an outer membrane (OM), which protects them against a harsh environment. This outer membrane consists of embedded proteins that perform a number of crucial tasks in the bacterial cell. These include the translocation of solutes, proteins, and signal transduction (Beis et al., 2006). Some organic acids are known to function by means of disrupting this membrane organization as well as oxidative stress (Hazan, Levine, and Abeliovich, 2004). Although alcohols, ethanol in particular, are known to disrupt membrane structure, and as a result alter membrane fluidity and dissipate proton gradients, a similar degree of inhibition is shown with sorbic acid and sorbic alcohol. This will ultimately result in ethanol tolerance in preservative-resistant bacteria as well as yeasts, suggesting the common mechanisms of action (Stratford and Anslow, 1998).

Together with sorbic acid, benzoic acid also acts as a membrane perturbing agent (Hazan, Levine, and Abeliovich, 2004). Disruption of the OM by organic acids involves the action of dissociated as well as undissociated forms (Alakomi et al., 2000). High lipoid solubility and the ability to form membrane polar/hydrogen unions are responsible for benzoic interaction with cell membranes and the modification of membrane properties (Otero-Losada, 2003).

Lactic acid is another potent OM-disintegrating agent, and may also cause LPS release which will sensitize bacteria to detergents or lysozyme. This disintegration of the LPS layer can also be caused by a fully dissociated acid, whereas undissociated lactic acid molecules may cause additional disintegrating of the outer membrane. It is known that at pH 4 ~40% and at pH 3.6 ~60% lactic acid is present in an undissociated form. Dissociated potassium lactate at neutral conditions does not have any permeabilizing activity. Lactic acid is water soluble and gains access to the periplasm through water-filled porin proteins of the OM and is, therefore, capable of causing sublethal injury to Gram-negative bacteria, including *E. coli*. Acetic acid possesses similar characteristics, and direct evidence has indicated that such injury involves the disruption of the LPS layer (Alakomi et al., 2000).

5.3.3 Accumulation of toxic anions

A very important inhibitory mechanism of organic acids is that a higher internal pH than the external environment will result in dissociation and accumulation of acid anion in the interior of bacterial cells (Breidt, Jr., Hayes, and McFeeters, 2004). Depending on the difference between internal and external pH, the molar quantities of the acid anion might be predicted inside cells and greatly increase the internal ionic strength

(Russell, 1992). This anion is not able to diffuse through the bacterial membrane (Breidt, Jr., Hayes, and McFeeters, 2004).

It is now established that once inside the cell, an organic acid will encounter a higher pH, dissociate to produce protons and anions, and also acidify the cytoplasm (Papadimitriou et al., 2007). The anionic part of the organic acids that cannot exit the microbial cell in its dissociated form will, therefore, accumulate within the cell and disrupt many metabolic functions. This will also lead to an osmotic pressure increase, which the organism will not be able to handle (Gauthier, 2005).

5.3.4   Inhibition of essential metabolic reactions

The higher pH environment of the cell cytosol (e.g., in *Aspergillus niger* this pH may be as high as 7.8) would promote rapid dissociation of the acid molecules into charged protons and anions, which cannot diffuse back across the cytoplasmic membrane (Plumridge et al., 2004). After dissociation of an organic acid inside the microbial cell, situations will develop to impair growth. The anionic part of the organic acids that cannot escape the bacteria in its dissociated form will accumulate within the bacteria and disrupt many metabolic functions and lead to an osmotic pressure increase, incompatible with the survival of the bacteria (Gauthier, 2005). Intracellular acidification of the cell cytosol as a result of the accumulation of protons generally inhibits the key metabolic functions associated with glycolysis and consequently inhibits ATP yields. This reduction in proton motive force (Δp) also leads to reduced uptake of amino acids by the cell (Plumridge et al., 2004). Release of protons destroys the buffering capacity of the cell and the cytoplasmic pH falls, leading to growth inhibition and ultimately death (Hill, O'Driscoll, and Booth, 1995).

It is also possible that organic acids or a lower pH will denature proteins, or sensitize them to oxidative damage (carbonylation). This is supported by the observation that organic acids induce the synthesis of chaperones, peroxidases, and catalases. Induction of peroxidases (such as KatE and Ahp) will increase the vulnerability of a microbial cell to oxidative damage (Price-Carter et al., 2005).

5.3.5   Stress on intracellular pH homeostasis

pH homeostasis is described as the ability of an organism to maintain its cytoplasmic pH at a value close to neutrality despite fluctuations in the external pH (Hill, O'Driscoll, and Booth, 1995). As a result of organic acids diffusing into the bacterial cell, where the pH is near or above neutrality, the acids will dissociate and lower the bacterial internal pH. This would then inevitably have an effect on the ability of an organism to maintain stable intracellular pH, needed for most normal cellular functions (Gauthier, 2005).

Adaptation time to an organic acid may be directly linked to the cell's ability to maintain its internal pH (pH_i). The cell may be able to maintain its pH_i and also adapt to mildly acidic conditions but only for a short period. After this period the pH_i protection system fails and the cells become sensitive to the toxic effects of the acid. It is known that certain organic acids enter the cell much more easily than others and as a result more readily alter the pH_i of the cell. It may, therefore, be possible to explain the differences in adaptation times between acidulants by the varying abilities of organic acids to alter the pH_i of the cell (Greenacre et al., 2003).

In bacterial cells homeostasis is achieved by a combination of passive and active mechanisms (Hill, O'Driscoll, and Booth, 1995).

Passive homeostasis: The very low permeability of the membrane to protons and other ions. This plays a major role in preventing large pH_i fluctuations, which follows variations in pH of the environment. Another factor in preventing serious disruption of the pH_i is the high buffering capacity of the cell, due to the protein content of the cytoplasm, as well as the presence of glutamate and polyamines.

Active pH homeostasis: Depends primarily on the potassium ion and the proton circuits.

In a low external pH a cell must extrude those protons associated with weak acids that are entering the cell. However, such translocation of protons across the membrane produces a membrane potential that then limits further proton extrusion. Major proton movement can only be achieved if the membrane potential is degenerated as a result of cations entering the cell. This purpose is fulfilled by potassium ion entering the cell and leading to the generation of a transmembrane pH gradient. As a bacterial cell divides and grows, bacteria must accumulate sufficient potassium to maintain constant cation concentrations. The net extrusion will consequently follow this potassium uptake and the cytoplasmic pH become more alkaline.

5.4   Factors that influence organic acid activity

Antimicrobial action of organic acids is dependent on various factors. The following are discussed (Chaveerach et al., 2002): number of undissociated organic acids that enter the bacterial cell, acidity constant (pK_a value), water activity (a_w), temperature, and production of H_2O_2.

Bactericidal activity of different acids on a bacterial cell has been found to depend on (1) the presence of organic compounds, (2) acid concentration, (3) structure of the acid, and (4) capacity of the cell to alkalinize cytoplasm.

5.4.1   Number of undissociated organic acids
that enter the bacterial cell

All three acids reduced pH to ±3 in unbuffered H_2O, even at concentrations as low as 10 mM (Lind, Jonsson, and Schnurer, 2005).

Organic acids can enter the microbial cell only in their undissociated forms, which diffuse across the microbial cell membrane. This entrance of the acid molecule then lowers the intracellular pH (pH_i) of the cell (Carrasco et al., 2006). The concentration of the undissociated form of an organic acid and the pH of the environment are interdependent variables, linked by the Henderson–Hasselbach equation (Breidt, Jr., Hayes, and McFeeters, 2004). As the extracellular pH decreases, the number of undissociated organic acids increases, and so do their activities toward the microbial cells (Kwon and Ricke, 1998). This undissociated state of the acid molecule is primarily responsible for any antimicrobial activity and effectiveness is dependent on the dissociation constants (pK_a) of the acid (Barbosa-Cánovas et al., 2003). This undissociated state of the organic acid is extremely important in the capacity to inhibit a microbial cell (Gauthier, 2005). More than 10–20 × the level of dissociated acid is needed to achieve the same inhibition as for undissociated acid (Presser, Ratkowsky, and Ross, 1997). However, although the undissociated molecules generally have a greater inhibitory effect, studies have also shown that the total inhibitory action of the weak acid is dependent upon a combination of the effects that are produced by undissociated acid molecules as well as the dissociated ions (Ray, 1992; Lück and Jager, 1997; Taniguchi et al., 1998).

5.4.2   Acidity constant (pK_a value)

The strengths of weak acids are measured on the pK_a scale. The smaller the number on this scale is, the stronger the acid (Namazian and Halvani, 2006). Published pK_a values for acids are generally measured at 25°C and under conditions of zero ionic strength in water. The pK_a values of organic acids are known to change in response to a change in temperature and ionic strength (Bjornsdottir, Breidt, Jr., and McFeeters, 2006). Different pK_a values and lipid solubility of an organic acid have both been suggested as reasons for the level of activity of an organic acid (Lambert and Bidlas, 2007). The closer the pH of a medium is to the pK_a of an acid, the greater the lag phase of microbial growth is expected to be (Livingston et al., 2004). Organic acids used for food preservation are weak acids with the pK_a range between pH 3 and 5 which possess some levels of buffer activity (Doores, 1993; Fang and Tsai, 2003).

5.4.3 Water activity (a_w)

For more than half a century it has been known that water activity (a_w) could be much more important to the quality and stability of food than the total amount of water present in any food (Maltini et al., 2003). This is based on several effects, which include (1) a_w is a determinant for the growth of microorganisms, (2) a_w is related to most degradation reactions, (3) moisture migration in multidomain foods obeys a_w and not moisture content, and (4) a_w is easier to measure than moisture content as this measure is nondestructive. The strong action of acid preservatives on membranes is apparent from the knowledge that monocarboxylic acids become more inhibitory as they become more lipophilic (Piper et al., 2001). Water activity is the most important environmental factor that will determine whether a microorganism will grow on food with intermediate moisture substance. However, water activity is not constant and undergoes various changes during food production and storage (Guynot et al., 2005). It is, therefore, of utmost importance in effective food preservation to define the moisture conditions in which pathogenic or spoilage microorganisms will be able to grow (Barbosa-Cánovas et al., 2003). It is known that a few fungi are actually capable of growing in environments with low water activity. Another factor to contend with is that water activity and pH are very tightly linked to the organoleptic qualities of food products and it is not easy to change these without altering them (Marín et al., 2003). For example, processed fruits and vegetables are often low-moisture foods that are sugar-rich and are characterized by color, flavor, and structural properties (Maltini et al., 2003).

5.4.4 Temperature

Although organic acids have been described as strong antimicrobial agents against psychrophilic and mesophilic microorganisms (Rico et al., 2007), temperature is a primary factor in the activity of organic acids, where a higher temperature typically causes increased effectiveness (Lin et al., 1996; Brudzinski and Harrison, 1998; Breidt, Jr., Hayes, and McFeeters, 2004). For example, it is known that efficacy of preservatives is decreased at refrigeration temperature and may take several days (Zhao, Doyle, and Besser, 1993) or even weeks (Uljas and Ingham, 1999) before a 5-log reduction in hazardous pathogens such as *E. coli* O157:H7 populations in cider is achieved. At 25°C *Escherichia coli* O157:H7 populations can be reduced by >5 logs after 6 h storage and at 35°C by the same after 3 h storage.

A higher temperature may cause an increase in fluidity of cell membranes, allowing more rapid diffusion of preservatives into the cytoplasm (Chikthimmah, LaBorde, and Beelman, 2003). For example, sodium lactate is primarily used in meat products to extend cold storage and its effects are bacteriostatic rather than bactericidal (De Wit and Rombouts, 1990; Brewer et al., 1991). Another interesting fact is that in general, microorganisms survive refrigeration best at lower pH (~pH 4.5) (Buchanan et al., 2004). However, in culture media, inhibitory effects are reported to be greater at lower temperatures (Houtsma, De Wit, and Rombouts, 1996; Buchanan, Golden, and Phillips, 1997; Gill, Greer, and Dilts, 1997; Doyle, 1999).

Pathogenic bacteria, such as *Listeria monocytogenes*, are capable of surviving refrigerated storage in the presence of additives such as organic acids. It is known that a storage temperature lower than 4.4°C is necessary for preservation of frankfurters, even when surface treatments utilizing organic acid salts are used (Lu et al., 2005). Some fungi, such as many *Aspergillus* species, are xerophilic and generally capable of growing on media containing high concentrations of salt or sugar. These xerophilic fungi show no special requirements with regard to growth temperature (Marín et al., 2003).

5.4.5   Production of H_2O_2

The presence of organic acids can lead to intracellular production of H_2O_2 and is toxic for a bacteria cell. Resistance to H_2O_2 can be induced by ethanol and could explain antagonistic effects that may result when using ethanol in combination with organic acids (Barker and Park, 2001).

5.5   The role of pH

Organic acids are most active at an environmental pH equal to or lower than their pK_a value (ionization constant) (Brul et al., 2002). Bacteriostatic effects of acids vary, and the type of acid, as well as the pH, are of utmost importance in controlling pathogens in foods. Antibacterial activity of organic acids is related to the reduction in pH, as well as its ability to dissociate, which is determined by the pK_a value and pH of the surrounding area, and as a result activity increases with a decrease in pH (Jung and Beuchat, 2000).

To demonstrate the organic acid activities at various pH values, a study done on the salts calcium propionate, sodium benzoate, and potassium sorbate is herewith discussed. These salts were found to be effective and to inhibit some isolates from bakery products at pH 4.5 when applied at concentration of 0.3%. In a sponge cake (an analogue of pH 6) addition of these same weak organic acids salts appeared to be effective only at low a_w levels (Guynot et al., 2005).

An important factor in food safety is that bacteria can only survive in acidic environments by regulating their cytoplasmic pH (pH_i). This is primarily driven by controlled movement of cations across the cell membrane. However, this ability to maintain the pH_i close to neutrality can be destroyed at decreased extracellular pH values, which could also lead to cell death (Hill, O'Driscoll, and Booth, 1995). Some organisms are, however, able to grow at lower pH. For example, the pH range for growth of *Campylobacter jejuni* is between 5.5 and 5.7, but this organism is not able to survive in extremely acidic conditions (below pH 3.0). At low pH (4.0 and 4.5) organic acids have a strong bactericidal affect on *C. jejuni* and *C. coli*, causing rapid death (Chaveerach et al., 2002). In Table 5.1 the pH limits for the growth of some foodborne microorganisms are shown.

At neutral pH many organic acids are essentially completely dissociated. These molecules will not actually be active against microorganisms and may even provide a potential carbon source (Piper et al., 2001). This is not true for sorbic acid, as at neutral pH, high sorbate levels still exert some inhibitory effects. This has especially been found in fungi, such as *Saccharomyces cerevisiae*. Unlike other preservative acids, sorbic acid can also be employed for preserving foodstuffs with a relatively high pH. This is due to its low dissociation constant of 1.73×10^{-5} (Marín et al., 2003). As a result, the activity of sorbic acid is less linked to pH than benzoic acid, and can still be effective at pH levels of 6.5, where in theory, weak-acid preservatives have little activity, although their activity also increases as pH decreases (Stratford and Anslow, 1998). Maximum pH for activity for sorbate is approximately 6.0–6.5, 5.0–5.5 for propionate and 4.0–4.5 for

Table 5.1 Growth pH Limits for Some Foodborne Microorganisms

Organism	Minimum pH	Optimum pH	Maximum pH
Bacteria (most)	4.5	6.5–7.5	9.0
Acetobacter spp.	3.0	5–6	6.0
Campylobacter jejuni	4.9	6.5–7.5	8.5
Escherichia coli	4.0	7.0	9.0
Lactobacillus brevis	3.2	4.6–5.0	6.3
Lactobacillus plantarum	<3	5.0–5.5	6.0
Bacillus cereus	5	6.5	8.8
Listeria monocytogenes	4.3	7.0	9.4
Salmonella spp.	3.8	7.0–7.5	9.5
Salmonella typhimurium	<4.0	6.6–8.2	9.0
Shigella flexnerii	4.5	6.0–6.5	9.19
Staphylococcus aureus	4.0	6.0–7.0	10.0
Vibrio parahaemolyticus	4.8	7.8–8.6	11.0
Yersinia enterocolitica	4.2	7.2	9.6–10.0
Yeast (in general)	1.5–3.5	4.5–6.8	8.0–8.9
Molds (in general)	1.5–3.5	4.5–6.8	8.0–11

Source: International Commission of Microbiological Specifications for Foods (ICMSF) *Microorganisms in Food,* Vol. 5. New York: Blackie Academic and Professional, 1996.

benzoate (Suhr and Nielsen, 2004). Of interest is an acid such as gluconic acid, which has a highly polar molecule and is seemingly unable to penetrate the cell membrane. The apparent antimicrobial activity of gluconic acid solutions between pH 3.0 and 4.0 is more likely to be due to pH effect alone (Bjornsdottir, Breidt, Jr., and McFeeters, 2006). When used in combination with other antimicrobial agents such as ethanol, an increase from pH 3.0 to 4.0 is known to reduce the ability of ethanol to stimulate cell death (Barker and Park, 2001).

In vitro assays are often conducted at low pH to show the antibacterial effectiveness of organic acids, ensuring that the acids are not dissociated, as it is known that at pH below 3–3.5, almost all organic acids are very efficacious in controlling bacterial growth. However, it is often necessary to simulate the "real" environment, such as the GIT of poultry or pigs (Gauthier, 2005).

Additives or preservatives have also been found to have varying effects on the surface tension of aqueous food and beverages, which usually results from a decrease in pH. For example, acetic acid can reduce the surface tension more efficiently than citric acid, as it has a lower ratio of hydrophobic group per hydrophilic group. Citric acid has twice as many hydrophilic groups as hydrophobic groups, and as a result has a strong affinity with water. It cannot, therefore, be adsorbed at the surface. Citric acid has no significant effect on the surface tension. The type of acid used to adjust the pH also has a different effect on surface tension (Permprasert and Devahastin, 2005).

Weak organic acid food preservatives, notably sorbic acid and benzoic acid, also exert pH-dependent effects on: (1) the heat-shock response and (2) the thermotolerance of fungi such as *S. cerevisiae*, specifically by lowering the pH_i. This decrease in pH_i caused by the organic acids, inhibits both the heat-shock protein (Hsp) and thermotolerance induction during sublethal (39°C) heat shock. However, at higher pH at 25°C sorbic acid treatment causes strong induction of thermotolerance, in the absence of sublethal heat treatment. At pH values above pH 5 sorbate acts as a potent inducer of thermotolerance, but no longer inhibits the capacity of cells to mount a heat-shock response. Around neutral pH, cells became more resistant to thermal killing at 52°C after preservative treatment. At very low pH (~pH 3.5), preservative treatment is known to cause a reduction in thermotolerance (Cheng, Moghraby, and Piper, 1999).

5.6 Antibacterial action

The accepted bactericidal actions of organic acids have long been the acidification of the bacterial cell cytoplasm following the diffusion of uncharged protonated acid across the membrane (Breidt, Jr., Hayes, and McFeeters, 2004). However, additional specific effects of organic acid on the bacterial cell have since been identified and vary with the type and concentration of acid. These include (Diez-Gonzalez and Russell, 1997; Krebs, Wiggins, and Stubbs, 1983; Shelef, 1994): (1) inhibition of cellular metabolic enzymes such as lactate dehydrogenase, (2) transport proteins, (3) membrane function, and (4) chelating metal ions.

Despite numerous investigations and the identification of inhibitory actions, antibacterial mechanism(s) for organic acids are still not fully understood. It is, however, recognized that they are capable of exerting bacteriostatic and bactericidal effects, depending on the physiological status of the organism as well as the external environmental conditions (Ricke, 2003). Other antibacterial activities, that may be less direct, have also been assigned to the organic acids. These include the inhibition of nutrient transport, damage to the cytoplasmic membrane, disruption of outer membrane permeability, and an influence on the macromolecular synthesis (Alakomi et al., 2000; Davidson, 2001).

Bacteria are known to react differently to antibacterial agents. This is due either to inherent differences, which may be their unique cell envelope composition or proteins not being susceptible or as a result of resistance development, by adaptation exchange of genetic information. For organic acids to be effective as an antibacterial, they must be able to penetrate the cell envelope and attain a sufficiently high concentration at the target site where they will exert antibacterial action (Cloete, 2003).

Lactic acid may pose as an example inasmuch as it can act either in a bactericidic or bacteriostatic manner, depending on the specific inhibitory conditions. This has been noted specifically for inhibition of *L. monocytogenes* where lactic acid is able to either kill or limit growth of the organism. Lactic acid is now included in growth prediction models for this pathogen (Gravesen et al., 2004).

Not much is known about the effect of organic acids on another important pathogen, *Salmonella* spp., but *E. coli* and *Salmonella* are both enteric bacteria and seem to have a similar physiology. It is, however, known that medium chain fatty acids (C6–C12, caproic, caprylic, capric, and lauric) are much more effective against *Salmonella* than the short-chain fatty acids (formic, acetic, propionic, and butyric). Another important finding was that the short-chain fatty acid butyrate specifically down-regulates the expression of invasion genes in *Salmonella* spp. at low doses. Medium-chain fatty acids and propionate also decrease the ability of *Salmonella* spp. to invade epithelial cells. This has been found to be in contrast to acetic acid (Van Immerseel et al., 2006).

5.7   Antifungal action

Many organic acids behave primarily as fungicides or fungistats, as fungi are known to be capable of growing at low-pH environments where the organic acids are most effective as antimicrobials. The use of sodium benzoate has been limited to products that are acidic in nature and is mainly used as an antimycotic agent. Most yeasts and molds are inhibited by concentrations as low as 0.05–0.1% (Barbosa-Cánovas et al., 2003). Even distribution of organic acids is essential for effective inhibition of spoilage molds (Arroyo, Aldred, and Magan, 2005).

Sorbic and benzoic acids are the two most popular organic acids in food preservation and are widely regarded as most active against yeasts and molds, and least active against bacteria. The "parabens" is a name given to the alkyl esters (methyl-, ethyl-, and propyl-) of *p*-hydroxybenzoic acid and are widely used because of their antifungal properties (Gonzalez, Gallego, and Valcarcel, 1998). Methyl esters are more effective against molds, whereas the propyl esters demonstrate better activity against yeasts. Inhibition of yeasts by sorbic acid is inconsistent as a classic weak-acid preservative (Stratford and Anslow, 1998). Sorbic acid delays spore germination and mycelial growth, causes intracellular acidification, collapses the pH gradient over the vacuole and rapidly disrupts pH homeostasis, and also reduces intracellular ATP pools and levels of sugar-phosphomonoesters and -phosphodiesters (Plumridge et al., 2004). Because of its lipophilicity, sorbic acid acts as a membrane-disrupting molecule and allows protons to enter the cytoplasm (Leyva and Peinado, 2005). Understanding the effect of sorbic acid on the molecular physiology of yeast cells will allow the food industry to develop knowledge-based strategies to make more optimal use of its preservative action as resistance rapidly develops in fungi (Papadimitriou et al., 2007).

It was initially speculated that the antifungal action of organic acids is based on intracellular acidification, because this inhibits glycolysis (Leyva and Peinado, 2005). However, results obtained with exponentially growing cells, both in batch (Warth, 1991) as well as chemostat cultures (Verduyn et al., 1990), showed that glycolysis was enhanced when cells were cultured in the presence of sublethal concentrations of organic acids. It was also detected that phosphofructokinase activity was not inhibited (Warth, 1991).

Although yeasts have a well-developed system for intracellular pH homeostasis, they are dependent upon a proton-translocating plasma membrane protein H^+-ATPase. Charged anions can only readily diffuse out of the cell with an active efflux process. Without this, these charged acid anions will accumulate in acid-stressed cells. The inhibitory activity of lipophilic and water-soluble monocarboxylic acids on the growth of yeasts such as *S. cerevisiae* is somewhat different and also has a different effect on the behavior of intracellular ATP concentration. Lipophilic acids are known to concentrate in membranes, exerting a membrane-disturbing effect and increasing the permeability of membranes by the release of protons. Octanic (caprylic) and decanoic (capric) acids are more toxic to yeast cells, showing destructive effects at very small concentrations (Kasemets et al., 2006). Studies with *S. cerevisiae* have shown that, in response to acidification, the plasma membrane H^+-ATPase pumps protons out of the cell in an ATP-dependent mechanism (Eraso and Gancedo, 1987; Ramos et al., 1989; Leyva and Peinado, 2005).

Citric acid is known to shift the primary energy metabolism in *S. cerevisiae* toward lower ethanol production and higher glycerol production. This results in lower ATP production. Citric acid also causes an increased glycerol production in *S. cerevisiae* (Lawrence et al., 2004; Nielsen and Arneborg, 2007). Citric acid inhibits growth of *S. cerevisiae* and *Zygosaccharomyces bailii* to a similar extent, but the growth-inhibitory mechanism of citric acid on yeasts is different from that of classic weak-acid preservatives, because the latter compounds exhibit their highest antimicrobial activity in their undissociated form at low pH values (Brul and Coote, 1999; Stratford, 1999). Citric acid is rather a chelator, that is, a lipophobic dissociated acid inhibiting growth by chelating metal ions from the medium.

5.8   Antiviral action

Little information exists on the activity of organic acids against viruses. However, enteric viruses have been shown not to possess similar susceptibility to weak organic acids compared to bacteria and consequently survive well in organic acid preserved foodstuffs such as lactate-fermented products (Adams and Nicolaides, 1997).

5.9   Acidified foods

Acid and acidified foods are defined as foods having a pH of 4.6 or lower, by the U.S. Code of Federal Regulations. In Table 5.2 some common foodstuffs are given together with their pH. Acid foods naturally have a pH below 4.6. Acidified foods have a low pH as a result of the addition of acid or acid food ingredients. In the current FDA regulations for acidified foods the amount or type of organic acid needed to lower pH is not specified. Although acidified foods have an excellent safety record, acid or fermented foods such as apple cider, salami, and apple juice have recently been associated with outbreaks of disease caused by *E. coli* O157:H7 and have raised concern about the general safety of acidified food. A better understanding of the microbial response to organic acids in foods is, however, evident for improvement of effective preservation (Bjornsdottir, Breidt, Jr., and McFeeters, 2006). The range of acidified food products is continuously expanding. Numerous acidified foods are currently produced without heat processing, which may pose the threat of acid-resistant food pathogens emerging (Breidt, Jr., Hayes, and McFeeters,

Table 5.2 Some Common Foodstuffs and pH

Food product	Approximate pH	Food product	Approximate pH
Limes	1.8–2.0	Molasses	4.9–5.4
Lemon juice	2.0–2.6	Bread (white)	5.0–6.2
Lime juice	2.0–2.4	Cabbage	5.2–5.4
Soft drinks	2.0–4.0	Beetroot	5.3–6.6
Lemons	2.2–2.4	Sweet potatoes	5.3–5.6
Vinegar	2.4–3.4	Coconut	5.5–7.8
Marinade	2.4–6.9	Flour (wheat)	5.5–6.5
Chili sauce	2.8–3.7	Beans	5.6–6.5
Wine	2.8–3.8	Potatoes	5.6–6.0
Cider	2.9–3.3	Celery	5.7–6.0
Grapefruit	3.0–3.7	Oysters	5.7–6.2
Pickles	3.0–3.4	Sardines	5.7–6.6
Apples	3.3–3.9	Calamari	5.8
Apricots	3.3–4.8	Mangos	5.8–6.0
Sauerkraut	3.4–3.6	Carrots	5.9–6.3
Sherry	3.4	Corn	5.9–7.3
Grapes	3.5–4.5	Tuna	5.9–6.1
Jam	3.5–4.0	Asparagus	6.0–6.7
Mustard	3.5–6.0	Capers	6.0
Fruit cocktail	3.6–4.0	Cranberry juice	6.0
Olives (green)	3.6–5.6	Melons	6.0–6.7
Mayonnaise	3.7–4.7	Olives (black)	6.0–7.0
Tomato sauce	3.9	Butter	6.1–6.4
Beer	4.0–5.0	Salmon	6.1–6.3
Yogurt	4.0–4.5	Cod liver	6.2
Tomato juice	4.1–4.6	Avocados	6.3–6.6
Tomatoes	4.3–4.9	Peanut butter	6.3
Tomato puree	4.3–4.5	Milk	6.4–6.8
Buttermilk	4.4–4.8	Crackers	6.5–8.5
Soy sauce	4.4–5.4	Dates	6.5–8.5
Bananas	4.5–5.2	Water	6.5–8.0
Maple syrup	4.6–5.5	Shrimp	6.8–7.0
Pimento	4.6–5.2	Tea	7.2
Cheese	4.8–6.4	Eggs (fresh)	7.6–8.0
Pumpkin	4.8–5.2		

2004). Home-made mayonnaise has been implicated in various incidences of food poisoning (Weagant, Bryant, and Bark, 1994; Erickson et al., 1995; Al-Ahmadi, El Bushra, and Al-Zahrani, 1998). The use of vinegar (acetic acid) as an acidulant, at pH 3.6–4.0 has been suggested for prevention of contamination with pathogens, particularly *Salmonella*.

Optimum growth of bacteria is generally at pH 6–7. A decrease in pH will inevitably cause a fall in growth rate, until it eventually reaches zero. In acidic foodstuffs it is important that the following factors are kept in mind: (1) weak carboxylic acids will only naturally dissociate in aqueous solution, and (2) the undissociated form is very lipid soluble which allows them to diffuse freely through the microbial cell membrane into the cytoplasm. It is, however, also possible that the low pH of acidic food will cause an increase in the amount of undissociated acid (Adams and Nicolaides, 1997).

5.10 Comparing effectiveness of organic acids with inorganic acids

Lowering the pH of a food product or beverage has long been used as a method of preserving food. In some foodstuffs it is a natural defense mechanism (Brown and Mayes, 1980), such as in fruit and fruit juices which are protected against microbial contamination by their organic acid content as well as inherent acidity (Fielding, Cook, and Grandison, 1997). Organic acids are considered to be more effective against foodborne pathogens than hydrochloric acid, because these molecules can easily penetrate bacterial cell membranes to cause a decrease in intracellular pH by dissociating. At a given environmental pH, organic acids will, therefore, cause more inactivation of cells than inorganic acids (such as HCl) (Deng, Ryu, and Beuchat, 2001; Buchanan et al., 2004). This increased antimicrobial activity will naturally vary among the organic acids as it is associated with the anion portion of the molecule (Buchanan et al., 2004).

In a study of four different organic acids and HCl, in the pH range between 4.0 and 5.5 it was found that among lactic, acetic, citric, malic, and hydrochloric acids, lactic acid consistently had the greatest activity against enterohemorrhagic *E. coli*, and HCl had the least (Buchanan et al., 2004). In another study of the inhibitory effect on *Listeria*, acetic acid was found to be more potent than lactic acid, whereas lactic acid was again more inhibitory than HCl (Doyle, 1999). In the control of *Campylobacter* contamination of poultry carcasses during slaughter, the effect of organic acids is well established. Contrary to this, at low pH levels (4.0 and 4.5) HCl has only a slight inhibitory effect on the *Campylobacter* population and has no effect at high pH levels (5.0 and 5.5). With organic acids, at low pH (4.0 and 4.5) *C. jejuni* and *C. coli* will rapidly die due to a strong bactericidal affect. The reduction rate is, therefore, much higher, compared to HCl. But at pH 5.0 and 5.5 *Campylobacter* can survive (Chaveerach et al., 2002).

It has been found in studies on *E. coli* O157:H7, that at both 22°C and 7°C, the addition of lactic acid instead of HCl (in reducing the pH to 4.5), resulted in a more rapid decrease, which may be the result of the additional antimicrobial action of lactic acid on the organism, apart from the lower pH (Uyttendaele, Taverniers, and Debevere, 2001).

5.11   Spectra of inhibition

Naturally occurring organic acids, such as sorbic acid, benzoic acid, and acetic acid, are the most commonly used chemical preservatives in food and all have a broad antimicrobial spectrum (Plumridge et al., 2004). Benzoic acid, in particular, although mainly associated with fruit preservation, is used in many types of acidic food products (Suhr and Nielsen, 2004). Sorbic acid inhibits both molds and yeasts, and is used in a broad variety of food products (Sofos and Busta, 1981), including fine bakery products, confectionery, and bread (Suhr and Nielsen, 2004).

Contrary to these organic acids, the spectrum of activity of other organic acids, such as sodium lactate, is narrow or may also be more specific to a certain group of organisms. The disadvantage of using this agent, is that with time, other organisms may take over (Lemay et al., 2002).

5.12   Improving effectiveness

It is apparent that any means of improving efficacy of organic acids should be made on the protective effect of various factors on microbial cells. Primary protective factors include the high lipid content and topography of foodstuffs such as chicken skin. Use of "transdermal synergists" has been proposed and may increase the killing effect by enhancing delivery of the acids to attached or embedded cells. Such transdermal compounds are commonly used in the pharmaceutical industry. Various pharmaceutical transdermal agents are emulsifiers that allow mixing of lipid and water phases. Emulsifiers are widely used in food processing and many pharmaceutical emulsifiers are GRAS (generally recognized as safe) in food applications. These agents enhance activity of an agent via (1) interaction of the synergist with intercellular lipids and intracellular proteins which promotes permeation or diffusion through layers of the skin, (2) solvent action of the transdermal agent to solubilize skin tissue components, and (3) increased partitioning of an acid into the membrane. The application of transdermal agents to enhance antimicrobial activity in the food industry is a fairly new concept, but has promising advantages (Tamblyn and Conner, 1997).

Improving effectiveness of organic acids will require a better understanding of the stress response capabilities of foodborne pathogens. These may include general and specific stress responses. This would entail the application of molecular tools to be used for studying pathogen behavior in microbial ecosystems in all the various food production environments. It is also essential to distinguish specific organic acid toxic mechanisms in foodborne pathogens. The process, however, is continuously difficult to achieve and will increasingly require novel genetic techniques that can be applied in screening for differential physiological responses when selected mutants are exposed to the numerous potential environmental influences. Genetic techniques are, however, becoming more readily available to enable the study of the regulations and are involved in the sensing and response of a pathogen to the environment and applied preservative actions (Ricke, 2003).

Direct assessment of the antimicrobial potency of an organic acid can be elusive, and is influenced by several factors. These include (1) the physical chemistry of individual acids, (2) specific type of organism, (3) growth conditions and culture, and (4) growth phase. A problem arises when conducting these investigations under *in vitro* conditions, as such specific quantitative *in vitro* responses may not necessarily be justifiable in translating to an *in vivo* condition of a particular organism. For example, many of the common foodborne pathogens, such as *Salmonella* spp. can grow in a wide range of ecosystems and *in vivo* growth conditions during organic acid exposure can vary enormously. It has also been described that, during a life cycle, *Salmonella* can grow aerobically and within a short time frame be forced to change to an anaerobic metabolism to survive. In so doing, the organisms can potentially colonize a highly fermentative environment, such as the gastrointestinal tract. Under such conditions, *Salmonella* would be required to survive high concentrations of short chain fatty acids produced by other organisms. It would also be producing organic acids of its own as a result of fermentation (Ricke, 2003).

Comparison of different inhibitory effects of an antimicrobial agent on microbial activity necessitates the determination of parameters that could express the influence of a toxic compound (Banerjee and Sarkar, 2004). Two such parameters have been identified: k_i, which is the exponential inhibition constant, and is inversely correlated with tolerance to antimicrobials, and $C_{50\%}$, which refers to the toxic concentration that would inhibit 50% of any pysiological mechanism (Liewen and Marth, 1985; Banerjee and Sarkar, 2004). To form a good picture of the action of organic acids k_i should be taken into account together with x_{min} values, in order to evaluate the inhibition of fermentation. However, the value of the reciprocal of k_i plus the value of x_{min} could be adapted if needed, for example, for a more practical industrial approach. There is a fair correlation between k_i and $C_{50\%}$. It is, however, easier to determine the latter parameter, and it would be adequate in expression of the inhibitory effect of an organic acid (Banerjee and Sarkar, 2004).

5.13   (Physical) factors that will enhance effectiveness

5.13.1   Ozone

In a study done on the antimicrobial effects of ozone alone or combined with organic acids in the control of *E. coli* O157:H7 and *L. monocytogenes* inoculated on mushrooms, it was found that ozone treatment alone had minimal influence on the numbers of *E. coli* O157:H7 and *L. monocytogenes,* whereas a combined treatment of 3 ppm ozone with 1% citric acid significantly reduced the population numbers (Yuk et al., 2007).

5.13.2   Ultrasound

In another study pH and organic acids were found to have a significant effect on ultrasound inactivation in solutions with soluble solids 12 g/100 ml in the preservation processes of fruit juices (Salleh-Mack and Roberts, 2007).

5.13.3   Ionizing radiation

Organic acid and ionizing radiation are both known to disrupt the membranes of bacterial cells. The effect of irradiation on *E. coli* (at all doses) is enhanced by the presence of acetic acid (0.02–1%), which can be attributed to at least three possibilities: (1) the organism may be sensitized to irradiation by the disruption of metabolic functions (energy production or enzyme activity), (2) it may result in an increased energy demand of homeostasis which may inhibit DNA repair, or (3) alternatively, the radiation process may sensitize the organism to acetic acid by damaging the cell membrane, which would allow uncontrolled influx of the acid molecules. However, the undissociated acetic acid has the ability to permeate the cell membrane freely. Another theory proposes that the effect of irradiation on the acetic acid molecule renders it a lethal free radical which then acts on the cell. However, neither irradiation nor the acetic acid molecule itself is producing the effect in this case and the synergism would then be an indirect effect. During irradiation the concentration of acetic acid has minimal effect on the extent of the synergistic response, but *E. coli* was found to be more susceptible to a higher concentration of acid after irradiation. More membrane damage is achieved with a combination of acetic acid and irradiation (Fielding, Cook, and Grandison, 1997).

5.13.4   Heat treatment

Inhibitory effects of organic acids can be adjusted by factors other than pH (Breidt, Jr., Hayes, and McFeeters, 2004). Temperature is a primary factor influencing organic acid activity as increasing temperature typically results in increased effectiveness (Krebs, Wiggins, and Stubbs, 1983; Brudzinski and Harrison, 1998; Uljas and Ingham, 1999).

For example, heat treatment, combined with the application of organic acids, has been successfully used to control *Salmonella* in feed. The use of organic acids was shown to enhance the destruction of *Salmonella* during heat treatment. Increasing the duration of a heat treatment is, however, more effective compared to the addition of organic acids for an equal heat treatment time. With organic acids feed is protected against *Salmonella* outgrowth, should a postprocess recontamination of the feed occur on the farm or during storage. Increased inactivation of *E. coli* O157:H7 has also been found at higher temperatures such as 35°C relative to that observed at 25°C, which can be attributed to two factors: (1) increased transport of organic acids into the cytoplasm of cells as a result of increased activity of the inner membrane acetoacetyl-CoA transferase, or (2) increased metabolic activity or increased cellular membrane fluidity, to allow more diffusion of the acids into the cytoplasm (Comes and Beelman, 2002).

Elevated temperature and citric acid have been found to increase the susceptibility to nisin, both in culture and in foods. Exposure to a temperature of 60°C for 10 min, followed by 24 h incubation by 30°C showed no survival of bacterial cells. However, at a temperature of 50°C some viable cells were found with the application of nisin, but survivor numbers were significantly reduced. Stationary phase cells are more resistant to both organic acids alone, as well as in combination with nisin, than log-phase cells. It is speculated that Gram-negative bacterial cells are protected from the effects of nisin by the composition of their outer layers. When these are weakened by elevated temperature, they become sensitive by similar mechanisms as Gram-positive bacteria (Phillips and Duggan, 2002).

5.13.5   Steam washing

Beef carcasses have been effectively treated using a steam vacuum (Kočevar et al., 1997). The organic acids, lactic or acetic acid, are usually applied by using a spray cabinet (Bolton, Doherty, and Sheridan, 2001). However, a number of problems have been identified with the application of the steam vacuum, such as:

- Fecal pathogens not completely removed.
- Temperature of meat surface should not exceed 34–49°C during treatment.

- Insufficient time for online workers to complete pasteurization, as this process requires at least ten seconds.
- Surface curving may cause problems with proper contact of vacuum head and limit effective treatment.
- Feces distributed and not necessarily removed.
- Unsuitable for decontamination of large areas on the carcass.

(Gill and Bryant, 1997; Phebus et al., 1997; Castillo et al., 1999; Bolton, Doherty, and Sheridan, 2001).

5.13.6   Vacuum

Vacuum-packed food allows even distribution of an antimicrobial solution in the food package, and small amounts of antimicrobial solution can efficiently cover food surfaces, so that large amounts of liquid are not needed in the package (Murphy et al., 2006).

5.13.7   Freezing

Freezing on its own does not provide a significant interference approach to reduce the risk of infection with *E. coli* O157:H7 after consumption of contaminated beef burgers. Application of a combination of organic acids, freezing, and pulsed electric fields did not deliver significant reductions in *E. coli* O157:H7 (Bolton, Doherty, and Sheridan, 2002).

5.13.8   Storage temperature

In a storage treatment of apple cider with a combination of 0.15% (w/v) fumaric acid and 0.05% (w/v) sodium benzoate, increasing storage temperature to 15 and 25°C significantly increased the destruction of *E. coli* O157:H7 populations (Chikthimmah, LaBorde, and Beelman, 2003). Opposed to this, the efficacy of organic acids decreases at refrigeration temperatures, and it may take several days (Zhao, Doyle, and Besser, 1993) or even weeks (Uljas and Ingham, 1999) before a 5-log reduction in *E. coli* O157:H7 populations in cider is achieved. Higher temperature causes increased fluidity of cell membranes, which allows more rapid diffusion of preservatives into the cytoplasm (Chikthimmah, LaBorde, and Beelman, 2003).

5.13.9   Do interactions exist?

There is confusion as to whether interactive effects exist and under which circumstances they occur, as individual environmental effects act independently. Two strong views have been suggested with regard to the effect of combined environmental factors: (1) interactions exist or (2) inter-actions do not. However, it has also been proposed that at the growth/no growth interface, interactions do exist, but elsewhere they are additive (McMeekin et al., 2000; Le Marc et al., 2002).

5.13.10 Buffering

Any buffering or resistance to changes in medium pH would limit the effects of organic acids. Proteinaceous material has also been suggested as major buffering material (Thomas, Hynes, and Ingledew, 2002; Abbott and Ingledew, 2004). Contrary to the previous finding, immobiliza-tion significantly enhances the inhibition of *E. coli* O157:H7 by organic acids, and combinations of immobilized nisin and antimicrobials at low temperatures increase the effectiveness of reducing *E. coli* O157:H7 on ground beef (Fang and Tsai, 2003).

5.14 Comparisons among organic acids

Studies have been conducted and investigations are still underway to determine the comparative effectiveness of the various organic acids com-monly employed as antimicrobials. Effectiveness has been shown to be dependent on several factors. In order to compare the inhibitory effects of organic acids on microbial activity, determination of the parameters that express the influence of toxic compounds is required (Malfeito Ferreira, Loureiro-Dias, and Loureiro, 1997). Although much of the inhibitory effect can be accounted for by pH, the various organic acids vary considerably in their inhibitory effects. Because organic acids are not members of a homogene series but vary in numbers of carboxy groups, hydroxy groups, and carbon–carbon double bonds in the molecule, it has typically not been possible to predict the magnitude, or in some cases even the direction, of the change in inhibitory effect upon substituting one acid for another or to predict the net result in food systems containing more than one acid. The mode of action may well differ in bacteria, yeasts, and molds (Hsiao and Siebert, 1999). A general rule that is sometimes applied to indicate activ-ity is the amount of weak acid in solution is determined by the ionization constant (K_a). A weaker acid would generally have a lower K_a as it yields a lower amount of hydrogen ions in solution. Acetic acid, for example, has a K_a of 1.753×10^{-5} whereas citric acid has 3 K_a (7.447×10^{-4}, 1.733×10^{-5}, and 4.018×10^{-6}); thus acetic acid may be considered "weaker" than citric (Permprasert and Devahastin, 2005).

Despite their various chemical structures, all organic acids inhibit microorganisms in a similar fashion. For example, all show increased inhi-bition at low pH, microbes are inhibited rather than killed, lag phases are prolonged, yields reduced, and active transport prevented (Stratford and

Anslow, 1998). Direct assessment and comparison of antibacterial potency of individual acids can be elusive because of the influence of variables such as the physical chemistry of respective acids, the bacterial species in question, the growth conditions or media composition, and the growth phase. This has been augmented in part due to complexities involved in designing and conducting the experiments that precisely delineate all of the contributing factors. Specific quantitative *in vitro* responses also do not necessarily translate to all *in vivo* possibilities for a particular organism. Furthermore, common foodborne pathogens, such as *Salmonella* spp. can grow in a multitude of ecosystems and, thus, *in vivo* growth conditions (when organic acid exposure occurs) can vary widely. It is, for example, possible that during their life cycle *Salmonella* can grow aerobically and within a short time frame switch to an anaerobic metabolism to survive and potentially colonize a highly fermentative environment, such as the gastrointestinal tract. Under these conditions, not only would *Salmonella* be required to survive high concentrations of short chain fatty acids (SCFA) produced by other organisms, but would also be generating and excreting organic acids of its own (Ricke, 2003).

In the pH range between 4.0 and 5.5 it was found that among five organic acids (lactic, acetic, citric, malic, and hydrochloric acids), lactic acid consistently had the greatest activity against enterohemorrhagic *E. coli* and HCl had the least (Buchanan, 2004). Tests on *Salmonella typhimurium* in stomach contents at pH 4 found the killing effect of organic acids to range from low to high in the following order: acetic < formic < propionic < lactic < sorbic < benzoic. From a purely biochemical angle, formic acid (the shortest-chain organic acid) should find it easier to diffuse into a cell and to cause acidification of cytoplasm (Chaveerach et al., 2002) compared to longer-chain molecules. In culture media, workers have shown that acetic acid exhibits a more potent antilisterial effect than lactic acid whereas the latter is more inhibitory than HCl. Malic acid (the predominant organic acid in apples) has been reported not to be as effective as lactic acid in suppressing growth of *L. monocytogenes* (Doyle, 1999). In another report acetic acid wash has been recommended over lactic acid due to its availability and safety (Kerth and Braden, 2006), whereas citric acid (present in citrus fruits) is more effective than acetic acid and lactic acids for inhibiting growth of thermophilic bacteria (Barbosa-Cánovas et al., 2003). In the case of comminuted sausages, sodium lactate has been suggested as a potential replacement for potassium sorbate or sodium benzoate (Diez-Gonzalez and Russell, 1997).

In the case of the organism *Y. enteric*, at high pH (5.8) the order of inhibition was reported to be formic > acetic > propionic > lactic, whereas at lower pH values it became formic > lactic > acetic > propionic (El-Ziney, De Meyer, and Debevere, 1997). In another study sodium benzoate was found to be more effective at inactivating *E. coli* O157:H7 than potassium sorbate (Comes and Beelman, 2002). Lactic acid was found to be more effective than citric acid and a 9% concentration was more effective than a 1% concentration in decontamination of chicken skin by immersion in acid. Ascorbic acid dipping was found to reduce psychrotrophic microbial counts whereas ascorbic and citric acid improved lipid stability (Huang, Ho, and McMillin, 2005). Sodium citrate showed more potent antibacterial activity compared with sodium lactate. The mode of action of these two acids has been reported to act through chelation mechanisms. In this case sodium citrate, which has three carboxylic groups, is likely to form more stable complexes than sodium lactate which has only one (Juneja and Thippareddi, 2004).

The alkyl esters (methyl, ethyl, and propyl) of *p*-hydroxybenzoic acid (the parabens) are widely used for their antifungal properties. The preservative effect of parabens tends to increase with increasing molecular mass. The methyl ester appears to be more effective against molds, whereas the propyl ester is more effective against yeasts (favored for oils and fats, for solubility reasons) (Gonzalez, Gallego, and Valcarcel, 1998).

Other studies have indicated that acetate is a weaker acid than lactate, and that tryptic soy broth yeast extract (TSBYE) has a higher buffering capacity when applied in meat washings. TSBYE acidified to pH 2.5 with acetic acid, was of immediate lethality (<1.0 log CFU/ml at time zero on all recovery media) to all *L. monocytogenes* strains and TSBYE acidified (pH 2.5) with lactic acid resulted in no survival (<1.0 log CFU at 60 min. In contrast, the 2% lactate washings of pH 2.5 permitted survival for 30 to 60 min. Accordingly TSBYE, acidified to pH 3.5 with acetate was more lethal to *L. monocytogenes* than 2% acetate washings (although of higher pH) (Samelis et al., 2001). The growth-inhibitory mechanism of citric acid on yeasts appears to be different from that of classic weak-acid preservatives in that the latter compounds exhibit their highest antimicrobial activity in their undissociated form at low pH values (Nielsen and Arneborg, 2007).

Lactic acid alone was unable to prevent growth of spoilage yeasts, even at high lactic acid concentration (pH 3.74) in olives, whereas potassium sorbate, at concentrations of 0.025% (as acid) showed good inactivation effect on yeasts in seasoned olives, regardless of its use in combination with 0.2% citric, lactic, or acetic acids, although it did not have an appreciable influence on LAB. Among the three acids, lactic acid was slightly less effective and showed a yeast population tail on both FF and SF in olives. Benzoate efficiently inactivated microbial growth in FF, and regardless of type of acid, the yeast population was rapidly destroyed by the presence of this preservative and disappeared completely after 25 h (citric and lactic acids) or 50 h (acetic acid) (Bravo et al., 2007). Lack of inhibitory action of sorbate and benzoate against LAB was also observed (Lues, 2000) and a favorable effect on LAB growth was even reported. Benzoate, in combination with acetic acid had a lower effect although it was consistent and progressive during the storage period. This acid would, therefore, be a better choice when benzoate is used to preserve seasoned olives from SF (Bravo et al., 2007).

References

Abbott, D.A. and Ingledew, W.M. 2004. Buffering capacity of whole corn mash alters concentrations of organic acids required to inhibit growth of *Saccharomyces cerevisiae* and ethanol production. *Biotechnology Letters* 26:1313–1316.

Adams, M.R. and Nicolaides, L. 1997. Review of the sensitivity of different food borne pathogens to fermentation. *Food Control* 8:227–239.

Al-Ahmadi, K.S., El Bushra, H.E., and Al-Zahrani, A.S. 1998. An outbreak of food poisoning associated with restaurant-made mayonnaise in Abha, Saudi Arabia. *Journal of Diarrhoeal Diseases Research* 16:201–204.

Alakomi, H.L., Skyttä, E., Saarela, M., Mattila-Sandholm, T., Latva-Kala, K., and Helander, I.M. 2000. Lactic acid permeabilizes gram-negative bacteria by disrupting the outer membrane. *Applied and Environmental Microbiology* 66:2001–2005.

Arroyo, M., Aldred, D., and Magan, N. 2005. Environmental factors and weak organic acid interactions have differential effects on control of growth and ochratoxin A production by *Penicillium verrucosum* isolates in bread. *International Journal of Food Microbiology* 98:223–231.

Banerjee, M. and Sarkar, P.K. 2004. Antibiotic resistance and susceptibility to some food preservative measures of spoilage and pathogenic micro-organisms from spices. *Food Microbiology* 21:335–342.

Barbosa-Cánovas, G.V., Fernández-Molina, J.J., Alzamora, S.M., Tapia, M.S., López-Malo, A., and Chanes, J.W. 2003. General considerations for preservation of fruits and vegetables. In: *Handling and Preservation of Fruits and Vegetables by Combined Methods for Rural Areas*. Rome: Food and Argriculture Organization of the United Nations.

Barker, C. and Park, S.F. 2001. Sensitization of *Listeria monocytogenes* to low pH, organic acids, and osmotic stress by ethanol. *Applied and Environmental Microbiology* 67:1594–1600.

Beis, K., Whitfield, C., Booth, I., and Naismith, J.H. 2006. Two-step purification of outer membrane proteins. *International Journal of Biological Macromolecules* 39:10–14.

Bjornsdottir, K., Breidt, F., Jr., and McFeeters, R.F. 2006. Protective effects of organic acids on survival of *Escherichia coli* O157:H7 in acidic environments. *Applied and Environmental Microbiology* 72:660–664.

Bolton, D.J., Catarame, T., Byrne, C., Sheridan, J.J., McDowell, D.A., and Blair, I.S. 2002. The ineffectiveness of organic acids, freezing and pulsed electric fields to control *Escherichia coli* O157:H7 in beef burgers. *Letters in Applied Microbiology* 34:139–143.

Bolton, D.J., Doherty, A.M., and Sheridan, J.J. 2001. Beef HACCP: Intervention and non-intervention systems. *International Journal of Food Microbiology* 66:119–129.

Bravo, K.A.S., López, F.N.A., Garcia, P.G., Quintana, M.C., and Fernandez, A.G. 2007. Treatment of green table olive solutions with ozone. Effect on their polyphenol content and on *Lactobacillus pentosus* and *Saccharomyces cerevisiae* growth. *International Journal of Food Microbiology* 114:60–68.

Breidt, F., Jr., Hayes, J.S., and McFeeters, R.F. 2004. Independent effects of acetic acid and pH on survival of *Escherichia coli* in simulated acidified pickle products. *Journal of Food Protection* 67:12–18.

Brewer, M.S., McKeith, F., Martin, S., Dallimer, M.A., and Meyer, J. 1991. Sodium lactate effects on shelf-life, sensory and physical characteristics of fresh pork sausage. *Journal of Food Science* 56:1176.

Brown, M.H. and Mayes, T. 1980. The growth of microbes at low pH values. In: G.W. Gould and J.E.L. Corry (Eds.), *Microbial Growth and Survival in Extremes of Environment*, SAB Technical Series No. 15, pp. 71–98. London: Academic Press.

Brudzinski, L. and Harrison, M.A. 1998. Influence of incubation conditions on survival and acid tolerance response of *Escherichia coli* O157:H7 and non-O157:H7 isolates exposed to acetic acid. *Journal of Food Protection* 61:542–546.

Brul, S. and Coote, P. 1999. Preservative agents in foods. Mode of action and microbial resistance mechanisms. *International Journal of Food Microbiology* 50:1–17.

Brul, S., Klis, F.M., Oomes, S.J.C.M., et al. 2002. Detailed process design based on genomics of survivors of food preservation processes. *Trends in Food Science and Technology* 13:325–333.

Buchanan, R.L., Edelson-Mammel, S.G., Boyd, G., and Marmer, B.S. 2004. Influence of acidulant identity on the effects of pH and acid resistance on the radiation resistance of *Escherichia coli* O157:H7. *Food Microbiology* 21:51–57.

Buchanan, R.L., Golden, M.H., and Phillips, J.G. 1997. Expanded models for the non-thermal inactivation of *Listeria monocytogenes*. *Journal of Applied Microbiology* 82:567–577.

Carrasco, E., Garcia-Gimeno, R., Seselovsky, R., et al. 2006. Predictive model of *Listeria Monocytogenes'* growth rate under different temperatures and acids. *Food Science and Technology International* 12:47–56.

Castillo, A., Lucia, L.M., Goodson, K.J., Savell, J.W., and Acuff, G.R. 1999. Decontamination of beef carcass tissue by steam vacuuming alone and combined with hot water and lactic acid sprays. *Journal of Food Protection* 62:146–151.

Chaveerach, P., Keuzenkamp, D.A., Urlings, H.A., Lipman, L.J., and Van, K.F. 2002. In vitro study on the effect of organic acids on *Campylobacter jejuni/coli* populations in mixtures of water and feed. *Poultry Science* 81:621–628.

Cheng, L., Moghraby, J., and Piper, P.W. 1999. Weak organic acid treatment causes a trehalose accumulation in low-pH cultures of *Saccharomyces cerevisiae*, not displayed by the more preservative-resistant *Zygosaccharomyces bailii*. *FEMS Microbiology Letters* 170:89–95.

Chikthimmah, N., LaBorde, L.F., and Beelman, R.B. 2003. Critical factors affecting the destruction of *Escherichia coli* O157:H7 in apple cider treated with fumaric acid and sodium benzoate. *Journal of Food Science* 68:1438–1442.

Choi, S.H. and Chin, K.B. 2003. Evaluation of sodium lactate as a replacement for conventional chemical preservatives in comminuted sausages inoculated with *Listeria monocytogenes*. *Meat Science* 65:531–537.

Cloete, T.E. 2003. Resistance mechanisms of bacteria to antimicrobial compounds. *International Biodeterioration and Biodegradation* 51:277–282.

Comes, J.E. and Beelman, R.B. 2002. Addition of fumaric acid and sodium benzoate as an alternative method to achieve a 5-log reduction of *Escherichia coli* O157:H7 populations in apple cider. *Journal of Food Protection* 65:476–483.

Davidson, P.M. 2001. Chemical preservatives and natural antimicrobial compounds. In: M.P. Doyle, L.R. Beuchat, and T.J. Montville (Eds.), *Food Microbiology—Fundamentals and Frontiers* 2nd ed., pp 593–627. Washington, DC: American Society for Microbiology.

Deng, Y., Ryu, J.-H., and Beuchat, L.R. 2001. Tolerance of acid-adapted and non-adapted *Escherichia coli* O157:H7 cells to reduced pH as affected by type of acidulant. *Journal of Applied MIcrobiology* 86:203–210.

De Wit, J.C. and Rombouts, F.M. 1990. Antimicrobial activity of sodium lactate. *Food Microbiology* 7:113–120.

Diez-Gonzalez, F. and Russell, J.B. 1997. The ability of *Escherichia coli* O157:H7 to decrease its intracellular pH and resist the toxicity of acetic acid. *Microbiology* 143:1175–1180.

Doores, S. 1993. Organic acids. In: A.L. Branen and P.M. Davidson (Eds.), *Antimicrobials in Foods* 2nd ed., pp. 95–136. New York: Marcel Dekker.

Doyle, E.M. 1999. *Use of organic acids to control Listeria in meat*. American Meat Institute Foundation. Washington, D.C.

Dziezak, J.D. 1990. Acidulants: Ingredients that do more than meet the acid test. *Food Technology* 44:78–83.

El-Ziney, M.G., De Meyer, H., and Debevere, J.M. 1997. Growth and survival kinetics of *Yersinia enterocolitica* IP 383 0:9 as affected by equimolar concentrations of un-dissociated short-chain organic acids. *International Journal of Food Microbiology* 34:233–247.

Eraso, P. and Gancedo, J.M. 1987. Activation of yeast plasma membrane ATPase by acid during growth. *FEBS Letters* 224:187–192.

Erickson, J.P., Stamer, J.W., Hayes, M., McKenna, D.N., and Van Alstine, L.A. 1995. An assessment of *Escherichia coli* O157:H7 contamination risks in commercial mayonnaise from pasteurized eggs and environmental sources, and behaviour in low-pH dressings. *Journal of Food Protection* 58:1059–1064.

Fang, T.J. and Tsai, H.C. 2003. Growth patterns of *Escherichia coli* O157:H7 in ground beef treated with nisin, chelators, organic acids and their combinations immobilized in calcium alginate gels. *Food Microbiology* 20:243–253.

Fernandes, A.R., Mira, N.P., Vargas, R.C., Canelhas, I., and Sa-Correia, I. 2005. *Saccharomyces cerevisiae* adaptation to weak acids involves the transcription factor Haa1p and Haa1p-regulated genes. *Biochemical and Biophysical Research Communications* 337:95–103.

Fielding, L.M., Cook, P.E., and Grandison, A.S. 1997. The effect of electron beam irradiation, combined with acetic acid, on the survival and recovery of *Escherichia coli* and *Lactobacillus curvatus*. *International Journal of Food Microbiology* 35:259–265.

Gauthier, R. 2005. Organic acids and essential oils, a realistic alternative to antibiotic growth promoters in poultry. *I Forum Internacional de avicultura* 17–19 August 2005, pp. 148–157.

Gill, C.O. and Bryant, J. 1997. Decontamination of carcasses by vacuum-hot water cleaning and steam pasteurization during routine operations at a beef packing plant. *Meat Science* 47:267–276.

Gill, C.O., Greer, G.G., and Dilts, B.D. 1997. The aerobic growth of *Aeromonas hydrophila* and *Listeria monocytogenes* in broths and on pork. *International Journal of Food Microbiology* 35:67–74.

Gonzalez, M., Gallego, M., and Valcarcel, M. 1998. Simultaneous gas chromatographic determination of food preservatives following solid-phase extraction. *Journal of Chromatography A* 823:321–329.

Gravesen, A., Diao, Z., Voss, J., Budde, B.B., and Knochel, S. 2004. Differential inactivation of *Listeria monocytogenes* by D- and L-lactic acid. *Letters in Applied Microbiology* 39:528–532.

Greenacre, E.J., Brocklehurst, T.F., Waspe, C.R., Wilson, D.R., and Wilson, P.D.G. 2003. *Salmonella enterica* Serovar Typhimurium and *Listeria monocytogenes* acid tolerance response induced by organic acids at 20°C: Optimization and modeling. *Applied and Environmental Microbiology* 69:3945–3951.

Guynot, M.E., Ramos, A.J., Sanchis, V., and Marín, S. 2005. Study of benzoate, propionate, and sorbate salts as mould spoilage inhibitors on intermediate moisture bakery products of low pH (4.5–5.5). *International Journal of Food Microbiology* 101:161–168.

Hazan, R., Levine, A., and Abeliovich, H. 2004. Benzoic acid, a weak organic acid food preservative, exerts specific effects on intracellular membrane trafficking pathways in *Saccharomyces cerevisiae*. *Applied and Environmental Microbiology* 70:4449–4457.

Hill, C., O'Driscoll, B., and Booth, I. 1995. Acid adaptation and food poisoning microorganisms. *International Journal of Food Microbiology* 28:245–254.

Hinton, A., Jr. 2006. Growth of *Campylobacter* in media supplemented with organic acids. *Journal of Food Protection* 69:34–38.

Hirshfield, I.N., Terzulli, S., and O'Byrne, C. 2003. Weak organic acids: A panoply of effects on bacteria. *Scientific Progress* 86:245–269.

Houtsma, P.C., De Wit, J.C., and Rombouts, F.M. 1996. Minimum inhibitory concentrations (MICs) of sodium lactate and sodium chloride for spoilage organisms and pathogens at different pH values and temperatures. *Journal of Food Protection* 59:1300–1304.

Hsiao, C.-P. and Siebert, K.J. 1999. Modeling the inhibitory effects of organic acids on bacteria. *International Journal of Food Microbiology* 47:189–201.

Huang, N., Ho, C., and McMillin, K.W. 2005. Retail shelf-life of pork dipped in organic acid before modified atmosphere or vacuum packaging. *Journal of Food Science* 70:M382–M387.

International Commission of Microbiological Specifications for Foods (ICMSF). 1996. Microbiological specifications of food pathogens. *Micro-organisms in Food*, Vol. 5. New York: Blackie Academic and Professional.

Juneja, V.K. and Thippareddi, H. 2004. Inhibitory effects of organic acid salts on growth of *Clostridium perfringens* from spore inocula during chilling of marinated ground turkey breast. *International Journal of Food Microbiology* 93:155–163.

Jung, Y.S. and Beuchat, L.R. 2000. Sensitivity of multidrug-resistant *Salmonella typhimurium* DT104 to organic acids and thermal inactivation in liquid egg products. *Food Microbiology* 17:63–71.

Kasemets, K., Kahru, A., Laht, T.M., and Paalme, T. 2006. Study of the toxic effect of short- and medium-chain monocarboxylic acids on the growth of *Saccharomyces cerevisiae* using the CO_2-auxo-accelerostat fermentation system. *International Journal of Food Microbiology* 111:206–215.

Kashket, E.R. 1987. Bioenergetics of lactic acid bacteria: Cytoplasmic pH and osmotolerance. *FEMS Microbiological Reviews* 46:233–244.

Kerth, C. and Braden, C. 2006. Using acetic acid rinse as a CCP for slaughter. Meat, Poultry & Egg Processors. *Virtual Library.* http://www.ag.auburn.edu/~curtipa/virtuallibrary/kerthaceticacid.html (accessed July 17, 2006).

Kocevar, S.L., Sofos, J.N., Bolin, R.B., O'Reagan, J.O., and Smith, G.C. 1997. Steam vacuuming as a pre-evisceration intervention to decontaminate beef carcasses. *Journal of Food Protection* 60:107–113.

Krebs, H.A., Wiggins, D., and Stubbs, M. 1983. Studies on the mechanism of the antifungal action of benzoate. *Biochemical Journal* 214:657–663.

Kwon, Y.M. and Ricke, S.C. 1998. Induction of acid resistance of *Salmonella typhimurium* by exposure to short-chain fatty acids. *Applied and Environmental Microbiology* 64:3458–3463.

Lambert, R.J. and Bidlas, E. 2007. An investigation of the Gamma hypothesis: A predictive modelling study of the effect of combined inhibitors (salt, pH and weak acids) on the growth of *Aeromonas hydrophila*. *International Journal of Food Microbiology* 115:12–28.

Lawrence, C.L., Botting, C.H., Antobus, R., and Coote, P.J. 2004. Evidence of a new role for the high-osmolarity glycerol mitogen-activated protein kinase pathway in yeast: Regulating adaptation to citric acid stress. *Molecular and Cellular Biology* 24:3307–3323.

Le Marc, Y., Huchet, V., Bourgeois, C.M., Guyonnet, J.P., Mafart, P., and Thuault, D. 2002. Modeling the growth kinetics of *Listeria* as a function of temperature, pH and organic acid concentration. *International Journal of Food Microbiology* 73:219–237.

Lemay, M.J., Choquette, J., Delaquis, P.J., Claude, G., Rodrigue, N., and Saucier, L. 2002. Antimicrobial effect of natural preservatives in a cooked and acidified chicken meat model. *International Journal of Food Microbiology* 78:217–226.

Leyva, J.S. and Peinado, J.M. 2005. ATP requirements for benzoic acid tolerance in *Zygosaccharomyces bailii*. *Journal of Applied Microbiology* 98:121–126.

Liewen, M.B. and Marth, E.H. 1985 Growth and inhibition of microoorganisms in the presence of sorbic acid. A review. *Journal of Food Protection* 48:364–375.

Lin, J., Smith, M.P., Chapin, K.C., Baik, H.S., Bennett, G.N., and Foster, J.W. 1996. Mechanisms of acid resistance in enterohemorrhagic *Escherichia coli*. *Applied and Environmental Microbiology* 62:3094–3100.

Lind, H., Jonsson, H., and Schnurer, J. 2005. Antifungal effect of dairy propionibacteria—Contribution of organic acids. *International Journal of Food Microbiology* 98:157–165.

Livingston, M., Brewer, M.S., Killifer, J., Bidner, B., and McKeith, F. 2004. Shelf life characteristics of enhanced modified atmosphere packaged pork. *Meat Science* 68:115–122.

Lu, Z., Sebranek, J.G., Dickson, J.S., Mendonca, A.F., and Baily, T.B. 2005. Inhibitory effects of organic acid salts for control of *Listeria monocytogenes* on frankfurters. *Journal of Food Protection* 68:499–506.

Lück, E. and Jager, M. 1997. Antimicrobial action of preservatives. In: *Antimicrobial Food Additives*, pp. 36–57. Berlin: Springer.

Lues, J.F.R. 2000. Organic acid and residual sugar variation in a South African cheddar cheese and possible relationships with uniformity. *Journal of Food Composition and Analytics* 13:819–825.

Lund, B.M., Baird-Parker, T.C., and Gould, G.W. (Eds.) 2000. *The Microbiological Safety and Quality of Food*, Vol. 1. Gaithersberg, MD: Aspen.

Malfeito Ferreira, M., Loureiro-Dias, M.C., and Loureiro, V. 1997. Weak acid inhibition of fermentation by *Zygosaccharomyces bailii* and *Saccharomyces cerevisiae*. *International Journal of Food Microbiology* 36:145–153.

Maltini, E., Torreggiani, D., Venir, E., and Bertolo, G. 2003. Water activity and the preservation of plant foods. *Food Chemistry* 82:79–86.

Marín, S., Abellana, M., Rubinat, M., Sanchis, V., and Ramos, A.J. 2003. Efficacy of sorbates on the control of the growth of *Eurotium* species in bakery products with near neutral pH. *International Journal of Food Microbiology* 87:251–258.

McMeekin, T.A., Presser, K., Ratkowsky, D., Ross, T., Salter, M., and Tienungoon, S. 2000. Quantifying the hurdle concept by modeling the bacterial growth/no growth interface. *International Journal of Food Microbiology* 55:93–98.

Murphy, R.Y., Hanson, R.E., Johnson, N.R., Chappa, K., and Berrang, M.E. 2006. Combining organic acid treatment with steam pasteurization to eliminate *Listeria monocytogenes* on fully cooked frankfurters. *Journal of Food Protection* 69:47–52.

Nakano, S., Fukaya, M., and Horinouchi, S. 2006. Putative ABC transporter responsible for acetic acid resistance in *Acetobacter aceti*. *Applied and Environmental Microbiology* 72:497–505.

Namazian, M. and Halvani, S. 2006. Calculations of pK_a values of carboxylic acids in aqueous solution using density functional theory. *The Journal of Chemical Functional Dynamics* 38:1495–1502.

Nielsen, M.K. and Arneborg, N. 2007. The effect of citric acid and pH on growth and metabolism of anaerobic *Saccharomyces cerevisiae* and *Zygosaccharomyces bailii* cultures. *Food Microbiology* 24:101–105.

Otero-Losada, M.E. 2003. Differential changes in taste perception induced by benzoic acid prickling. *Physiology and Behavior* 78:415–425.

Papadimitriou, M.N., Resende, C., Kuchler, K., and Brul, S. 2007. High Pdr12 levels in spoilage yeast (*Saccharomyces cerevisiae*) correlate directly with sorbic acid levels in the culture medium but are not sufficient to provide cells with acquired resistance to the food preservative. *International Journal of Food Microbiology* 113:173–179.

Permprasert, J. and Devahastin, S. 2005. Evaluation of the effects of some additives and pH on surface tension of aqueous solutions using a drop-weight method. *Journal of Food Engineering* 70:219–226.

Phebus, R.K., Nutsch, A.L., Schafer, D.E., et al. 1997. Comparison of steam pasteurization and other methods for reduction of pathogens on surfaces of freshly slaughtered beef. *Journal of Food Protection* 60:476–484.

Phillips, C.A. and Duggan, J. 2002. The effect of temperature and citric acid, alone, and in combination with nisin, on the growth of *Arcobacter butzleri* in culture. *Food Control* 13:463–468.

Piper, P., Calderon, C.O., Hatzixanthis, K., and Mollapour, M. 2001. Weak acid adaptation: The stress response that confers yeasts with resistance to organic acid food preservatives. *Microbiology* 147:2635–2642.

Plumridge, A., Hesse, S.J., Watson, A.J., Lowe, K.C., Stratford, M., and Archer, D.B. 2004. The weak acid preservative sorbic acid inhibits conidial germination and mycelial growth of *Aspergillus niger* through intracellular acidification. *Applied and Environmental Microbiology* 70:3506–3511.

Presser, K.A., Ratkowsky, D.A., and Ross, T. 1997. Modeling the growth rate of *Escherichia coli* as a function of pH and lactic acid concentration. *Applied and Environmental Microbiology* 63:2335–2360.

Price-Carter, M., Fazzio, T.G., Vallbona, E.I., and Roth, J.R. 2005. Polyphosphate kinase protects *Salmonella enterica* from weak organic acid stress. *Journal of Bacteriology* 187:3088–3099.

Ramos, S.M., Balbin, M., Raposo, E., and Pardo, L.A. 1989. The mechanism of intracellular acidification induced by glucose in *Saccharomyces cerevisiae*. *Journal of General Microbiology* 135:2413–2422.

Ray, B. 1992. Acetic, propionic, and lactic acids of starter culture bacteria as biopreservatives. In: *Food Preservatives of Microbial Origin*. B. Ray and M. Daeschel (Eds.), pp. 103–136. Boca Raton, FL: CRC Press.

Ricke, S.C. 2003. Perspectives on the use of organic acids and short chain fatty acids as antimicrobials. *Poultry Science* 82:632–639.

Rico, D., Martin-Diana, A.B., Barat, J.M., and Barry-Ryan, C. 2007. Extending and measuring the quality of fresh-cut fruit and vegetables: A review. *Trends in Food Science & Technology* 18:373–386.

Russell, J.B. 1992. Another explanation for the toxicity of fermentation acids at low pH—Anion accumulation versus uncoupling. *Journal of Applied Bacteriology* 73:363–370

Russell, J. 2003. Swiping pathogens. *National Provisioner* 217:63–69.

Salleh-Mack, S.Z. and Roberts, J.S. 2007. Ultrasound pasteurization: The effects of temperature, soluble solids, organic acids and pH on the inactivation of *Escherichia coli* ATCC 25922. *Ultrasonics Sonochemistry* 14:323–329.

Samelis, J., Sofos, J.N., Kain, M.L., Scanga, J.A., Belk, K.E., and Smith, G.C. 2001. Organic acids and their salts as dipping solutions to control *Listeria monocytogenes* inoculated following processing of sliced pork bologna stored at 4°C in vacuum packages. *Journal of Food Protection* 64:1722–1729.

Shelef, L.A. 1994. Antimicrobial effects of lactates: A review. *Journal of Food Protection* 57:4445–4450.

Skrivanova, E. and Marounek, M. 2007. Influence of pH on antimicrobial activity of organic acids against rabbit enteropathogenic strain of *Escherichia coli*. *Folia Microbiol (Praha)* 52:70–72.

Sofos, J.N. and Busta, F.F. 1981. Antimicrobial activity of sorbate. *Journal of Food Protection* 44:614–622.

Stratford, M. 1999. Weak acid and "weak-acid preservative" inhibition of yeasts. In: A.C.J. Tuijtelaars, R.A. Samson, F.M. Rombouts, and S. Notermans (Eds.), *Food Microbiology and Food Safety in the Next Millenium*, pp. 315–319. Foundation Food Micro '99. Veldhoven, The Netherlands.

Stratford, M. and Anslow, P.A. 1998. Evidence that sorbic acid does not inhibit yeast as a classic 'weak acid preservative'. *Letters in Applied Microbiology* 27:203–206.

Suhr, K.I. and Nielsen, P.V. 2004. Effect of weak acid preservatives on growth of bakery product spoilage fungi at different water activities and pH values. *International Journal of Food Microbiology* 95:67–78.

Tamblyn, K.C. and Conner, D.E. 1997. Bactericidal activity of organic acids in combination with transdermal compounds against *Salmonella typhimurium* attached to broiler skin. *Food Microbiology* 14:477–484.

Taniguchi, M., Nakazawa, H., Takeda, O., Kaneko, T., Hoshino, K., and Tanaka, T. 1998. Production of a mixture of antimicrobial organic acids from lactose by co-culture of *Bifidobacterium longum* and *Propionibacterium freudenreichii*. *Bioscience, Biotechnology and Biochemistry* 62:1522–1527.

Thomas, K.C., Hynes, S.H., and Ingledew, W.M. 2002. Influence of medium buffering capacity on inhibition of *Saccharomyces cerevisiae* growth by acetic and lactic acids. *Applied and Environmental Microbiology* 68:1616–1623.

Uljas, H.E. and Ingham, S.C. 1999. Combinations of intervention treatments resulting in 5-log10-unit reductions in numbers of *Escherichia coli* O157:H7 and *Salmonella typhimurium* DT104 organisms in apple cider. *Applied and Environmental Microbiology* 65:1924–1929.

Uyttendaele, M., Taverniers, I., and Debevere, J. 2001. Effect of stress induced by suboptimal growth factors on survival of Escherichia coli O157:H7. *International Journal of Food Microbiology* 66:31–37.

Van Immerseel, F., Russell, J.B., Flythe, M.D., et al. 2006. The use of organic acids to combat *Salmonella* in poultry: A mechanistic explanation of the efficacy. *Avian Pathology* 35:182–188.

Verduyn, C., Postma, E., Scheffers, W.A., and Van Dijken, J.P. 1990. Energetics of *Saccharomyces cerevisiae* in anaerobic glucose-limited chemostat cultures. *Journal of General Microbiology* 136:405–412.

Warth, A.D. 1991. Mechanism of action of benzoic acid on *Zygosaccharomyces bailii*: Effects on glycolytic metabolite levels, energy production, and intracellular pH. *Applied and Environmental Microbiology* 57:3410–3414.

Weagant, S.D., Bryant, J.L., and Bark, D.H. 1994. Survival of *Escherichia coli* O157:H7 in mayonnaise and mayonnaise-based sauces at room and refrigerated temperatures. *Journal of Food Protection* 57:629–631.

Yuk, H.G., Yoo, M.Y., Yoon, J.W., Marshall, D.L., and Oh, D.H. 2007. Effect of combined ozone and organic acid treatment for control of *Escherichia coli* O157:H7 and *Listeria monocytogenes* on enoki mushroom. *Food Control* 18:548–553.

Zhao, T., Doyle, M.P., and Besser, R.E. 1993. Fate of enterohemorrhagic *Escherichia coli* O157:H7 in apple cider with and without preservatives. *Applied and Environmental Microbiology* 59:2526–2530.

chapter six

Problems associated with organic acid preservation

6.1 Adverse effects on humans and animals

6.1.1 Chemical reactions in humans ("allergies")

Although only about 2% of adults and 5% of children have been estimated to be sensitive to foodborne allergens, the number of patients suffering from food allergies has increased dramatically in the last decades, rendering allergic diseases a serious health problem. Changes in eating habits and changes in the environment have been indicated as the main reasons (Hayakawa, Linko, and Linko, 1999). True food allergies are often difficult to diagnose, as similar symptoms are caused by several factors such as food intolerance, seafood toxins, or chemical additives (Furukawa, 1991; Hefle, Nordlee, and Taylor, 1996).

It is important to note at this stage, that preservatives and additives cause chemical and not allergic reactions (Motala and Steinman, 2008). Although it is now evident that benzoic acid and sorbic acid are the most commonly used additives and are applied to a wide variety of foods (Gould and Jones, 1989; Wen, Wang, and Feng, 2007), sodium benzoate is known to be the causative agent of hyperactivity or tight chests in individuals who have asthma, although only some asthmatics will react to these substances (Motala and Steinman, 2008; Poulter, 2007). Consumers also often complain of a scratchy feeling at the back of their throats, and other reactions, such as rashes have also been reported. There are no reliable tests to confirm sensitivity to these chemicals (Motala and Steinman, 2008).

Sorbic acid is preferred to other organic acids due to its harmlessness and organoleptic neutrality (Banerjee and Sarkar, 2004). However, concern has arisen that sorbic acid, together with benzoic acid, may cause oxidative stress in humans. Benzoic acid is also generally assumed to be safe, being conjugated in the liver to produce benzoylglycine (hippuric acid), which is then excreted in the urine. Sorbic acid is also largely excreted, as the oxidation product 2,4-hexadienedioxic acid. The problem is that before these acids are transported to the liver, they will come into contact with epithelial cells of the gastrointestinal tract. Further information is needed to determine if weak organic acid food preservatives exert pro-oxidant effects on these epithelia (Piper, 1999). There is also concern that excessive use of benzoic or sorbic acid may lead to serious side effects such as meta-bolic acidosis, convulsions, and even hyperpnea in humans (Tfouni and Toledo, 2002; Wen, Wang, and Feng, 2007).

The American Academy of Allergy and Immunology (AAAI) has defined a food allergy or hypersensitivity as an immunological reaction resulting from ingestion of a food or food additive (Hayakawa, Linko, and Linko, 1999). This reaction occurs only in some patients, may occur when only a small amount of the substance is ingested, and is related to any physiological effect of the food or the food additive (Anderson and Sogn, 1984; Hayakawa, Linko, and Linko, 1999). As a result of pressure from consumers in particular, interest in the reduction of allergenic activity of foods during processing has increased. This may involve enzymatic mod-ification or genetic engineering, and the development of either hypoaller-genic foods or the substitution of foods (Itoh, 1996).

6.1.2   Organic acids as pro-oxidants

The use of sulfite, one of the very first preservatives, has been greatly reduced because of reports of adverse effects on human health, especially in steroid-dependent asthmatics. It has been replaced in food preserva-tion by the use of organic acids. In various studies, however, it has been found that preservatives are also acting as pro-oxidants and can even be mutagenic toward the mitochondrial genome of yeasts in aerobic environ-ments. The potential for weak organic acid food preservatives to act as pro-oxidants in humans should receive much attention if only to reassure the public of the complete safety of these compounds (Piper, 1999).

A pro-oxidant is a chemical that induces oxidative stress, either through creating reactive oxygen or inhibiting antioxidant systems. Benzoic acid is one of the organic acids found to be a pro-oxidant. This is a rather sur-prising finding, as it is also known that the related 2-hydroxybenzoic acid (salicylic acid) acts as a scavenger of free radicals *in vivo*. Sorbic acid and benzoic acid were also found to be mutagenic toward the yeast mitochon-drial genome, but only in the presence of oxygen (Piper, 1999).

Ascorbic acid acts as both an antioxidant and a pro-oxidant. Appropriate levels for preventing muscle discoloration are, therefore, dependent on a number of factors, such as the presence and concentration of metals within a food (Mancini et al., 2007). Ascorbic acid's pro-oxidant action could also be attributed to the production of ferrous heme proteins, which may be more reactive and more oxidative than ferric derivatives (Yamamoto, Takahashi, and Niki, 1987). The addition of citric acid as che-lator may improve the efficacy of ascorbic acid in muscle food (Mancini et al., 2007).

6.2 Adverse effects on foodstuffs

Organic acids have a direct influence on the quality of many natural and processed foods. The primary purpose of added organic acids is often to provide a particular flavor, taste, or aroma supplementary to their role as stabilizers or preservatives (Gomis, 1992). Unfortunately, often these organic acids may have a negative effect on the appearance of food products. For example, after organic acid application, the surface of cut meat often appears deteriorated and undesirable to any consumer (Smulders and Greer, 1998). However, studies have shown that supplementation of beef trimmings or minced beef with lactic acid or citric acid to a final concentration of 2% (v/v) did not produce significant changes in the appearance or smell of these products (Bolton et al., 2002). When organic acids are used as marinades, it can have a negative effect on the quality of meat with regard to water-holding capacity, marinade retention, percentage loss during the cooking process, and also moisture and binding ability (Carroll et al., 2007). Many countries (including Canada) do not allow the use of preservatives in various foodstuffs, as it is believed that this would reduce the "natural" appeal of minimally processed food products, for example, fruit yogurts, despite the fact that organic acids are mainly natural products, commonly found in various foodstuffs (Penney et al., 2004).

Sorbic acid has many advantages over many other organic acids, such as being less linked to pH for optimum antimicrobial activity, low in price, and having very little residual taste. Unfortunately, a drawback associated with sorbic acid is the adverse effect on yeast activity and dough rheology in the production of bread. This inevitably results in production of reduced loaf volume and also renders the dough sticky and difficult to process (Guynot et al., 2005).

6.3 Protective effects on microorganisms

In fermented and minimally processed products background flora are often present and can have a notable influence on the behavior of contaminating microorganisms (Janssen et al., 2006). These microbial interactions are very important in determining the safety of such products (Breidt, Jr., and Fleming, 1998). Various factors may exert profound influences on the negative microbial interaction between microorganisms. These include organic acid production and a decrease in pH, as well as the production of bacteriocins (Janssen et al., 2006). *In vitro* studies have shown that the use of certain weak-acid preservatives may, under some conditions enhance the growth of fungal species isolates in food products (Marín et al., 2002, 2003).

The hurdle concept for effective preservation of foods may inhibit outgrowth, but has been found to induce prolonged survival of *Escherichia coli* O157:H7 in minimally processed foods (Uyttendaele, Taverniers, and Debevere, 2001). When compared to pH, acetic acid, malic acid, and L-lactic acid can provide protective effects on the survival of *E. coli* O157:H7 at concentrations between 5 and 10 mM. D-Lactic has a greater protective effect (approximately a 4-log increase in survival) over a wider range of concentration, from 1 to 20 mM. Citric acid, on the other hand, does not appear to exhibit any protective effect at similar concentrations (Bjornsdottir, Breidt, Jr., and McFeeters, 2006).

6.4   Sensorial effects and consumer perception

Food producers are increasingly faced with challenges as a result of demands and sometimes inconceivable expectations of the consumer. Although it is known that the color of food and beverages has a dramatic influence on consumer preferences (Calvo, Salvador, and Fiszman, 2001; Gamble, Jaeger, and Harker, 2006; Cortés, Esteve, and Frigola, 2008), consumers have also become more critical of the use of additives to preserve food or to improve color and flavor (Bruhn, 2000). Minimal processing techniques have been put in place to meet these challenges (Rico et al., 2007). The color of citric beverages, in particular, is related to the consumer's perception of flavor, sweetness, and other characteristics in relation to the quality of these products (Tepper, 1993; Cortés, Esteve, and Frigola, 2008). Consumer purchases of raw chilled meat are critically influenced by color, as appearance is the first impression of meat offered to the public. Other aspects of meat quality, such as composition, palatability, and safety, are of secondary concern (Huang, Ho, and McMillin, 2005). It must be kept in mind that consumer perceptions, related to human subjectivity, will ultimately determine purchasing and consumption decisions (Verbeke et al., 2007).

It is becoming increasingly obvious that what the public says it wants, and what it actually buys, are often not the same, which often indicates that people's desires and realities are at opposite ends of the spectrum. People want food that is fast and easy to prepare (such as that intended for microwave heating) to look, smell, and taste as good as food prepared in a conventional oven, and also to be as safe and nutritious. To accomplish these desires and demands, it is often necessary to add more additives not fewer! (Forman, 1998).

6.5   Recommended daily intake

Organic production methods lead to increases in nutrients, particularly organic acids and polyphenolic compounds, many of which are considered to have potential human health benefits as antioxidants. The impact on health when consuming great levels of organic acids and polyphenolics has not yet been determined (Winter and Davis, 2006).

Both the FDA and USDA provide extensive regulatory oversight in ensuring the safety of the food supply (Donoghue, 2003). Moreover, in current FDA regulations for acidified foods, no mention is made of the amount or type of organic acid needed to lower the pH (Bjornsdottir et al., 2006). The maximum permitted concentration in different foods should be regulated by legislation (EC, 1995; Wen, Wang, and Feng, 2007), and a mandatory step in routine food analyses should be the determination of this maximum permitted concentration (Gonzalez, Gallego, and Valcarcel, 1998).

6.6   Odors and palatability

Organic acids significantly affect the flavor and quality of food, although some acids generally have aggressive odors and may reduce the palatability of food. This is, however, more obvious in animal feed and diet (Tsiloyiannis et al., 2001). In addition to this, when using individual aromatic compounds in preservation, high concentrations are often needed to inhibit contaminating microorganisms. Such high concentrations often exceed the flavor threshold acceptable to consumers. Alternative ways of preventing growth, while keeping high flavor quality, are to combine different preservative techniques, so-called hurdle technology (Nazer et al., 2005).

Flavors of organic acids range from fresh acidic (citric acid) to salty and bitter (succinic acid). This characteristic plays an important role in the wine industry. The natural malolactic conversion causes major changes in fermented beverages, red wine, and cider, which results in a decrease in the acidity of red wine and cider, and the replacement of malic acid with lactic acid, which has a weaker taste (Gomis, 1992).

Sensory characteristics of benzoic acid are not commonly known, but relating to its chemical structure, benzoic acid might induce either sourness in taste or pungency (irritation) due to general chemical stimulation (Otero-Losada, 1999). The hydroxy derivatives of benzoic acid (salicylic [2-hydroxy], m-hydroxy [3-hydroxy], gentisic [2,5-digydroxy], protochatechuic [3,4-hydroxy], and gallic [3,4,5-trigydroxy] acids) also induce sourness, astringency, and prickling (pungency), although with different intensities. However, from previous evidence, it is known that benzoic stimulates the general chemical sense through activation of trigeminal nerve endings coexisting with taste cells, to induce a mild pungency, or a prickling taste (Peleg and Noble, 1995). Benzoic is, therefore, considered a prototypical mild pungent stimulus. It induces prickling and may be perceived as slightly sour (Otero-Losada, 1999).

The sharp, strong acidic taste of citric acid may overpower the flavor of sweeteners and flavorants in a product if additionally added. The use of certain synthetic high-intensity sweeteners in a foodstuff may result in a bitter aftertaste. To overcome this, it is used in conjunction with citric acid. However, citric acid also has a short-lived tartness flavor, which is rapidly lost, leaving the bitter aftertaste (Fowlds, 2002).

6.7   Cost

A major predicament in the application of organic acids is the high cost. The most expensive organic acids are known to be malic acid, citric acid, and tartaric acid (Castillo et al., 2004; Couto and Sanroman, 2006). The latter is generally the most expensive of the commonly used food acids (Fowlds, 2002). Citric acid is one of the most commonly used organic acids in both the food and pharmaceutical industries because of its pleasant taste and high solubility as well as its flavor-enhancing properties. Although citric acid can be obtained by chemical synthesis, the cost is much higher than using fermentation (Couto and Sanroman, 2006).

6.8   Application methods

Effective food processing entails overpowering the microbial cell's ability to survive in hostile environments, while taking great care to employ a minimum of energy-wasting processing steps (Hill, O'Driscoll, and Booth, 1995). In 1987 a group of international experts discussed the possibility of an integrated approach on how to assess the efficacies of various separate or combined production practices (Smulders, 1987). A consensus was reached, concluding that strict adherence to measures of good financial practice (GFP) and good manufacturing practice (GMP) during production and processing should be the basis of muscle food safety strategies. It was also decided that any decontamination steps should only serve as additional measures to assure safety of a food product (Smulders and Greer, 1998; Couto and Sanroman, 2006).

In carcass decontamination, dilute solutions of organic acids (1–3%) generally have no effect on desirable sensory properties of meat. However, this is dependent on treatment conditions, and some acids, in particular, lactic and acetic acid, can produce adverse sensory changes when applied to meat cuts. These changes are frequently irreversible. Problems are also often encountered after organic acid spray in carcass decontamination as a result of recontamination during carcass breaking and also further processing steps (Smulders and Greer, 1998). Various studies have reported on such sensory effects after application of the organic acids in meat decontamination. Some of the results are presented in Table 6.1.

Organic acids have to overcome the protective effect posed by some food products, such as the high lipid content and topography of chicken skin, which are primary protective factors (Tamblyn and Conner, 1997).

Table 6.1 Some Effects Detected on Sensory Properties of Subprimal or Retail
Cut Red Meat (Beef) after Application of Organic Acids

Meat portion	Organic acid treatment	Application method	Effect on sensory quality
Steaks	1% acetic acid or lactic acid	Spray	No effect on color or odor
Steaks	2% acetic acid or lactic acid	Dip	Discoloration
Loin strips	1% acetic acid or lactic acid	Spray	No effect on color or odor
Loin strips	2% acetic–lactic combination	Spray	No effect on color
Loin strips	2% acetic–lactic combination	Spray	No effect on color or odor
Cubes	1.2% acetic acid	Dip	Discoloration, off flavor

Sources: Data from Bell et al., *Journal of Food Protection*, 49:207–210, 1986; Acuff et al. *Meat Science* 19:217–226, 1987; Dixon et al., *International Journal of Food Microbiology* 5:181–186, 1987; Kotula and Thelappurate, *Journal of Food Protection* 57:665–670, 1994; Goddard et al., *Journal of Food Protection* 59:849–853, 1996; Mikel et al., *Journal of Food Science* 61:1058–1061, 1996.

However, the maximum concentration of preservatives permitted in each type of food is controlled by legislation (Gonzalez, Gallego, and Valcarcel, 1998). The ideal would be to increase the antimicrobial activity of lower concentrations of organic acids as food preservatives (≤ 1%). This would overcome the problems with cost and quality associated with high concentrations of acids (Tamblyn and Conner, 1997). There is, therefore, the increasing need in the regulation of food safety to be able to predict the safe limits of weak-acid preservation processes (Quintas et al., 2005). The application of inadequate concentrations may lead to induction of tolerance to organic acids, which in turn would affect the virulence of a pathogenic invasive organism (Hill, O'Driscoll, and Booth, 1995).

Commercial trials are continuously necessary to develop and evaluate organic acid systems designed for the commercial environment, in order to enhance meat safety without sacrificing desirable sensory properties. These trials are essential to provide regulatory agencies and meat processors with the information needed to redirect some of their decontamination efforts to postcarcass processing and storage (Smulders and Greer, 1998). Spraying with diluted solutions of organic acids may also cause a decrease in the pH value of the meat (Álvarez-Ordóñez et al., 2009).

Low concentrations of organic acids are not effective against *Salmonella enterica* sv. Typhimurium attached to broiler carcasses (Lillard et al., 1987; Tamblyn and Conner, 1994), and application could result in induction of tolerance, which in turn would have an impact on the virulence of a pathogenic invasive organism (Hill, O'Driscoll, and Booth, 1995).

However, higher concentrations, for example, 4%, which is usually effective against *S. enterica* sv. Typhimurium (attached to broiler skin) would produce adverse carcass characteristics, such as bleaching of the skin and would also be too costly for use in the processing plant (Tamblyn and Conner, 1997).

The possible long-term consequences of meat decontamination on the microbial ecology of plants and products is a cause for concern (Stopforth et al., 2003). The runoff and aerosol dispersion from spray-washing applications may provide a favorable environment for bacterial contamination from washed carcasses, leading to possible biofilm formation (Hood and Zottola, 1997; Wong, 1998). Organic acid meat runoff may also mix with the water meat runoff in packing plants and may exert a sublethal acid stressing effect on meatborne pathogens (Samelis et al., 2002). This may result in biofilms composed of acid stressed/adapted pathogenic strains (Stopforth et al., 2003). Extended acid stressing may, therefore, result in pathogen survivors adapted to acid, enhance virulence, and trigger adaptive mutations of permanent acid/stress resistance (Sheridan and MacDowell, 1998). These pathogens, that are now part of mixed microbial cultures in a food environment, may react differently to decontamination than pure cultures in controlled environments (Samelis et al., 2001).

6.9   Oxidation

Oxidation is a common problem associated with some organic acids, for example, ascorbic acid (vitamin C). When dissolved, ascorbic acid is readily oxidized to dehydroascorbic acid, catalyzed by air or light exposure (Wu et al., 1995). Ascorbic acid solutions should be freshly prepared and kept tightly closed and not exposed to light. Organic acids also promote lipid oxidation which increases with increasing acid concentration (Ogden et al., 1995).

6.10   Ineffectiveness

Several factors can alter the effective concentration of a preservative in food products or feed (Ricke, 2003). The spectrum of activity of some organic acids, in particular sodium lactate, is relatively narrow or may be specific to a certain group of organisms. This may pose a problem, as other mechanisms may take over after time (Lemay et al., 2002). Sodium benzoate or potassium sorbate are both chemical preservatives, approved for use in a wide range of foodstuffs, but have been shown to have only limited effect on *E. coli* O157:H7 (Comes and Beelman, 2002). This observation is augmented by the fact that various microbiota can utilize organic acids as both carbon and energy sources (Van Immerseel et al., 2006).

6.11   Influence on tolerance to other stresses

The organic acids in food preservation cause definite culture pH-dependent effects on both (1) the heat-shock response and (2) the thermotolerance of the two important food-associated fungi *Saccharomyces cerevisiae* and *Zygosaccharomyces bailli*. Sorbic and benzoic acids have definite influences on the heat-shock response and thermotolerance of *S. cerevisiae* and these effects are strongly dependent on pH. At low pH these acids will inhibit both the heat-shock protein (Hsp) and thermotolerance induction during sublethal (39°C) heat shock and also cause strong induction of respiratory-deficient petites among the survivors of lethal (50–52°C) treatment. At higher pH, sorbic acid causes strong induction of thermotolerance, but without inducing a heat-shock response (Cheng, Moghraby, and Piper, 1999). Undissociated acetate (but not dissociated acetate) induce the extreme acid resistance of *E. coli* O157:H7. It appears, therefore, that pH effects are mediated via acetate dissociation. Propionate and butyrate can encourage the development of extreme acid resistance in nonpathogenic *E. coli*, and are less effective than acetate in increasing the extreme acid resistance of *E. coli* O157:H7. Benzoic acid is ~100× less active than acetic acid in promoting the survival of *E. coli* O157:H7. However, acetic acid only promoted the survival of *E. coli* O157:H7 cells when added before they reached the stationary phase (Diez-Gonzalez and Russell, 1999).

Exposure to acid frequently induces stress responses in listeriae which make these bacteria more tolerant of acidity, ethanol, or hydrogen peroxide (Doyle, 1999). *L. monocytogenes* can grow in the presence of organic acids and salt, apart from its ability to grow at temperatures as low as −0.4°C (Sleator et al., 2001; Bereksi et al., 2002). It was found that exposure to salt and organic acids increased the ability of *L. monocytogenes* to invade Caco-2 cells (tumor-derived cells). However, exposure to organic acids and salt decreases the ability of *L. monocytogenes* to survive gastric stress (Garner et al., 2006).

References

Acuff, G., Vanderzant, C., Savell, J., Jones, D., Griffen, D., and Ehlers, J. 1987. Effect of acid decontamination of beef subprimal cuts on the microbiological and sensory characteristics of steaks. *Meat Science* 19:217–226.

Álvarez-Ordóñez, A., Fernandez, A., Bernardo, A., and Lopez, M. 2009. Comparison of acids on the induction of an acid tolerance response in *Salmonella typhimurium*, consequences for food safety. *Meat Science* 81:65–70.

Anderson, J.A. and Sogn, D.D. (Eds.). 1984. *Adverse Reactions to Foods*, PHS-NIH Publication No. 84-2442, pp. 1–6. Milwaukee, WI: American Academy of Allergy and Immunology.

Banerjee, M. and Sarkar, P.K. 2004. Antibiotic resistance and susceptibility to some food preservative measures of spoilage and pathogenic micro-organisms from spices. *Food Microbiology* 21:335–342.

Bell, M., Marshall, R., and Anderson, M. 1986. Microbiological and sensory tests of beef treated with acetic and formic acids. *Journal of Food Protection* 49:207–210.

Bereksi, N., Gavini, F., Benezech, T., and Faille, C. 2002. Growth, morphology and surface properties of *Listeria monocytogenes* Scott A and LO28 under saline and acid environments. *Journal of Applied Microbiology* 92:556–565.

Bjornsdottir, K., Breidt, F., Jr., and McFeeters, R.F. 2006. Protective effects of organic acids on survival of *Escherichia coli* O157:H7 in acidic environments. *Applied and Environmental Microbiology* 72:660–664.

Bolton, D.J., Catarame, T., Byrne, C., Sheridan, J.J., McDowell, D.A., and Blair, I.S. 2002. The ineffectiveness of organic acids, freezing and pulsed electric fields to control *Escherichia coli* O157:H7 in beef burgers. *Letters in Applied Microbiology* 34:139–143.

Breidt, F., Jr. and Fleming, H.P. 1998. Modeling of the competitive growth of *Listeria monocytogenes* and *Lcatococcus lactis* in vegetable broth. *Applied and Environmental Microbiology* 64:3159–3165.

Bruhn, C. 2000. Food labelling: Consumer needs. In: J.R. Blanchfield (Ed.), *Food Labelling*. Cambridge, UK: Woodhead,.

Calvo, C., Salvador, A., and Fiszman, S.M. 2001. Influence of colour intensity on the perception of colour and sweetness in various fruit-flavoured yoghurts. *European Food Research and Technology* 213:99–103.

Carroll, C.D., Alvarado, C.Z., Brashears, M.M., Thompson, L.D., and Boyce, J. 2007. Marination of turkey breast fillets to control the growth of *Listeria monocytogenes* and improve meat quality in deli loaves. *Poultry Science* 86:150–155.

Castillo, C., Benedito, J.L., Mendez, J., Pereira, V., Lopez-Alonso, M., Miranda, M., and Hernandez, J. 2004. Organic acids as a substitute for monensin in diets for beef cattle. *Animal Feed Science and Technology* 115:101–116.

Cheng, L., Moghraby, L., and Piper, P.W. 1999. Weak organic acid treatment causes a trehalose accumulation in low-pH cultures of Saccharomyces cerevisiae, not displayed by the more preservative-resistant Zygosaccharomyces bailii. *FEMS Microbiology Letters* 170:89–95.

Comes, J.E. and Beelman, R.B. 2002. Addition of fumaric acid and sodium benzoate as an alternative method to achieve a 5-log reduction of *Escherichia coli* O157:H7 populations in apple cider. *Journal of Food Protection* 65:476–483.

Cortés, C., Esteve, M.J., and Frigola, A. 2008. Color of orange juice treated by high intensity pulsed electric fields during refrigerated storage and comparison with pasteurized juice. *Food Control* 19:151–158.

Couto, S.R. and Sanroman, M.A. 2006. Application of solid-state fermentation to food industry—A review. *Journal of Food Engineering* 76:291–302.

Diez-Gonzalez, F. and Russell, J.B. 1999. Factors affecting the extreme acid resistance of *Escherichia coli* O157:H7. *Food Microbiology* 16:367–374.

Dixon, Z., Vanderzant, C., Acuff, G., Savell, J., and Jones, D. 1987. Effect of acid treatment of beef strip loin steaks on microbiological and sensory characteristics. *International Journal of Food Microbiology* 5:181–186.

Donoghue, D.J. 2003. Antibiotic residues in poultry tissues and eggs: Human health concerns? *Poultry Science* 82:618–621.

Doyle, E.M. 1999. *Use of organic acids to control Listeria in meat.* American Meat Institute Foundation. Washington, DC.

EC. 1995. European Parliament and Council Directive No. 95/2/EC. European Communities, Brussels.

Forman, C. 1998. The food additives business: GA-040N. Business Communications Company Inc. http://www.bccresearch.com. (Accessed November 15, 2006.)

Fowlds, R.W.R. 2002. Production of a food acid mixture containing fumaric acid. http://www.freepatentsonline.com. (Accessed June 18, 2007.)

Furukawa, C.T. 1991. Non-immunogologic food reactions that can be confused with allergy. *Immunology and Allergy Clinics of North America* 11:815–818.

Gamble, J., Jaeger, S.R., and Harker, F.R. 2006. Preferences in pear appearance and response to novelty among Australian and New Zealand consumers. *Postharvest Biology and Technology* 41:38–47.

Garner, M.R., James, K.E., Callahan, M.C., Wiedmann, M., and Boor, K.J. 2006. Exosure to salt and organic acids increases the ability of *Listeria monocytogenes* to invade Caco-cells but decreases its ability to survive gastric stress. *Applied and Environmental Microbiology* 72:5384–5395.

Goddard, B., Mikel, W., Conner, D., and Jones, W. 1996. Use of organic acids to improve the chemical, physical and microbiological attributes of beef strip loins stored at −1°C for 112 days. *Journal of Food Protection* 59:849–853.

Gomis, D.B. 1992. HPLC analysis of organic acids. In: L.M.L. Nollet (Ed.), *Food Analysis by HPLC*, pp. 371–385. New York: Marcel Dekker.

Gonzalez, M., Gallego, M., and Valcarcel, M. 1998. Simultaneous gas chromatographic determination of food preservatives following solid-phase extraction. *Journal of Chromatography A* 823:321–329.

Gould, G.W. and Jones, M.V. 1989. Combination and synergistic effects. In: *Mechanisms of Action of Food Preservation Procedures.* G.W. Gould (Ed.) London: Elsevier Science.

Guynot, M.E., Ramos, A.J., Sanchis, V., and Marín, S. 2005. Study of benzoate, propionate, and sorbate salts as mould spoilage inhibitors on intermediate moisture bakery products of low pH (4.5–5.5). *International Journal of Food Microbiology* 101:161–168.

Hayakawa, K., Linko, Y.Y., and Linko, P. 1999. Mechanism and control of food allergy. *Lebensmittel-Wissenschaft und-Technologie* 32:1–11.

Hefle, S.L., Nordlee, J.A., and Taylor, S.L. 1996. Allergenic foods. *Critical Review of Food Science and Nutrition* 36:S69–S89.

Hill, C., O'Driscoll, B., and Booth, I. 1995. Acid adaptation and food poisoning microorganisms. *International Journal of Food Microbiology* 28:245–254.

Hood, S.K. and Zottolla, E.A. 1997. Growth media and surface conditioning influence the adherence of *Pseudomonas fragi, Salmonella typhimurium* and *Listeria monocytogenes* cells to stainless steel. *Journal of Food Protection* 60:1034–1037.

Huang, N., Ho, C., and McMillin, K.W. 2005. Retail shelf-life of pork dipped in organic acid before modified atmosphere or vacuum packaging. *Journal of Food Science* 70:M382–M387.

Itoh, Y. 1996. Current status of development of substitute, hypoallergenic and anti-allergenic foods for food allergy. *Bio Industry* 13:36–43.

Janssen, M., Geeraerd, A.H., Logist, F., et al. 2006. Modelling *Yersinia enterocolitica* inactivation in coculture experiments with *Lactobacillus sakei* as based on pH and lactic acid profiles. *International Journal of Food Microbiology* 111:59–72.

Kotula, K. and Thelappurate, R. 1994. Microbiological and sensory attributes of retail cuts of beef treated with acetic and lactic acid solutions. *Journal of Food Protection* 57:665–670.

Lemay, M.J., Choquette, J., Delaquis, P.J., Claude, G., Rodrigue, N., and Saucier, L. 2002. Antimicrobial effect of natural preservatives in a cooked and acidified chicken meat model. *International Journal of Food Microbiology* 78:217–226.

Lillard, H.S., Blankenship, L.C., Dickens, J.A., Craven, S.E., and Schakelford, A.D. 1987. Effect of acetic acid on the microbiological quality of scalded picked and unpicked broiler carcasses. *Journal of Food Protection* 50:112–114.

Mancini, R.A., Hunt, M.C., Seyfert, M., Kropf, D.H., Hachmeister, K.A., Herald, T.J., and Johnson, D.E. 2007. Effects of ascorbic and citric acid on beef lumbar vertebrae marrow colour. *Meat Science* 76:568–573.

Marín, S., Abellana, M., Rubinat, M., Sanchis, V., and Ramos, A.J. 2003. Efficacy of sorbates on the control of the growth of *Eurotium* species in bakery products with near neutral pH. *International Journal of Food Microbiology* 87:251–258.

Marín, S., Guynot, M.E., Neira, P., Bernadó, M., Sanchis, V., and Ramos, A.J. 2002. Risk assessment of the use of sub-optimal levels of weak-acid preservatives in the control of mould growth on bakery products. *International Journal of Food Microbiology* 79:203–211.

Mikel, W.B., Goddard, B.L., and Bradford, D.D. 1996. Muscle microstructure and sensory attributes of organic acid-treated beef strip loins. *Journal of Food Science* 61:1058–1061.

Motala, C. and Steinman, H. 2008. Food allergy, preservatives and asthma. *http://www.asthma.co.za/*. (Accessed September 15, 2009.)

Nazer, A.I., Kobilinsky, A., Tholozan, J.L., and Dubois-Brissonnet, F. 2005. Combinations of food antimicrobials at low levels to inhibit the growth of *Salmonella* sv. *Typhimurium*: A synergistic effect? *Food Microbiology* 22:391–398.

Ogden, S.K., Guerrero, I., Taylor, A.J., Escalona Buendia, H., and Gallardo, F. 1995. Changes in odour, colour and texture during the storage of acid preserved meat. *Lebensmittel-Wissenschaft und-Technologie* 28:521–527.

Otero-Losada, M.E. 1999. A kinetic study on benzoic acid pungency and sensory attributes of benzoic acid. *Chemical Senses* 24:245–253.

Peleg, H. and Noble, A.C. 1995. Perceptual properties of benzoic acid derivatives. *Chemical Senses* 20:393–400.

Penney, V., Henderson, G., Blum, C., and Johnson-Green, P. 2004. The potential of phytopreservatives and nisin to control microbial spoilage of minimally processed fruit yogurts. *Innovative Food Science and Emerging Technologies* 5:369–375.

Piper, P.W. 1999. Yeast superoxide dismutase mutants reveal a pro-oxidant action of weak organic acid food preservatives. *Free Radical Biology and Medicine* 27:11–12.

Poulter, S. 2007. The proof food additives ARE as bad as we feared. *Daily Mail* MailOnline, 18 May 2007.

Quintas, C., Leyva, J.S., Sotoca, R., Loureiro-Dias, M.C., and Peinado, J.M. 2005. A model of the specific growth rate inhibition by weak acids in yeasts based on energy requirements. *International Journal of Food Microbiology* 100:125–130.

Ricke, S.C. 2003. Perspectives on the use of organic acids and short chain fatty acids as antimicrobials. *Poultry Science* 82:632–639.

Rico, D., Martin-Diana, A.B., Barat, J.M., and Barry-Ryan, C. 2007. Extending and measuring the quality of fresh-cut fruit and vegetables: A review. *Trends in Food Science and Technology* 18:373–386.

Samelis, J., Sofos, J.N., Kendall, P.A., and Smith, G.C. 2001. Influence of the natural microbial flora on the acid tolerance response of *Listeria monocytogenes* in a model system of fresh meat decontamination fluids. *Applied and Environmental Microbiology* 67:2410–2420.

Samelis, J., Sofos, J.N., Kendall, P.A., and Smith, G.C. 2002. Effect of acid adaptation on survival of *Escherichia coli* O157:H7 in meat decontamination washing fluids and potential effects of organic acid interventions on the microbial ecology of the meat plant environment. *Journal of Food Protection* 65:33–40.

Sheridan, J.J. and McDowell, D.A. 1998. Factors affecting the emergence of pathogens on foods. *Meat Science* 49:S151–S167.

Sleator, R.D., Wouters, J., Gahan, C.G., Abee, T., and Hill, C. 2001. Analysis of the role of OpuC, an osmolyte transport system, in salt tolerance and virulence potential of *Listeria monocytogenes*. *Applied and Environmental Microbiology* 67:2692–2698.

Smulders, F. 1987. Prospectives for microbiological decontamination of meat and poultry by organic acids with special reference to lactic acid. In: F.J.M. Smulders (Ed.), *Elimination of Pathogenic Organisms from Meat and Poultry*, pp. 319–344. London: Elsevier Science,.

Smulders, F.J. and Greer, G.G. 1998. Integrating microbial decontamination with organic acids in HACCP programmes for muscle foods: Prospects and controversies. *International Journal of Food Microbiology* 44:149–169.

State Bureau for Quality Supervision, Inspection and Quarantine. 1996. Hygienic standards for the use of food additives (National Standard of the People's Republic of China GB 2760). *Standards Press of China*, Beijing, p. 290.

Stopforth, J.D., Samelis, J., Sofos, J.N., Kendall, P.A., and Smith, G.C. 2003. Influence of organic acid concentration on survival of *Listeria monocytogenes* and *Escherichia coli* O157:H7 in beef carcass wash water and on model equipment surfaces. *Food Microbiology* 20:651–660.

Tamblyn, K.C. and Conner, D.E. 1994. Antibacterial activity of citric, malic, and propionic acids against *Salmonella typhimurium* attached to broiler skin. *Poultry Science* 73:23.

Tamblyn, K.C. and Conner, D.E. 1997. Bactericidal activity of organic acids in combination with transdermal compounds against *Salmonella typhimurium* attached to broiler skin. *Food Microbiology* 14:477–484.

Tepper, B.J. 1993. Effects of a slight color variation on consumer acceptance of orange juice. *Journal of Sensory Study* 8:145–154.

Tfouni, S.A.V. and Toledo, M.C.F. 2002. Determination of benzoic and sorbic acids in Brazilian food. *Food Control* 13:117–123.

Tsiloyiannis, V.K., Kyriakis, S.C., Vlemmas, J., and Sarris, K. 2001. The effect of organic acids on the control of porcine post-weaning diarrhoea. *Research in Veterinary Sciences* 70:287–293.

Uyttendaele, M., Taverniers, I., and Debevere, J. 2001. Effect of stress induced by suboptimal growth factors on survival of Escherichia coli O157:H7. *International Journal of Food Microbiology* 66:31–37.

Van Immerseel, F., Russell, J.B., Flythe, M.D., et al. 2006. The use of organic acids to combat *Salmonella* in poultry: A mechanistic explanation of the efficacy. *Avian Pathology* 35:182–188.

Verbeke, W., Frewer, L.J., Scholderer, J., and De Brabander, H.F. 2007. Why consumers behave as they do with respect to food safety and risk information. *Analytica Chimica Acta* 586:2–7.

Wen, Y., Wang, Y., and Feng, Y.Q. 2007. A simple and rapid method for simultaneous determination of benzoic and sorbic acids in food using in-tube solid-phase microextraction coupled with high-performance liquid chromatography. *Analytical and Bioanalytical Chemistry* 388:1779–1787.

Winter, C.K. and Davis, S.F. 2006. Organic foods. *Journal of Food Science* 71:R117–R124.

Wong, A.C.L. 1998. Biofilms in food processing environments. *Journal of Dairy Science* 81:2765–2770.

Wu, C.H., Lo, Y.S., Lee, Y.-H., and Lin, T.-I. 1995. Capillary electrophoretic determination of organic acids with indirect detection. *Journal of Chromatography A* 716:291–301.

Yamamoto, K., Takahashi, M., and Niki, E. 1987. Role of iron and ascorbic acid in the oxidation of methyl linoleate micelles. *Chemical Letter* 6:1149–1152.

chapter seven

Large-scale organic acid production

7.1 Introduction

Organic acids are widely distributed in nature as normal constituents of plants or animal tissues and for their various industrial applications are commonly derived from food plants and food grade microbes (Gauthier, 2005; Penney et al., 2004). Mycelial fungi and certain groups of bacteria (e.g., lactic acid and acetic acid bacteria) have for many years been used as conventional producers of organic acids. In the 1960s the ability of yeasts to synthesize the carboxylic acids was stumbled upon unexpectedly. This was during the early stages of worldwide research on the use of oil hydrocarbons as raw materials for the production of microbial protein. The first notable reports of success came from Japan and the Soviet Union. From Russia, it was reported that alkaline-assimilating yeasts were able to produce considerable amounts of organic acids (the Krebs cycle intermediates, α-ketoglutaric, citric acid, and *treo-D$_s$-*(+)-isocitric) from petrolatum (Finogenova et al., 2005).

7.2 Naturally occurring weak organic acids

Organic acids occur naturally in a variety of foods as a result of (1) normal biochemical metabolic processes, (2) direct addition as acidulants, (3) hydrolysis, or (4) bacterial growth (Gomis, 1992). In Table 7.1 some of the most abundantly found organic acids are described.

7.3 Microbial physiology and organic acids

Production of organic acids is found among various bacterial and fungal species. This is particularly common among all lactic acid bacteria (LAB) (Vesterlund et al., 2004). Heterofermentative LAB are able to ferment various organic acids, predominantly citrate, malate, and pyruvate (Zaunmuller et al., 2006). In various species and strains of LAB, organic acid production may, however, vary. For example, in *Lactococcus lactus* pyruvate is partially converted to α-acetolactate when electron acceptors (such as citrate) are present, whereas *Lactobacillus sanfranciscensis*

Table 7.1 Natural Occurrence of Organic Acids

Organic acid	Source found	Natural function	Reference
Lactic acid	Not naturally in food, formed during fermentation Abundant in fermented food	Natural antimicrobial	Alakomi et al. 2000; Barbosa-Cánovas et al. 2003
Succinic acid	Plant and animal tissues	Role in intermediary metabolism of human body	Barbosa-Cánovas et al. 2003; Food-Info 2009
Citric acid	Citrus fruits	Essential for citric acid cycle in respiration of plants	Barbosa-Cánovas et al. 2003; Food-Info 2009
Malic acid	Fruit and vegetables	Part of metabolic pathway of every living cell	Barbosa-Cánovas et al. 2003; Food-Info 2009
Tartaric acid	Fruits, such as grapes and pineapples	Regulates acidity, enhances taste	Barbosa-Cánovas et al. 2003; Food-Info 2009
Benzoic acid	Various fruits, especially berries (cranberries, raspberries), plums, prunes, also cinnamon and cloves	Antifungal action, specifically in acidic substrates	Barbosa-Cánovas et al. 2003; Chen et al. 2001; Penney et al. 2004
Propionic acid	In foods by natural processing	Product of fermented foods	Barbosa-Cánovas et al. 2003; Food-Info 2009
Cinnamic acid	Spices, including cinnamon plants	Phenolic component	Roller 1995

strains use a pyruvate branch when converting citrate to lactate and acetate. Malate and fumarate are converted to lactate by *L. sanfranciscensis*, whereas *Lactobacillus reuteri* and *Lactobacillus pontis* do not utilize citrate, but convert malate and fumarate to succinate. Citrate conversion to succinate is more commonly found in lactobacilli (Ganzle, Vermeulen, and Vogel, 2007). Accelerated acetic acid production is typical for citrate utilization. *Lactobacillus* strains produce acetic acid quite late in fermentation and may appear low when compared to lactic acid (Rossland et al., 2005).

Efficient lactic acid production from cane sugar molasses is achieved by *Lactobacillus delbrueckii* in batch fermentation. Fermentative production of lactic acid is very effective in producing optically pure L- or D-lactic and also DL-lactic acid, depending on the strain (Dumbrepatil et al., 2008). *Lactobacillus plantarum* cells are homofermentative, often used for production of lactic acid from glucose fermentation (Krishnan et al., 2001).

The acetic acid bacteria (AAB) play an important role in the production of vinegar (acetic acid) from fruit juices and alcohols as they are able to oxidize sugars, sugar alcohols, and ethanol. Acetic acid is the major end product of this unique metabolism. These bacteria, therefore, often occur in sugar–alcohol-enriched environments. The production of acetic acid is in essence a two-stage process, with yeasts converting sugars to alcohol ($C_6H_{12}O_6 \rightarrow 2CH_3CH_2OH + CO_2$), followed by oxidation of ethanol to acetic acid by AAB ($C_2H_5O + O_2 \rightarrow CH_3COOH + H_2O$). Vinegar is the traditional product of acetous fermentation of natural alcoholic substrates, and is classified as fruit, starch, or spirit substrate vinegar, depending on the substrate (Beuchat, 1995; Trček, 2005).

Various genera of the AAB are classified in the family *Acetobacteriaceae*. AAB that are most commonly used for commercial production of vinegar are members of *Acetobacter*, *Gluconobacter*, and *Gluconoacetobacter*. Industrial manufacturing of vinegar is categorized into: (1) slow processes, (2) fast processes, and (3) submerged processes. Other products of the AAB include 2-keto-L-gulonic acid, D-tagatose, and shikimate, all products used in various other food and pharmaceutical industries.

Propionibacteria are able to utilize lactate as a substrate much more rapidly than glucose. An alternative production of propionic acid by species of propionibacteria, is achieved from sugars through lactate as an intermediate (Tyree, Clausen, and Gaddy, 1991). Overall production of propionic and acetic acid from glucose by *Propionibacterium* is given by:

$$1.5C_6H_{12}O_6 \rightarrow 2C_2H_5COOH + CH_3COOH + CO_2 + H_2O \qquad (7.1)$$

The sugars are first fermented to lactic acid by a species of *Lactobacillus*. Lactic acid is then converted to propionic and acetic acids by propionibacteria (Tyree, Clausen, and Gaddy, 1991). Lactic acid is rapidly formed from glucose using lactobacilli:

$$C_6H_{12}O_6 \rightarrow 2CH_3CHOHCOOH \qquad (7.2a)$$

Propionic and acetic acids are then produced from lactic acid:

$$3CH_3CHOHCOOH \rightarrow 2CH_3CH_2COOH + CH_3COOH + CO_2 + H_2O \quad (7.2b)$$

Both of these fermentation processes theoretically give the same yields of propionic and acetic acids from glucose (Tyree, Clausen, and Gaddy, 1991).

The most widely used fermentation process utilizes the fungus *Aspergillus niger*. Microbial production of many organic acids is the preferred method and dates back several decades (Ramanchandran et al., 2006). Members of the fungal group *Aspergillus* section *Nigri* (formerly *A. niger* group) are considered common food spoilage fungi and are distributed worldwide. However, some are widely used and studied for industrial purposes as they are common sources of organic acids and also extracellular enzymes for application in food processing (Abarca et al., 2004). Although being a recognized opportunistic pathogen, *A. niger* holds the GRAS (generally recognized as safe) status from the FDA (Bigelis and Lasure, 1987). Despite the fact that this organism is also notorious for its ability to produce ochratoxin A (OTA), it has long been used in the food industry, without any apparent adverse effects on human health (Abarca et al., 2004).

7.4   Substrates and yields

The world's food supply strongly depends on access to fermented food. In turn, successful fermentation depends on the production of antimicrobial metabolites by starter cultures (Nes and Johnsborg, 2004). Fermentation of a wide range of dairy products, meat, and vegetable foods, commonly involves the use of bacteria such as LAB as starter cultures. During the fermentation of dairy products, these cultures metabolize lactose to lactic acid, which lowers the pH to create an environment unfavorable to pathogens and spoilage organisms, and also acts as a powerful inhibitor of the growth of these organisms (Davidson et al., 1995; Aslim et al., 2005). Low pH also assists in producing the desired organoleptic properties of a fermented foodstuff. Although other compounds are also produced, the production of sufficient quantities of an organic acid, in particular lactic acid, is essential for the success of the fermentation (Davidson et al., 1995). Lactic acid is not naturally present in foods, but is commonly formed during fermentation of foods such as sauerkraut, pickles, olives, some meats, and cheeses. Propionic acid also occurs in foods as a result of natural processing. An example is Swiss cheese at concentrations up to 1%, where propionic acid may be found, as produced by *Propionibacterium shermanii* (Barbosa-Cánovas et al., 2003).

In vegetables, natural microflora may consist of various genera of bacteria, yeasts, and molds. The organic acid content of a tomato is responsible for the low pH of between 4.0 and 4.6 (Gutheil, Price, and Swanson, 1980; Sajur, Saguir, and Manca de Nadra, 2007). Sorbic acid and perbenzoic acid are found in particularly high levels in several wild berries (Piper, 1999).

Organic acids are also general compounds in wine-derived products. Some of them originate from the grape, whereas others may first appear during various kinds of fermentation. Tartaric acid is a major acid in wine, but is also prevalent in the grape. It is, therefore, commonly found in grape-derived products, and is also present in appreciable amounts in wine vinegar and in low concentration in cider vinegars. Total acidity of vinegars is expressed as acetic acid because it is the major organic acid in vinegars. The level and character of organic acids in vinegars may actually provide information concerning (1) the origin of the raw material, (2) the nature of microbiological growth, and even (3) the processing techniques used (Morales, Gonzalez, and Troncoso, 1998).

Malic acid is present in low amounts in wine vinegars, but may be very variable, depending on the origin of the wine and also on the different treatments. In cider vinegars and sherry vinegars, malic acid is the main representative. High concentrations are, for example, found in balsamic vinegar. As a result of malolactic fermentation, malic acid is converted to lactic acid and generally produced after alcoholic fermentation. The high malic acid content of apples is commonly reduced by this fermentation. A high amount of lactic acid is formed in cider vinegars, and partially oxidized by acetic bacteria, whereas citric acid is not abundantly found in cider vinegars. Acetic acid levels are usually very high in sherry vinegars (Morales, Gonzalez, and Troncoso, 1998). The natural process of malolactic conversion causes considerable changes in fermented beverages, red wine, and cider. Such changes may include a decrease in the acidity of red wine and cider, and also the replacement of malic acid with lactic acid. The latter has a weaker taste and ultimately improves the flavor of these two types of drinks as well as increases their microbial safety (Gomis, 1992).

Microbiological changes in acidic food and beverages at low pH (pH < 4.5), such as fruit juices and purées generally do not involve toxin production (except mycotoxin formation as a result of some mold types), but can cause a degradation in quality and consequent commercial damages. The presence of fumaric acid is often the result of spoilage of fruit juices by molds, such as *Rhizopus stolonifer*. However, fumaric formation can also arise from the addition of synthetic malic acid, which is considered as an index of adulteration, when confirmed by detection of D-malic acid not present in malic acid from natural sources (Trifiro et al., 1997). The capacity to produce organic acids by spoilage microbiota may, however, pose a further spoilage factor. Citrate is, for example, produced in huge amounts by various organisms and has distinctive acidifying power (Guerzoni, Lanciotti, and Marchetti, 1993).

Acetic acid is one of the main products of AAB metabolism and is found in many foods as the result of the presence and activity of these bacteria. Acetic acid is also a major volatile acid in wine but also one of the main reasons for wine spoilage; for example, when reaching levels above 1.2–1.4 g/l it becomes a problem. AAB are, however, able to survive in high acidic and high ethanol conditions, such as wine, where they may be the main oxidative microorganisms (Gonzalez et al., 2005).

7.5 Industrial fermentation

Organic acids are mainly produced by fermentation (Bailly et al., 2001). More recently the term "biopreservation" has been employed to link fermentation and preservation as it refers to increasing the shelf life and improving the safety of food by using microorganisms or their metabolites or both (Ross et al., 2002). Citric acid is a typical fermentation product and was originally produced by large fermentation companies in Western countries (Marz, 2002). However, the types of organic acids and also the levels that are produced during a fermentation process mostly depend on various factors such as microbial species or strains, culture composition, and growth conditions (Lindgren and Dobrogosz, 1990). For example, an organic acid such as lactic acid may be in equilibrium with its undissociated forms during fermentation and the extent of this dissociation usually depends on the pH of the culture (Ammor et al., 2006).

Lactic acid is such an important factor in the whole process of fermentation that, although other compounds made by starter bacteria can prevent the growth of unwanted microorganisms, the production of adequate quantities of lactic acid is fundamental to the success of the fermentation. This dependence on the low pH that results from synthesis of organic acids acts as a powerful inhibitor of the growth of food spoilage organisms and human pathogens, but also adds to the organoleptic properties of the fermented food product (Davidson et al., 1995).

Currently the major production of organic acids is still in place for traditional applications, which entails food additives and pharmaceutical components as organic acids are traditionally used in food and pharmaceutical industries as (1) preservatives, (2) chemical intermediates, or (3) buffer media. However, this is also true for new applications, such as the biodegradable plastics industry as the demand for organic acids for production of polymeric materials is constantly growing, due to their biodegradable properties, which has resulted in the development of new plants (Bailly, 2002). The fermentation industry comprises five major ingredient categories: (1) antibiotics, (2) organic acids, (3) amino acids, (4) enzymes, and (5) vitamins (Marz, 2000).

Electrodialysis (ED) has been applied in the recovery and production of a variety of acids (Schugerl, 2000). These including lactate (Boyaval, Corre, and Terre, 1987; De Raucourt et al., 1989), lactic acid (Nomura, Iwahara, and Hongo, 1987), malic acid and acetic acid (Eystmondt, Vasic-Racki, and Wandrey, 1989; Weier, Glatz, and Glatz, 1992), propionic acid (Weier,

Glatz, and Glatz, 1992; Boyaval, Stata, and Gavach, 1993), and amino acids (Tichy, Vasic-Racki, and Wandrey, 1990). An important feature of ED is that it increases the volumetric productivity of an organic acid and eliminates microbial growth inhibition, resulting in an excessive concentration of acid. Organic acids are prepared by fermentation and then separated from the fermentation broth by means of electrodialysis (Bazinet, 2004). In Figure 7.1 electrolytic production of organic acids is schematically illustrated (Bailly, 2002).

7.5.1 Monopolar

The use of ED with a monopolar membrane for protein separation and acid caseinate production and in bioreactors for organic acid production is a well-proven technology with huge operating systems worldwide. Such ED is applied to different food systems, which include (Bazinet, 2004)

1. Whey demineralization
2. Organic acids
3. Sugar
4. Wine stabilization
5. Fruit juice deacidification
6. Separation of proteins
7. Separation of amino acids
8. Blood treatments

These applications use the dilution–concentration feature of monopolar ion-exchange membranes.

7.5.2 Bipolar

ED with bipolar membranes is increasingly being applied to the production of (Bazinet, 2004)

1. Mineral and organic acids
2. Inhibition of enzymatic browning
3. Separation of protein

These applications are based on water dissociation at the interface of a bipolar membrane and are coupled with the action of the monopolar membrane action. Deacidification and acid production, however, entail conventional ED. In the recovery of organic acids from fermentation broths the elimination of cations has often been a major problem, as fermentation typically performs better in pHs significantly above the pK_a of the acid produced. Bipolar membranes offer a solution to the elimination

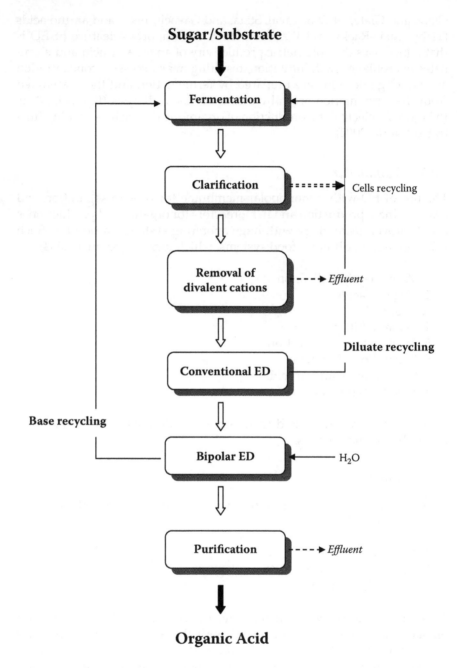

Figure 7.1 Electrodialytic production of fermented organic acids (Adapted from M. Bailly (2002). *Desalination* 144:157–162). With permission.

of cations by allowing a salt to be split into the corresponding alkali and acid solutions. When bipolar membranes are used in water splitting processes, it is possible to recover organic acids and also to control the pH of liquids (Bazinet, 2004).

Continuous lactic acid production from whey permeate is carried out in a process that consists of three separate operations in (1) a bioreactor, (2) an ultrafiltered (UF) model, and (3) an ED cell. With the UF process, recycling of all or part of the biomass is achieved. It is also possible to separate low molecular weight metabolites, such as sodium lactate, resulting from lactose fermentation. This product can then be extracted and concentrated continuously by ED. A disadvantage of continuous lactic acid production is, however, that it tends to clog the ultrafiltration membranes, which restricts permeate flow (Bazinet, 2004).

Rhizopus oryzae is an indispensable microorganism in industrial fermentation, as it is widely employed to produce L-lactic acid as well as other organic acids. This organism is able to produce only one stereospecific product (L-lactic acid), rather than a racemic mixture and can, therefore, fulfill the need for producing a food additive to be used as both acidulant and preservative. During L-lactic acid fermentation many other metabolites can be produced as by-products. These include fumaric acid, malic acid, ethanol, and the like. However, these metabolites can greatly influence the downstream process and the quality of the L(+)-lactic acid produced. Fumaric acid is the main by-product, as a result of a special metabolic pathway in L-lactic acid production by *R. oryzae* (Wang et al., 2005).

Various residues from agriculture and industry are, however, also employed by different microorganisms. Such residues include cassava bagasse, coffee husk and pulp, apple pomace, and soybean and potato residues. A great advantage of the production of various organic acids by *A. niger*, in particular citric acid, is the use of agricultural products and their wastes such as date, maize, citrus and kiwifruit peel, apple and grape pomace, pineapple, mandarin orange, carob pod, and brewery wastes (Roukas, 1999).

7.6 Organic acid demand

Markets constantly undergo changes in response to changes in marketing environments and political frameworks (Marz, 2002). Organic acids have been estimated to represent the third largest category after antibiotics and amino acids in the global market of fermentation. Organic acids that dominate the market are benzoic, sorbic, lactic, and citric acids. Benzoic acid is widely used, primarily because of its low price (Banerjee and Sarkar, 2004), and citric acid dominates the market of organic acids due to its versatility (Ramanchandran et al., 2006). The total market value of organic acid is estimated to have risen to $3 million in 2009 (Marz, 2005).

Information on the main ingredients of feed compounds, as well as on volumes of compound feed produced is public, but is scattered in many different journals and publications. For example, market information on the ingredients of feed is rarely published and there is no readily available comprehensive overview on the subject. Direct contact with the industry is necessary to provide reliable and precise quantitative information (Marz, 2002; Marz, 2005).

7.7   Lactic acid production

Lactic acid fermentation has for many years received extensive attention because lactic acid has wide applications in the food, pharmaceutical, leather, and textile industries as well as being a chemical additive in feedstock (Couto and Sanroman, 2006). The acid exists in two enantiomers (L(+) and (D–)) of which L(+) is used by the human metabolism due to the presence of L-lactate dehydrogenase and is preferred for food. In addition, elevated levels of the D-isomer are harmful to humans. Lactic acid can be produced commercially by fermentation or by means of chemical synthesis (Zhang, Jin, and Kelly, 2007). Chemical synthesis requires several steps and produces a racemic mixture of the two isomers, whereas the carbohydrate fermentation process can produce an optically pure form of lactic acid or isomer and is less expensive (Valli et al., 2006). This is, however, dependent on the microorganisms, substrates, and the fermentation conditions employed in the production process (Zhang, Jin, and Kelly, 2007). In Figures 7.2 and 7.3 the conventional method of lactic acid production from raw materials such as starch is compared with a method where the enzymatic hydrolysis of carbohydrate substrates is coupled with the microbial fermentation of the sugars (Zhang, Jin, and Kelly, 2007). This single step is referred to as "simultaneous saccharification and fermentation" (SSF).

Current industrial lactic acid fermentations are widely performed with lactic acid bacteria (Valli et al., 2006). LAB have received wide interest because of their high growth rate and product yield (Zhang, Jin, and Kelly, 2007). For example, *Lactococcus* and *Carnobacterium* produce L-lactic acid, and *Leuconostoc* produces D-lactic acid (Liu, 2003; Gravesen et al., 2004). However, lactic acid can be produced using bacteria as well as fungi (Zhang, Jin, and Kelly, 2007). Natural substrates such as starch (Altaf et al., 2006; Ohkouchi and Inoue, 2006) and cellulose (Chen and Lee, 1997; Venkatesh, 1997) are not economically feasible to use, because of being very expensive and also requiring pretreatment for the release of fermentable sugars. The manufacturing cost of lactic acid is notably reduced when waste products such as whey or molasses

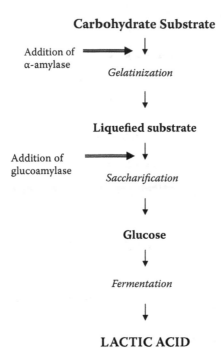

Figure 7.2 Conventional production of lactic acid (Adapted from Zhang, Jin, and Kelly (2007). *Biochemical Engineering Journal* 35:251–263). With permission.

containing fermentable sugars are used for the production of lactic acid (Dumbrepatil et al., 2008).

Lactic acid can be produced from renewable materials using various species of the fungus *Rhizopus*. Renewable material includes molasses (corn starch, wheat starch, and potato starch) and lignocellulose (corn cobs and woody materials) (Zhang, Jin, and Kelly, 2007). Molasses is an economically feasible raw material for industrial production of lactic acid. However, lactic acid productivity and cell growth in molasses fermentation can be severely affected by the concentration of yeast extract added to the medium (Wee et al., 2004). Production by fungi has advantages over bacteria, such as their amylolytic characteristics, low nutrient requirements, and, very important, the valuable fermentation by-product fungal biomass (Zhang, Jin, and Kelly, 2007). Various methods have been adapted to increase productivity of LAB in lactic acid production, but their pH sensitivity is still proving to be problematic (Valli et al., 2006). Genetic analysis of lactic acid synthesis in LAB is providing a potential of genetically modifying production rates of industrially important strains (Wang et al., 2005).

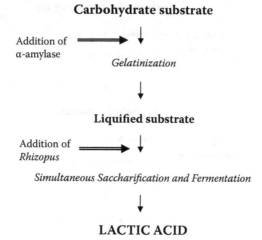

Figure 7.3 Production of lactic acid by *Rhizopus* and simultaneous saccharification and fermentation (SSF) (Adapted from Zhang, Jin, and Kelly (2007). *Biochemical Engineering Journal* 35:251–263. With permission.)

7.7.1   Factors affecting production of lactic acid

Effect of pH: pH is one of the most crucial factors that influence the production of lactic acid. The pH range for optimal production is 5.0–6.0.

Oxygen supply: In fungal fermentation oxygen plays an important role in lactic acid production as this fermentation is an aerobic process.

Temperature: Temperature also has an important impact on lactic acid production. (Zhang, Jin, and Kelly, 2007 p. 154)

Strains of *Rhizopus* species have been divided into two groups, this being the lactic acid producers and fumaric acid (or malic acid) producers, based on the yield and type of organic acids and the proportion of metabolites produced. Strains of *Rhizopus* species that produce predominantly lactic acid possess two *ldh* genes, *ldhA* and *ldhB* (Skory, 2000), whereas strains of *Rhizopus* species that produce fumaric and malic acids contain a single ORF of *ldhB* (Saito et al., 2004). Three lactate dehydrogenase enzymes have been described in R. *oryzae*. These include one NAD-independent LDH and two NAD-dependent LDH isozymes (Pritchard, 1971; Zhang, Jin, and Kelly, 2007). One NAD^+-dependent LDH (EC1.1.1.27, LdhA) converts pyruvate to lactate, and is produced during early growth and synthesis of lactic acid. The other NAD-dependent (LdhB), converts L-lactate to pyruvate, and is produced when cultures are grown on nonfermentable substrates, such as gycerol, ethanol, and lactate. When glucose is exhausted, a slight decrease in lactic acid production can be detected. New kinetic models for lactic acid production using bacteria and, to a lesser extent, *Rhizopus* species are being researched and developed. Existing kinetics models relating to production with *Rhizopus* species are built on immobilized cultures, but this is, however, not an economical process. Efforts have been made to minimize formation of fumaric acid and ethanol, the two major products through molecular genetic technologies, and an alternative way to enhance lactic acid production has been proposed to minimize accumulation of ethanol in the *Rhizopus* genus via pyruvate decarboxylase and alcohol dehydrogenase (ADH) (Zhang, Jin, and Kelly, 2007).

More and relatively new applications of organic acids such as biodegradable plastic (polylactic acid) are also becoming essential in further promoting the application of L-lactic acid (Wang et al., 2005). It was estimated that by 2010 the L-lactic acid market could exceed 10^6 tons (Datta, Tsai, and Bonsignore, 1995). In a more recent report, the estimation was that the lactic acid market would reach 259,000 metric tons by 2012. According to the Global Industry Analysts, Inc., the United States, Asia-Pacific and Europe dominate the world lactic acid market. In Europe, Germany is ranked the largest individual market for lactic acid, followed by France and Italy (press release, 2008).

7.8   Citric acid production

Citric acid is one of the most frequently used organic acids in the food and pharmaceutical industries and production still follows a conventional procedure where citric acid is produced by *A. niger* from molasses as substrate (Finogenova et al., 2005). The food industry is the largest consumer of citric acid, using approximately 70% of the total production. The pharmaceutical industry follows with 12%, and 18% is used in other applications (Shah et al., 1993; Couto and Sanromán, 2006). Citric acid has a pleasant taste, is highly soluble, has characteristic flavor-enhancing properties, and, therefore, dominates the market. Citric acid can also be produced by chemical synthesis, but due to high cost, the preferred method is through fermentation. Citric acid is mainly produced by submerged fermentation (SmF), by the fungus *A. niger*, but the solid-state fermentation (SsF) method is increasingly considered as a potential alternative to SmF in increasing the efficiency of citric acid production. SsF has been found to be a very effective method of using nutrient-rich solid waste as a substrate (Finogenova et al., 2005).

> SmF: Industrial production of organic acids by growing the microorganisms that produce the product in a submerged culture

SsF: Growth of microorganisms on moist solid particles, where the spaces between the particles contain a continuous gas phase and a minimum of visible water

A. niger has been used for the production of organic acids since the beginning of the last century (Abarca et al., 2004). Two acids, citric acid and gluconic acid, are the only organic acids still produced by mycological processes that are used in significant quantities as food additives (Bigelis and Lasure, 1987). Citric acid and its production have not received much consideration with regard to modern technology and molecular biology. This may be the result of its production being considered a mature area (Abarca et al., 2004). The conventional method of citric acid production by *A. niger* from molasses is summarized and illustrated in Figure 7.4 (Finogenova et al., 2005).

Production of citric acid is strictly dependent on the specific strain and operational conditions. The oxygen level, for example, is an important parameter for citric acid fermentation. Potential substrates that have been proposed and used for citric acid production include residues such as apple pomace, coffee husk, wheat straw, pineapple waste, mixed fruit, maosmi waste, cassava bagasse, banana, sugar beet, and kiwi fruit. Using fruit waste for citric acid production is advantageous as it utilizes low-grade waste, and at the same time produces a commercially valuable product (Kumar et al., 2003). Molasses is also a popular substrate, but because of difficulties in purification, beet sugar is often used as a reliable alternative (Brul et al., 2002; Bizukojc and Ledakowicz, 2004). High CO_2 concentrations are beneficial for citric acid synthesis (Pintado, 1998), whereas a low O_2 environment has a direct influence on growth inhibition, a factor that is crucial for production (Soccol and Vandenberghe, 2003). Certain metal ions (such as Fe^{++}, MN^{++}, and Zn^{++}) are inhibitory to citric acid production in SmF, whereas SsF gives high citric acid yield without inhibition by the presence of metals, even at high concentrations (Kumar et al., 2003).

Global production of citric acid is estimated to be around 500,000 tons annually, with the entire production process conducted by fermentation (Soccol and Vandenberghe, 2003), and there is a constant increase each year in the consumption of citric acid. As a result, production is increasing at a rate of 5% per year (Finogenova et al., 2005) and there is a serious need for finding new manufacturing alternatives (Soccol and Vandenberghe, 2003). A contributing factor to the annual increase in the production of citric acid is also attributable to the use of trisodium citrate. This salt is used to replace tripolyphosphates in the production of ecologically pure synthetic detergents (Finogenova et al., 2005).

Conventional production of citric acid is complex and ecologically unsafe, as a result of the characteristics of the raw material used and concentrated acids and alkali being used throughout the entire process. The

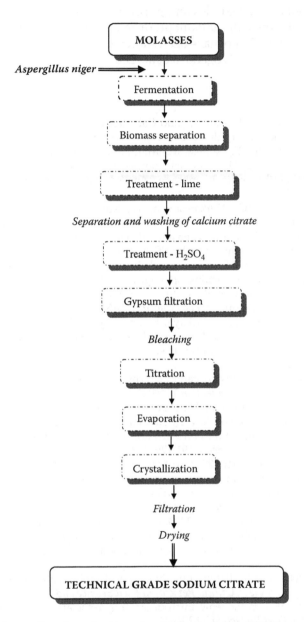

Figure 7.4 Conventional production of sodium citrate by utilization of the fungus *Aspergillus niger* (Adapted from Finogenova, et al. (2005). *Applied Biochemistry and Microbiology* 41:418–425).

effluent (containing cyanides and gypsum) into the environment is comparable in relation to the amount produced. However, new technologies of production from ethanol have been shown to be ecologically pure, simple, and readily scalable (Finogenova et al., 2005).

References

Abarca, M.L., Accensi, F., Cano, J., and Cabanes, F.J. 2004. Taxonomy and significance of black aspergilli. *Antonie Van Leeuwenhoek* 86:33–49.

Alakomi, H.L., Skyttä, E., Saarela, M., Mattila-Sandholm, T., Latva-Kala, K., and Helander, I.M. 2000. Lactic acid permeabilizes Gram-negative bacteria by disrupting the outer membrane. *Applied and Environmental Microbiology* 66:2001–2005.

Altaf, M., Naveena, B.J., Venkateshwar, M., Kumar, E.V., and Reddy, G. 2006. Single step fermentation of starch to l(+) lactic acid by *Lactobacillus amylophilus* GV6 in SSF using inexpensive nitrogen sources to replace peptone and yeast extract—Optimization by RSM. *Proceedings in Biochemistry* 41:465–472.

Ammor, S., Tauveron, G., Dufour, E., and Chevallier, I. 2006. Antibacterial activity of lactic acid bacteria against spoilage and pathogenic bacteria isolated from the same meat small-scale facility: 1—Screening and characterization of the antibacterial compounds. *Food Control* 17:454–461.

Aslim, B., Yuksekdag, Z.N., Sarikaya, E., and Beyatli, Y. 2005. Determination of the bacteriocin-like substances produced by some lactic acid bacteria isolated from Turkish dairy products. *Swiss Society of Food Science and Technology* 38:691–694.

Bailly, M. 2002. Production of organic acids by bipolar electrodialysis: Realizations and perspectives. *Desalination* 144:157–162.

Bailly, M., Roux-de Balmann, H., Aimar, P., Lutin, F., and Cheryan, M. 2001. Production processes of fermented organic acids targeted around membrane operations: Design of the concentration step by conventional electrodialysis. *Journal of Membrane Science* 191:129–142.

Banerjee, M. and Sarkar, P.K. 2004. Antibiotic resistance and susceptibility to some food preservative measures of spoilage and pathogenic micro-organisms from spices. *Food Microbiology* 21:335–342.

Barbosa-Cánovas, G.V., Fernández-Molina, J.J., Alzamora, S.M., Tapia, M.S., López-Malo, A., and Chanes, J.W. 2003. General considerations for preservation of fruits and vegetables. In: *Handling and Preservation of Fruits and Vegetables by Combined Methods for Rural Areas*. Rome: Food and Argriculture Organization of the United Nations.

Bazinet, L. 2004. Electrodialytic phenomena and their applications in the dairy industry: A review. *Critical Reviews in Food Science and Nutrition* 44:525–544.

Beuchat, L.R. 1995. Pathogenic microorganisms associated with fresh produce, *Journal of Food Protection* 59:204–216.

Bigelis, R. and Lasure, L.L. 1987. Fungal enzymes and primary metabolites used in food processing. In: L.R. Beuchat (Ed.), *Food and Beverage Mycology*, pp. 473–516. New York: Van Nostrand Reinhold.

Bizukojc, M. and Ledakowicz, S. 2004. The kinetics of simultaneous glucose and fructose uptake and product formation by *Aspergillus niger* in citric acid fermentation. *Process Biochemistry* 39:2261–2268.

Boyaval, P., Corre, C., and Terre, S. 1987: Continuous lactic acid fermentation with concentrated product recovery by ultra-filtration and electrodialysis. *Biotechnology Letters* 9:207–212.

Boyaval, P., Stata, J., and Gavach, C. 1993. Concentrated propionic acid production by electrodialysis. *Enzyme and Microbial Technology* 15:683–686.

Brul, S., Coote, P., Oomes, S., Mensonides, F., Hellingwerf, K., and Klis, F. 2002. Physiological actions of preservative agents: Prospective of use of modern microbiological techniques in assessing microbial behaviour in food preservation. *International Journal of Food Microbiology* 79:55–64.

Chen, H., Zuo, Y., and Deng, Y. 2001. Separation and determination of flavonoids and other compounds in cranberry juice by high-performance liquid chromatography. *Journal of Chromatography A* 913:387–395.

Chen, R. and Lee, Y.Y. 1997. Membrane-mediated extractive fermentation for lactic acid production from cellulosic biomass. *Applied Biochemistry and Biotechnology* 63:435–448.

Couto, S.R. and Sanromán, M.A. 2006. Application of solid-state fermentation to food industry—A review. *Journal of Food Engineering* 76:291–302.

Datta, R., Tsai, S.P., and Bonsignore, P. 1995. Technological and economic potential of poly(lactic acid) and lactic acid derivatives. *FEMS Microbiological Reviews* 16:221–231.

Davidson, B.E., Llanos, R.M., Cancilla, M.R., Redman, N.C., and Hillier, A.J. 1995. Current research on the genetics of lactic acid production in lactic acid bacteria. *International Dairy Journal* 5:763–784.

De Raucourt, A., Girard, D., Prigent, Y., and Boyaval, P. 1989. Lactose fermentation with cells recycled by ultra-filtration and lactate separation by electrodialysis: Modeling and simulation. *Applied Microbiology and Biotechnology* 30:521–527.

Dumbrepatil, A., Adsul, M., Chaudhari, S., Khire, J., and Gokhale, D. 2008. Utilization of molasses sugar for lactic acid production by *Lactobacillus delbrueckii* subsp. *delbrueckii* mutant Uc-3 in batch fermentation. *Applied and Environmental Microbiology* 74:333–335.

Eysmondt, V.J., Vasic-Racki, D., and Wandrey, C. 1993. The continuous production of acetic acid by electrodialysis integrated fermentation. Modeling and computer simulation. *Chemical and Biochemical Engineering Quarterly* 7:139–148.

Finogenova, T.V., Morgunov, I.G., Kamsolova, S.V., and Chernyavskaya, O.G. 2005. Organic acid production by the yeast *Yarrowia lipolytica*: A review of prospects. *Applied Biochemistry and Microbiology* 41:418–425.

Food-Info. 2009. Information on food production, food safety, additives, ingredients and other topics. http://www.food-info.net.

Ganzle, M.G., Vermeulen, N., and Vogel, R.F. 2007. Carbohydrate, peptide and lipid metabolism of lactic acid bacteria in sourdough. *Food Microbiology* 24:128–138.

Gauthier, R. 2005. Organic acids and essential oils, a realistic alternative to antibiotic growth promoters in poultry. I Forum Internacional de avicultura 17–19 August 2005, pp. 148–157.

Gomis, D.B. 1992. HPLC analysis of organic acids. In: L.M.L. Nollet (Ed.), *Food Analysis by HPLC*, pp. 371–385. New York: Marcel Dekker.

182   *Organic Acids and Food Preservation*

Gonzalez, A., Hierro, N., Poblet, M., Mas, A., and Guillamon, J.M. 2005. Application of molecular methods to demonstrate species and strain evolution of acetic acid bacteria population during wine production. *International Journal of Food Microbiology* 102:295–304.

Gravesen, A., Diao, Z., Voss, J., Budde, B.B., and Knochel, S. 2004. Differential inactivation of *Listeria monocytogenes* by d- and l-lactic acid. *Letters in Applied Microbiology* 39:528–532.

Guerzoni, M.E., Lanciotti, R., and Marchetti, R. 1993. Survey of the physiological properties of the most frequent yeasts associated with commercial chilled foods. *International Journal of Food Microbiology* 17:329–341.

Gutheil, R.A., Price, L.G., and Swanson, B.G. 1980. pH, acidity and vitamin C content of fresh and canned homegrown Washington tomatoes. *Journal of Food Protection* 43:366–369.

Krishnan, S., Gowda, L.R., Misra, M.C., and Karanth, N.G. 2001. Physiological and morphological changes in immobilized *L. plantarum* NCIM 2084 cells during repeated batch fermentation for production of lactic acid. *Food Biotechnology* 15:193–202.

Kumar, D., Jain, V.K., Shanker, G., and Srivastava, A. 2003. Utilisation of fruits waste for citric acid production by solid state fermentation. *Process Biochemistry* 38:1725–1729.

Lindgren, S.E. and Dobrogosz, W.J. 1990. Antagonistic activities of lactic acid bacteria in food and feed fermentation. *FEMS Microbiology Reviews* 7:149–163.

Liu, S.Q. 2003. Practical implications of lactate and pyruvate metabolism by lactic acid bacteria in food and beverage fermentations. *International Journal of Food Microbiology* 83:115–131.

Marz, U. 2000. World markets for fermentation ingredients. Business Communications Company (Inc.): *Food and Beverage Publications*.

Marz, U. 2002. World markets for citric, ascorbic, isoascorbic acids: Highlighting antioxidants in food. Business Communications Company (Inc.): *Food & Beverage Publications*. Report Code FOD017B, Published June 2002. http://www.bccresearch.com (accessed July 29, 2006).

Marz, U. 2005. GA-103R world markets for fermentation ingredients. http://www.bccresearch.com (accessed July 29, 2006.)

Morales, M.L., Gonzalez, A.G., and Troncoso, A.M. 1998. Ion-exclusion chromatographic determination of organic acids in vinegars. *Journal of Chromatography A* 822:45–51.

Nes, I.F. and Johnsborg, O. 2004. Exploration of antimicrobial potential in LAB by genomics. *Current Opinion in Biotechnology* 15:100–104.

Nomura, Y., Iwahara, M., and Hongo, M. 1987. Lactic acid production by electrodialysis fermentation using immobilized growing cells. *Biotechnology and Bioengineering* 30:788–793.

Ohkouchi, Y. and Inoue, Y. 2006. Direct production of L(+)lactic acid from starch and food wastes using *Lactobacillus manivotivorans* LMG18011. *Bioresource Technology* 97:1554–1562.

Penney, V., Henderson, G., Blum, C., and Johnson-Green, P. 2004. The potential of phytopreservatives and nisin to control microbial spoilage of minimally processed fruit yogurts. *Innovative Food Science & Emerging Technologies* 5:369–375.

Pintado, J., Lonsane, B.K., Gaime-Perraud, I., and Roussos, S. 1998. On-line monitoring of citric acid production in solid-state culture by respirometry. *Process Biochemistry* 33:513–518.

Piper, P.W. 1999. Yeast superoxide dismutase mutants reveal a pro-oxidant action of weak organic acid food preservatives. *Free Radical Biology & Medicine* 27:1219–1227.

Press Release 2008. World lactic acid market to reach 259 000 metric tons by 2012. *Veille Bioplastiques*. http://www.bioplastique.wordpress.com/2008/10/29/ (accessed April 22, 2010).

Pritchard, G.G. 1971. An NAD-independent l-lactate dehydrogenase from *Rhizopus oryzae*. *Biochimica et Biophysica Acta* 250:25–34.

Ramanchandran, S., Fontanille, P., Pandey, A., and Larroche, C. 2006. Gluconic acid: Properties, applications and microbial production. *Food Technology and Biotechnology* 44:185–195.

Roller, S. 1995. The quest for natural antimicrobials as novel means of food preservation: Status report on a European research project. *International Biodeterioration & Biodegradation* 36:333–345.

Ross, R.P., Morgan, S., and Hill, C. 2002. Preservation and fermentation: Past, present and future. *International Journal of Food Microbiology* 79:3–16.

Rossland, E., Langsrud, T., Granum, P.E., and Sorhaug, T. 2005. Production of antimicrobial metabolites by strains of *Lactobacillus* or *Lactococcus* co-cultured with *Bacillus cereus* in milk. *International Journal of Food Microbiology* 98:193–200.

Roukas, T. 1999. Citric acid production from carob pod by solid-state fermentation. *Enzyme and Microbial Technology* 24:54–59.

Saito, K., Saito, A., Ohnishi, M., and Oda, Y. 2004. Genetic diversity in *Rhizopus oryzae* strains as revealed by the sequence of lactate dehydrogenase genes. *Archives of Microbiology* 182:30–36.

Sajur, S.A., Saguir, F.M., and Manca de Nadra, M.C. 2007. Effect of dominant specie of lactic acid bacteria from tomato on natural microflora development in tomato purée. *Food Control* 18:594–600.

Shah, N.D., Chattoo, B.B., Baroda, R.M., and Patiala, V.M. 1993. Starch hydrolysate, an optimal and economical source of carbon for the secretion of citric acid by *Yarrowia lipolytica*. *Starch* 45:104–109.

Schügerl, K. 2000. Integrated processing of biotechnology products. *Biotechnology Advances* 18:581–599.

Skory, C.D. 2000. Isolation and expression of lactate dehydrogenase genes from *Rhizopus oryzae*. *Applied and Environmental Microbiology* 66:2343–2348.

Soccol, C.R. and Vandenberghe, L.P.S. 2003. Overview of applied solid-state fermentation in Brazil. *Biochemical Engineering Journal* 13:205–218.

Tichy, S., Vasic-Racki, D., and Wandrey, C. 1990. Electrodialysis as an integrated downstream process in amino acid production. *Chemical and Biochemical Engineering Quarterly* 7L127–145.

Trček, J. 2005. Quick identification of acetic acid bacteria based on nucleotide sequences of the 16S–23S rDNA internal transcribed spacer region and of the PQQ-dependent alcohol dehydrogenase gene. *Systematic and Applied Microbiology* 28:735–745.

Trifiro, A., Saccani, G., Gherardi, S., et al. 1997. Use of ion chromatography for monitoring microbial spoilage in the fruit juice industry. *Journal of Chromatography A* 770:243–252.

Tyree, R.W., Clausen, E.C., and Gaddy, J.L. 1991. The production of propionic acid from sugars by fermentation through lactic acid as an intermediate. *Journal of Chemical Technology and Biotechnology* 50:157–166.

Valli, M., Sauer, M., Branduardi, P., Borth, N., Porro, D., and Mattanovich, D. 2006. Improvement of lactic acid production in *Saccharomyces cerevisiae* by cell sorting for high intracellular pH. *Applied and Environmental Microbiology* 72:5492–5499.

Venkatesh, K.V. 1997. Simultaneous saccharification and fermentation of cellulose to lactic acid. *Bioresource Technology* 62:91–98.

Vesterlund, S., Paltta, J., Laukova, A., Karp, M., and Ouwehand, A.C. 2004. Rapid screening method for the detection of antimicrobial substances. *Journal of Microbiological Methods* 57:23–31.

Wang, X., Sun, L., Wei, D., and Wang, R. 2005. Reducing by-product formation in l-lactic acid fermentation by *Rhizopus oryzae. Journal of Industrial Microbiology and Biotechnology* 32:38–40.

Wee, Y-J., Kim, J-M., Yun, J-S., and Ryu, H-W. 2004. Utilization of sugar molasses for economical l(+)-lactic acid production by batch fermentation of *Enterococcus faecalis. Enzme and Microbial Technology* 35:568–573.

Weier, A.J., Glatz, B.A., and Glatz, C.E. 1992. Recovery of propionic and acetic acids from fermentation broth by electrodialysis. *Biotechnology Progress* 8:479–485.

Zaunmuller, T., Eichert, M., Richter, H., and Unden, G. 2006. Variations in the energy metabolism of biotechnologically relevant heterofermentative lactic acid bacteria during growth on sugars and organic acids. *Applied Microbiology and Biotechnology* 72:421–429.

Zhang, Z.Y., Jin, B., and Kelly, J.M. 2007. Production of lactic acid from renewable materials by *Rhizopus* fungi. *Biochemical Engineering Journal* 35:251–263.

chapter eight

Resistance to organic acids

8.1   Introduction

Antibiotics have for many years received the majority of the focus in the war against their vanishing effectiveness. However, preservative compounds that are used to control microbial contamination in food production are increasingly receiving more attention, as they are progressively more reported to elicit resistance responses in microorganisms (Chapman, 1998). These responses of organisms may be a deciding factor in their survival of a variety of environmental stressors in food production and processing (Ricke, 2003). Stresses encountered by foodborne pathogens may include heat- and cold-shock, oxidative stress, osmotic stress, extreme pH, and nutrient deficiencies (Archer, 1996).

Although organic acids have for many years been proven to be effective under a wide variety of food processing conditions, there is increasing indication that some foodborne pathogens are intrinsically tolerant, whereas others are becoming tolerant to the organic acids. It may be possible that this tolerance is only true under particular circumstances, for example, exposure to organic acids at less than maximum concentration which provides the organism the opportunity to induce specific stress mechanisms that allow for resistance, and also resistance to more severe acidic conditions. Determining the efficacy of organic acids in specific applications will require a better understanding of the general and specific stress response potential of foodborne pathogens (Ricke, 2003).

8.2   Intrinsic (natural) resistance

Much evidence has indicated the ability of food-associated microorganisms to adapt to various environmental factors and should not be underestimated (Suhr and Nielsen, 2004). There may also be a greater variety between isolates of the same species than generally accepted. Some meatborne pathogens are particularly sensitive to organic acids (e.g., *Yersinia enterocolitica*), whereas others are highly resistant, particularly *Escherichia coli* O157:H7 (Smulders and Greer, 1998). Lactic acid bacteria (LAB) are normally resistant to the majority of antibiotics and consequently also much more resistant to the activity of organic acids (Fielding, Cook, and Grandison, 1997).

The proportion of undissociated constituent can be calculated from the dissociation constant by the following formula (Figure 8.1).

$$a = \frac{[H^+]}{[H^+] + D}$$

Where a   = amount of undissociated acid

[H+]   = Hydrogen concentration

D   = Dissociation constant

Figure 8.1 The content of undissociated acid declines with an increase in pH value.

Although much of the inhibition by organic acids is accounted for by pH, they vary considerably in inhibitory effects (Hsiao and Siebert, 1999). Different organisms have also shown variations in susceptibility to inhibition by organic acids (Matsuda et al., 1994). For example, lactobacilli may be much more resistant to acetic, benzoic, butyric, and lactic acids, whereas *E. coli* has been found to be most resistant toward citric, malic, and tartaric acids (Hsiao and Siebert, 1999).

8.2.1   Bacteria

Organic acids differ in their ability to enter a cell, and because of this, adaptation time to these acidulants may vary between organisms. Optimum adaptation time, therefore, may differ according to organic acid used and because of differences in the membrane structure (Greenacre et al., 2003). It is a well-known fact that organic acids enter bacterial cells only in the undissociated form, as the extracellular pH decreases. This then causes the portion of undissociated organic acid to increase and subsequently increase their action on the bacterial cells (Kwon and Ricke, 1998). The proportion of undissociated constituent can be calculated from the dissociation constant by the following formula:

$$a = \frac{[H^+]}{[H^+]+D}$$

where

a = Amount of undissociated acid

$[H^+]$ = Hydrogen concentration

D = Dissociation constant

The content of undissociated acid declines with an increase in pH value.

Acetic acid bacteria are naturally highly resistant to acetic acid, with notable differences in tolerance between species (Trček et al., 2006). The enzyme AatA plays an important role in acetic acid resistance in these bacteria, with the *aatA* gene distributed in the genera *Acetobacter* and *Gluconobacter* (Nakano, Fukaya, and Horinouchi, 2006). The pathogen *Listeria monocytogenes* has become a major concern worldwide in the food manufacturing industry due to its resistance to various food preservation practices, and as a result being well adapted to food environments. The organism is also able to grow at refrigeration temperatures and on dry surfaces, and can tolerate acidic conditions, all factors that normally restrict bacterial growth (Barker and Park, 2001). Listeriae exposed to organic acids are known to be able to repair themselves during storage at low temperature and eventually begin to multiply if no other hurdles are present (Doyle, 1999).

8.2.2   Fungi

Certain yeasts are able to grow at low ambient pH and also in the presence of organic acids which is an important cause of food spoilage and inflicts heavy losses on the food industry (Thomas and Davenport, 1985; Deak, 1991; Cheng, Moghraby, and Piper, 1999). Yeasts possess mechanisms to maintain their intracellular pH values such that would enable them to maintain their glycolytic rate, and even enhance it. Differences found in yeast tolerance to organic acids can be attributed to the efficiency of these mechanisms. At sublethal concentrations organic acids do not inhibit the glycolytic rate in exponentially growing yeasts; they actually enhance the fermentation rate. At acid pH values the activity of several yeast glycolytic enzymes will be inhibited in cell extracts or in permeabilized cells, but the glycolytic rate is not inhibited in growing cells (Leyva and Peinado, 2005).

Saccharomyces cerevisiae and *Zygosaccharomyces bailii* are both resistant to benzoic and sorbic acids and able to spoil low pH foods (pH < 4.5) (Stratford and Lambert, 1999; Nielsen and Arneborg, 2007). Inhibition of yeasts such as *Z. bailii* and *S. cerevisiae* often requires levels of organic acids greater than the legal limits (Hazan, Levine, and Abeliovich, 2004). The fungus *Zygosaccharomyces* is generally associated with the more extreme spoilage yeasts, being osmotolerant, fructophilic (preferring fructose), highly fermentative, and also highly resistant to preservatives. *Z. bailii* can grow in the presence of commonly used organic acids at concentrations much higher than the legally permitted or organoleptically acceptable concentrations (Steels et al., 2000). *Z. bailii* is, therefore, notably more resistant to preservatives compared to *S. cerevisiae*, as a result of its ability to more effectively limit entry of undissociated acid and greater ability to degrade preservatives (Piper et al., 2001). These adaptations prevent cytosolic acidification by reducing the amount of preservative that would eventually reach the cytosol. Stimulation of the net H^+ efflux in the presence or absence of an organic acid is accompanied by the net K^+

influx. This mechanism is similar in *S. cerevisiae*, where the net K^+ influx initially serves to counter the loss of positive charge from the cell, due to H^+-ATPase activity (Macpherson et al., 2005). However, the two yeasts do not seem to have different citric acid tolerances, although in *Z. bailii* the primary energy metabolism is not as affected by citric acid as in *S. cerevisiae*. This is an indication that glycerol production is less enhanced and production of ethanol and ATP production is not affected (Nielsen and Arneborg, 2007).

A. niger is able to metabolize or degrade sorbic acid, although the precise mechanism involved and the nature of the breakdown products are still unclear. Sorbic acid is particularly an antifungal agent, known to delay spore germination and also to retard mycelial growth. It is, however, a common feature of filamentous fungi to metabolize sorbic acid and it has been suggested that degradation occurs via β-oxidation, with the resulting production of CO_2 and H_2O. However, the exact metabolic pathways and enzymes involved in converting sorbic acid into its metabolic products are only speculative, as sorbic acid is thought to undergo auto-oxidative degradation in aqueous systems to form malonaldehyde and other carbonyls (Plumridge et al., 2004).

8.3   Development of resistance

There are clear indications that microorganisms will increasingly become resistant to preservatives such as the organic acids, as is the case in human pathogens becoming increasingly resistant to antibiotics. Some organisms have already acquired resistance to organic acids and various mechanisms have been described for the development of this resistance (Theron and Lues, 2007).

Concern has been expressed that decontamination with organic acids could result in the emergence of acid-tolerant foodborne pathogens (Quintavilla and Vicini, 2002) and also in the risk of acid resistance spreading, as bacteria have the ability to transfer characteristics between and across species genetically encoded for by genes carried on extrachromosomal DNA (plasmids, transposons, etc.) (Mellor, 2005). Microorganisms exposed to a moderately acidic environment have, in response to this, been reported to develop cells with increased acid resistance, known as acid tolerance response (ATR). These acid-adapted cells appear to be more resistant under stress conditions, which may cause them to be more infective than the unadapted cells (Cheng, Yu, and Chou, 2003). In addition to this, some bacterial cells have become acid-tolerant or have undergone acid shock or have developed cross-resistance to other environmental stresses such as heat (Ryu, Deng, and Beuchat, 1999), osmosis, and salt (Leyer and Johnson, 1993).

Weak-acid adaptation may have evolved to facilitate growth at low pH in the presence of weak organic acids. Unfortunately this poses problems for the food industry where substantial increases in resistance to major organic acid food preservatives are obvious (Piper et al., 2001). Inducible resistance to organic acids (or acid tolerance), in previously susceptible organisms, has recently been more extensively studied and it is known that resistance to weak organic acids can be attributed to various mechanisms (Brul and Coote, 1999). Such inducible resistance is regarded as an important survival strategy of many microorganisms. The organism senses that the environment is becoming unsuitable and undergoes a programmed molecular response, which involves the synthesis of specific stress-inducible proteins. These proteins are then possibly responsible for repairing macromolecular damage caused by the stress. Of importance in future research is the knowledge that some of these stress proteins are induced by a specific stress factor, whereas others may require a range of stress conditions (Bearson, Bearson, and Foster, 1997).

For any food spoilage microorganism to become resistant to sorbate or benzoate, these organisms should be able to catalyze degradation of these compounds. *Z. bailii* has become an important spoilage agent of wine, because of being highly tolerant to ethanol and able to metabolize acetic acid in wine fermentations. However, individual cells differ in resistance, and only a small fraction are highly resistant. *S. cerevisiae* can also frequently adapt to grow in the presence of sorbate and benzoate, but at levels lower than those inhibitory to *Z. bailii*. *Z. bailii* is able to grow on sorbate or benzoate only at pH $\geq$ 5.5. At lower pH values these acids are in an undissociated state, causing severe weak-acid stress (Mollapour and Piper, 2001).

Quite a high concentration of acetic acid (80–150 mM) is needed to totally inhibit growth of *S. cerevisiae* at pH 4.5 (Piper et al., 2001). Increased plasma membrane H^+-ATPase activity is common to both *S. cerevisiae* and *Z. bailii* and this enzyme is identified as a target preventing spoilage. Short-term K^+ influx and long-term K^+ accumulation are common responses to stress caused by preservatives and suggest that K^+ transporters may also serve as control targets (Macpherson et al., 2005). However, restricting K^+ availability in food and beverages may not be a realistic control strategy, particularly for fruit-derived drinks such as wines that tend to have high (mM) and variable K^+ content, set by K^+ accumulation of the fruit (Nuñez et al., 2000).

8.4   Inducible resistance

It is constantly found that microorganisms can often be induced to become resistant when exposed to sublethal concentrations of an antimicrobial or preservative. For example, *S. cerevisiae* can be induced to adapt to weak-acid stress by addition of 0.5–2.5 mM sorbate or benzoate. The organism cannot, however, degrade either of the compounds and their presence provides continuous weak-acid stress (Piper et al., 2001). After exposure to the acids, cells become static for a long period and generally resume growth after several hours. In yeasts, exposure to organic acids results in induction of two additional proteins (plasma membrane proteins), a smaller heat-shock protein Hsp30, and a larger one, ATP-binding cassette (ABC) transporter Pdr12, a homolog of Snq2 and Pdr5 ABC drug efflux pumps (Holyoak et al., 1999). Hsp30 assists in adaptation to weak acids by regulating activity of the membrane H$^+$-ATPase. In the presence of organic acids, growth of a yeast will depend upon optimal glycolytic flux. Glycolysis is inhibited by organic acids, as a result of acidification of the cytosol (Kasemets et al., 2006). The PDR12 gene is induced very strongly by weak organic acid stress, activating Pdr12, a 171-kDa plasma membrane ATP-binding cassette transporter, suggesting a specific system (that exists) for the sensing of weak-acid stress (Piper et al., 2001). Pdr12 is essential for yeast cells to adapt and grow in the presence of weak-acid preservatives (Holyoak et al., 1999). The cell may also respond by inducing the activity of a proton-pumping membrane protein H$^+$-ATPase. However, Pdr12 is not induced by acetic acid (Papadimitriou et al., 2007).

Long-term adaptation to organic acids is the result of increased expression of the ATP-binding cassette transporter Pdr12p, which catalyzes the active efflux of acids from the cytosol (Mollapour et al., 2006). Chronic exposure and adaptation to benzoate causes accumulation of K$^+$, whereas an increase in the availability of K$^+$ could enhance growth in benzoate and acute exposure to butyric or propionic acids stimulates H$^+$ efflux and K$^+$ influx in *S. cerevisiae*. K$^+$ countertransport is probably acting to regulate charge balance (Boxman et al., 1985). Growth of *S. cerevisiae* in benzoate also appears to enhance basal levels of H$^+$ extrusion, as a result of (1) increased availability of H$^+$-ATPase, (2) increased turnover, or (3) more efficient coupling ratio (Macpherson et al., 2005). Furthermore, this basal rate increase is similar to that found in late-exponential-phase *S. cerevisiae* grown in sorbic acid (Holyoak et al., 1996). In *Z. bailii*, such H$^+$ extrusion is accompanied by net K$^+$ influx (Macpherson et al., 2005).

In *S. cerevisiae* sorbic acid evokes more stress reactions than just the induction of acid and anion extrusion pumps (De Nobel et al., 2001). Often long-term effective adaptation to sorbic acid also involves changes in the cell envelope, both the plasma membrane and cell wall (Henriques, Quintas, and Loureiro-Dias, 1997). These changes will eventually cause a decrease in organic acid influx to provide yeast cells with long-term sorbic acid resistance (Brul et al., 2002). Sorbic and benzoic acids also have profound influences on heat-shock response as well as thermotolerance of *S. cerevisiae*, although these reactions are highly dependent on pH. At low pH these acids inhibit both the heat-shock protein and thermotolerance induction during sublethal (39°C) heat shock, and at pH values above pH 5 sorbate acts as a potent chemical inducer of thermotolerance (in the absence of sublethal heat treatment), and no longer inhibits the cell's ability to mount a heat-shock response. Effects of sorbic acid on thermal inactivation of *Z. bailii* at 52°C are also pH dependent. At neutral pH at 25°C sorbic acid treatment rendered cells more resistant to thermal killing at 52°C, whereas at lower pH (pH 3.5) sorbic acid causes a reduction in thermotolerance. Another factor in resistance development may be the accumulation of trehalose. Trehalose is known as a stress protectant, and at low pH, accumulates rapidly in *S. cerevisiae* exposed to sorbate (Cheng, Moghraby, and Piper, 1999).

In exponentially growing cells glycolysis may be enhanced in the presence of sublethal concentrations of organic acids, and phosphofructokinase (PF1K) activity is not inhibited (Warth 1991), indicating that may also play some role in organic acid tolerance. Increasing PF1K levels by genetic manipulation has shown improvement of resistance to benzoic acid in *S. cerevisiae* (Pearce, Booth, and Brown, 2001; Leyva and Peinado, 2005).

Some organic acids have no inhibitory effect on microorganisms; for example, gluconic acid is a highly polar molecule, and is unable to penetrate the cell membrane of *E. coli* O157:H7 (Bjornsdottir, Breidt, Jr., and McFeeters, 2006). A tolerance response in *Salmonella enterica* sv. Typhimurium, induced by lactic acid, will, however, induce sensitivity to hydrogen peroxide (Greenacre et al., 2006).

At low extracellular pH, organic acids cause a reduction in PtdIns(3,5) P_2 accumulation, but a rise in PtdIns(4,5)P_2 levels. Changes in PtdInsP_2 levels are, however, independent of weak-acid-induced Pdr12p expression. PtdIns(3,5)P_2 synthesis is required for correct vacuole acidification. Levels of this molecule may be adjusted to maintain intracellular pH homoeostasis in response to stresses imposed by organic acids (Mollapour et al., 2006).

8.5 Mechanisms of resistance

8.5.1 Bacteria

Survival of bacteria against organic and inorganic challenges differs with regard to physiological requirements (Ferreira et al., 2003). As a result of inherent differences, bacteria react differently to antimicrobial agents. These differences include a unique composition of cell envelope or proteins. Differences may also be attributed to the development of resistance by adaptation, induction, or genetic exchange. An antimicrobial agent, or in this context a preservative, must be able to penetrate the cell envelope. To exert antibacterial action, this should be followed by attainment of a sufficiently high concentration at the target site (Cloete, 2003). Resistance to antimicrobial agents is, therefore, primarily attributable to an adaptation mechanism of the cell envelope. LPS of sufficient length protects against the action of organic acids in some species of Enterobacteriaceae. Some types of O-polysaccharides are more protective than others. In the resistance development of STEC O157 to organic acids the O157 polysaccharide is involved (Barua et al., 2002).

Biofilm bacteria are also notorious for being highly resistant to antimicrobial agents. Some possible explanations are listed as follows (Cloete, 2003):

1. Limited diffusion of antimicrobial agents through the biofilm matrix
2. Outer membrane structure
3. Enzyme production
4. Metabolic activity within the biofilm
5. Efflux pumps
6. Genetic adaptation

Some microorganisms are able to degrade the added preservatives by producing specific enzymes (e.g., *Pseudomonas aeruginosa* can degrade methyl *para*(4)-hydroxybenzoate), whereas the intestinal pathogen *Salmonella enterica* sv. Typhimurium may react to many potential stress factors in its so-called natural habitat, conveniently conveying cellular resistance against the weak butyric, acetic, and propionic acids (Theron and Lues, 2007).

It is especially important to understand the effect of food-associated organic acids on *E. coli* O157:H7, as this pathogen is relatively tolerant to acid (Roering et al., 1999; Casey and Condon, 2002). Resistance in *E. coli* O157:H7 to benzoic acid was recently observed after induction of an acid-tolerance response with a strong acid (pH 2). In certain Gram-positive bacteria (e.g., *L. monocytogenes*), acid tolerance response has also been induced to pH 3 after prior exposure to a mild acid at pH 5. However, it is not yet certain if this response would give protection to microorganisms against weak acids used in the food industry (Brul and Coote, 1999). It has, however, been found that in *L. monocytogenes* δ^B may mediate protection against organic and inorganic acid stresses through different mechanisms (Ferreira et al., 2003).

Findings from a study done on acid-adapted *Salmonella* to lactic acid rinses from artificially inoculated beef muscle slices showed that acid-adapted strains were not any more resistant to acid decontamination than parental strains (Dickson and Kunduru, 1995). In a study done by Steiner and Sauer (2003), the overexpression of the ATP-dependent helicase RecG was found to increase resistance to weak organic acids in *E. coli*. This was achieved by reduction of the toxic effects of the organic acids, reduction of the effects of the synthetic uncouplers (CCCP and DNP), and a reduction of the ATPase and cytochrome *c* inhibitor azide as a result of a decrease in pH or available ATP. In LAB, resistance mechanisms to various substances have been identified as enzymic activation, restricted import, active export, or target modification of the antimicrobial agent (Herreros et al., 2005).

In AAB, for example, *Acetobacter acetii*, a defect in membrane-bound alcohol dehydrogenase has been associated with the reduction in resistance to acetic acid. A gene cluster involved in acetic acid resistance in *A. acetii* has been identified and includes (1) *aarA* encoding a citrate synthase, (2) *aarB* encoding a functionally unknown protein, and (3) *aarC* encoding a protein suspected of being involved in acetic acid assimilation. Protein profiles may also change in response to acetic acid, specifically with regard to the production of many proteins. One such protein is aconitase, whose production was enhanced in response to acetic acid (Nakano, Fukaya, and Horinouchi, 2006).

Other mechanisms conferring acetic acid resistance are also located in the cell membrane, because the toxicity of organic acids may be partly a result of a disruption of the proton motive force by acetic acid, acting as an uncoupling agent. A proton motive efflux system for acetic acid has also recently been described in *A. acetii* (Matsushita et al., 2005). Mutation and overexpression of the *aatA* gene also caused resistance to formic and propionic acids simultaneously to that of acetic acid. *aatA* functions as an efflux pump of acetic acid. Several transporters for monocarboxylic acids are known (Nakano, Fukaya, and Horinouchi, 2006):

1. Bacterial lactate permease LctP family
2. Eukaryotic proton-linked monocarboxylate transporter MCT family
3. Monocarboxylate permease having a sodium-binding *Rhizobium leguminosarum*

However, these transporters do not contain any ABC motifs and transport monocarboxylic acids via a proton-coupled reaction. Acetic acid resistance in *A. acetii* is, therefore, conferred by at least two mechanisms: assimilation of the weak acid by enzymes, such as citrate synthase or aconitase and export of acetic acid by ABC transporter. Both mechanisms are implicated in reducing intracellular acetic acid concentration (Nakano, Fukaya, and Horinouchi, 2006).

8.5.2   Fungi

In fungi, some resistance mechanisms may be similar to those of bacteria, for example, the enzymatic degradation of sorbic acid to pentadine by certain fungal species (Samson et al., 1995). Resistance development is attributed to (1) decreased plasma membrane permeability to diffusional preservative influx (Warth, 1989), (2) active extrusion of organic acid anion

(Holyoak et al., 1999), and (3) the ability to degrade these organic acids (Mollapour and Piper, 2001; Macpherson et al., 2005).

In yeasts associated with food spoilage, resistance to organic acids is often a multicomponent process and may vary between genera and even species (Macpherson et al., 2005). *Z. bailii* is generally more resistant to preservatives than any other spoilage yeast and consequently a greater economic threat. One reason for this results from a greater ability to limit the diffusional entry of an undissociated acid as well as a greater ability to degrade these preservatives (Piper et al., 2001). Adaptations may prevent acidification of the cytosol by limiting the amount of preservative that would reach the cytosol. However, resistance mechanisms of *S. cerevisiae* are more familiar (Macpherson et al., 2005). *S. cerevisiae* is more prone to perturbations in cytosolic pH caused by preservatives. The main adaptive strategy in *S. cerevisiae* to resist intracellular acidification is proposed to be H^+ extrusion by the plasma membrane H^+-ATPase coupled with acid anion extrusion by the ABC transporter Pdr12 (Holyoak et al., 1999; Piper et al., 2001).

In *S. cerevisiae* acetate resistance determinants differ from those of resistance to the more lipophilic acids sorbate and benzoate. Although acetic acid has the same pK_a (4.76) as sorbic acid, acetate is far less inhibitory to yeasts than the more lipophilic sorbate (*trans-trans*hexanedienoate), although they have the same pK_a. Sorbate also has a much higher capacity to dissolve in membranes and as a result, can easily disorder. *S. cerevisiae* is able to grow in the presence of sorbate, mediated due to War1p transcription factor-dependent induction of a single ATP-binding cassette (ABC) transporter (Pdr12p). Pdr12p induction in cells is very prominent in sorbate-stressed cells, reaching levels in the plasma membrane similar to the most abundant plasma membrane protein. Pdr12p catalyzes active efflux of weak organic carboxylate anions, which lowers intracellular levels. This causes resistance to sorbate, benzoate, and aliphatic short-chain (C_{3-8}) carboxylic acids that are relatively soluble in water (Bauer et al., 2003).

Increased sorbate resistance in *S. cerevisiae* is also associated with a loss in the Yap5 transcription factor (Mollapour et al., 2004). *S. cerevisiae* has adopted a strategy to confer resistance by extrusion of high levels of proton and acid. However, active anion extrusion, which is mainly Pdr12-catalyzed at the plasma membrane can export the acid only as far as the periplasm. The undissociated acid can then easily diffuse back into the cell (Piper et al., 2001). Pdr12 is essential for the adaptation of yeast to grow under weak-acid stress, inasmuch as the normal physiological function of Pdr12 is to protect against the toxicity of organic acids, produced by competitor organisms, as these acids may accumulate to inhibitory levels in cells at low pH (Piper et al., 1998). To counteract proton toxification as a result of the dissociation of organic acids in the cell cytosol, *S. cerevisiae* respond by activating H$^+$-ATPase (Plumridge et al., 2004).

Acetate uptake and utilization systems of *S. cerevisiae* are also glucose repressed, whereas *Z. bailii* can utilize acetate when cultured on fermentable sugars (Mollapour and Piper, 2001). Genes specific to *Z. bailii* enable it to degrade lipophilic weak organic acid preservatives (Mollapour and Piper, 2001). In cells adapting to sorbic acid, strong induction has been reported of a gene encoding a specific heat-shock protein, of genes encoding proteins involved in cell wall maintenance, and of genes involved in transposon activity (Brul et al., 2002). *Z. bailii* can also degrade sorbate and benzoate, but only in the presence of oxygen, whereas *S. cerevisiae* is unable to use these compounds as a sole carbon source. *ZbYME2* is the first *Zygosaccharomyces* gene of resistance to food preservatives to be characterized genetically (Mollapour and Piper, 2001). In *Z. bailii* initial diffusion of acid is often more restricted, causing extreme organic acid resistance (Piper et al., 2001).

Weak-acid resistance generally requires a high level of energy-dependent extrusion of both protons and anions from acid-adapted *S. cerevisiae*. Although weak organic acids as food preservatives may be pro-oxidants as well as mutagenic toward the mitochondrial genome in an aerobic yeast, at least one antioxidant activity (Sod2) may play a major role in causing weak-acid resistance of aerobic PDR12+ yeast (Piper, 1999).

It is proposed that anions are actively extruded from cells to account for lower intracellular concentration. In theory, if anions are extruded from the cell, they would reassociate when exposed to the lower external pH and freely diffuse back into the cell, resulting in a futile cycle without resistance development. However, adapted yeast cells can reduce the diffusion coefficient of preservatives across the plasma membrane, and the diffusion of weak acids into the cell is then reduced. This adaptive mechanism is based on the efflux of preservative anions by Pdr12. Efflux of anions together with a reduction in the diffusion coefficient of the membrane will result in the maintenance of cell homeostasis, which will enable the organism to survive and grow (Holyoak et al., 1999).

Transcription factor Haa1p was recently identified and is required for rapid adaptation of a yeast to several organic acids. Transcription of 9 Haa1p-target genes are activated under weak-acid stress. Haa1-regulated genes that may be required for more rapid organic acid adaptation include (1) *TPO2* and *TPO3*, encoding two plasma membrane multidrug transporters, and (2) *YGP1*, encoding a poorly characterized cell wall glycoprotein. Organic acids do not kill the microorganisms; they only inhibit growth and cause highly extended lag phases. The role of Haa1p is important during this adaptation period (Fernandes et al., 2005).

Genes upregulated in sorbate-adapted cells include many that are regulated by Msn2p and Msn4p, the main regulators of general stress response. However, both appear to be activated by exposure to sorbic acid (Görner et al., 1998). Pdr12p induction is specific for weak-acid stress and is independent of Msn2p/4p (Piper et al., 1998; Schüller et al., 2004).

8.6   Transmission of resistance

It is possible for resistance of food-associated bacteria treated with antibiotics to be transmitted to the human population through the food chain. For example, LAB may contaminate milk and meat from animals treated with antibiotics and consequently carry populations of resistant bacteria. Although many of these organisms are not pathogenic, they could constitute a reservoir of genes conferring resistance to antibiotics that might be transferred to pathogenic strains. There is also the possibility that cross-resistance to preservatives and organic acids may occur (Herreros et al., 2005).

8.7   Extent of the situation

The number of microbial species becoming resistant to antibiotics is constantly increasing, and fungi are no exception. Fungal human pathogens as well as spoilage molds in food and feed systems are becoming resistant. In fact, yeasts and molds are not only becoming resistant to antibiotics, but also to preservatives such as the organic acids (Schnürer and Magnusson, 2005).

Until recently organic acids such as sorbic acids, were still regarded as the standard approach by the food industry for microbial inhibition. In the past 30–40 years many studies have, however, focused on resistance development against this organic acid (Brul et al., 2002). Many studies have dealt with spoilage yeasts such as *Z. bailii* and *Zygosaccharomyces lentus*, notorious for being difficult to control (Mollapour and Piper, 2001). The higher fermentation rate of *S. cerevisiae* is one of the reasons why this organism should also be considered more dangerous than is generally expected in food stabilization (Guerzoni, Gardini, and Duan, 1990; Ferreira, Loureiro-Dias, and Loureiro, 1997).

There is not yet a clear understanding of the actual effectiveness of the use of classical preservatives such as organic acids in conjunction with other common components of food preservation systems. This entails the physiological and molecular mechanisms within cells and whether microorganisms die or adapt and survive as a result of this preservation. For example, which signal transduction systems are involved and which stress proteins are induced, how are these systems induced, and how much cellular energy (ATP) is involved in each system? The amount of available energy will determine to what extent a microbial cell will activate the various stress response pathways (Brul and Coote, 1999).

Pathogens in mixed microbial cultures in food or food environments may react differently to decontamination stresses than pure cultures in a laboratory (Samelis et al., 2001; Stopforth et al., 2003). Many species of spoilage yeasts and molds are able to adapt and grow in the presence of the maximum permitted levels of sorbic and benzoic acids used in manufactured foods and beverages, resulting in inconvenience to producer and consumer as well as significant economic loss (Holyoak et al., 1999). High levels of sorbic acid or benzoic acid are often needed to inhibit growth of spoilage molds and yeasts in preserved foods. Such high concentrations are routinely used in soft drink manufacturing, causing consumption of these compounds daily in possibly physiologically considerable amounts (Piper, 1999).

The potential long-term effects of organic acid decontamination on the microbial ecology of meat-packing plants and products and to manage potential safety risks must be taken seriously. Gram-negative bacteria that survive meat decontamination by organic acid rinses have been found to establish niches in these plants (Samelis et al., 2002).

It is speculated that the microbial growth phase, acid tolerance, and virulence may be linked, and it is, therefore, reasonable to suggest that organic acids may also influence virulence (El-Gedaily et al., 1997; Ricke, 2003). Another serious problem is that cross-resistance in *Salmonella* to other stress conditions, such as reactive oxygen and high osmolarity may be increased by exposure to a mixture of organic acids. This ability of an organic acid to induce cross-protection to other stress conditions may indicate that these acids could have a more profound effect on survival and also virulence of *S. typhimurium*, which may have serious practical implications for controlling this pathogen in food production systems (Kwon and Ricke, 1998).

8.8   E. coli O157:H7

E. coli O157:H7 is notorious for being extremely acid resistant, and seriously decreases the effectiveness of organic acid spray washings on carcasses. For example, the acid-resistant strain ATCC 43895 has been found to be able to survive in undiluted 2% lactate (pH 2.3–2.5) or 2% acetate (pH 3.0–3.2) spray washings on beef at 4 or 10°C for two to seven days. The survival rate of *E. coli* O157:H7 at 4°C is greater and longer lasting than most other pathogenic bacteria. This was specifically detected in comparison to *L. monocytogenes* and *Salmonella typhimurium* DT104 under various conditions. Survival of *E. coli* O157:H7 may become higher in number or last longer upon previous acid adaptation. *E. coli* O157:H7 has a wide spectrum of sensitivity to the organic acids, specifically in carcass decontamination. This is specifically true for acetic acid and lactic acid. It has not been elucidated which of these acids is the most effective against this pathogen (Samelis et al., 2002).

8.9   Protective effects of organic acids

Under certain conditions, the use of organic acids may enhance the growth of some fungal strains in bakery products (Marín et al., 2002). For example, a low concentration of sorbic acid (0.025 and 0.05%), almost without exception, has been shown to enhance fungal growth, regardless of a_w and temperature levels (Samelis et al., 2002). Potassium benzoate (at a concentration of 2 mM) has been found to produce an increase in growth rate of 52% in *Z. bailii* cells (Macpherson et al., 2005).

Organic acids such as acetic, malic, and L-lactic acids have been found to have protective effects (1–2 log increases in survival) on the survival of *E. coli* O157:H7, compared to organic acids, at concentrations between 5 and 10 mM. D-lactic has a greater protective effect (~4-log increase in survival) over a concentration range of 1 to 20 mM. However, citric acid has not yet shown any protective effect at these concentrations (Bjornsdottir, Breidt, Jr., and McFeeters, 2006).

8.10   Possible advantages of organic acid resistance

In bakery products, propionic acid is known to inhibit some spoilage agents such as molds and *Bacillus* spores, but not yeasts to the same extent. This acid is, therefore, the traditional choice for bread preservation (Ponte and Tsen, 1987; Suhr and Nielsen, 2004). In wine, *S. cerevisiae* may be inhibited by competitor microbes producing organic acids. A stress response is activated to prevent accumulation of these acids to potentially toxic levels (Piper et al., 2001).

8.11   Industry strategies

The quest to evaluate the effectiveness of organic acids for specific applications increasingly requires a better understanding of the general as well as specific stress-response capabilities of foodborne microorganisms and pathogens. Modern tools must be applied to study microbial behavior in all aspects of food production environments to more targeted strategies in controlling foodborne pathogens with organic acids (Ricke, 2003). Understanding resistance development is the next step in designing improved preservation strategies and is important in the development of more effective preservation protocols (Macpherson et al., 2005; Papadimitriou et al., 2007). Microorganisms adapt to survive in previously effective control methods and new techniques are essential for maintaining food quality, while at the same time inhibiting contamination in every production step and throughout the distribution chain (Rico et al., 2007).

8.11.1 Targets

Any protective effect that an organic acid may have on contaminating organisms must be addressed when improving the efficacy of organic acids (Tamblyn and Conner, 1994). Often, low concentrations of organic acids are ineffective against various pathogenic organisms, for example, *S. typhimurium*, attached to carcasses, especially broiler chicken carcasses (Lillard et al., 1987; Tamblyn and Conner, 1994). Future techniques in attending to such aspects may include the use of transdermal agents (Tamblyn and Conner, 1997). Long-term effects of organic treatments on microbial ecology of food products is an important aspect to consider in managing potential safety risks. One such aspect is spray washings used in carcass decontamination, which may select from acid-tolerant or organic acid resistant spoilage bacteria or yeasts (Samelis et al., 2002).

The ability of an organism to resist intracellular acidification is an essential survival mechanism, if pH homeostasis is critical to cellular function (Krebs et al., 1983; Holyoak et al., 1996). The plasma membrane H^+-ATPase plays an important role in resisting intracellular acidification and provides a common target in control strategies to be applied for both *Z. bailii* and *S. cerevisiae* (Macpherson et al., 2005). *Z. bailii* is known to limit the entry of weak-acid preservatives, and *S. cerevisiae* relies on plasma membrane efflux systems to remove H^+ and organic acid from a cell (Piper et al., 2001). If *Z. bailii* were also to rely on K^+ as counter-ion during stress response, plasma membrane K^+ transporters may form a common target for control strategy of yeasts (Macpherson et al., 2005).

References

Archer, D.L. 1996. Preservation microbiology and safety: Evidence that stress enhances virulence and triggers adaptive mutations. *Trends in Food Science and Technology* 7:91–95.

Barker, C. and Park, S.F. 2001. Sensitization of *Listeria monocytogenes* to low pH, organic acids, and osmotic stress by ethanol. *Applied and Environmental Microbiology* 67:1594–1600.

Barua, S., Yamashino, T., Hasegawa, T., Yokoyama, K., Torii, K., and Ohta, M. 2002. Involvement of surface polysaccharides in the organic acid resistance of Shiga toxin-producing *Escherichia coli* O157:H7. *Molecular Microbiology* 43:629–640.

Bauer, B.E., Rossington, D., Mollapour, M., Mamnun, Y., Kuchler, K., and Piper, P.W. 2003. Weak organic acid stress inhibits aromatic amino acid uptake by yeast, causing a strong influence of amino acid auxotrophies on the phenotypes of membrane transporter mutants. *European Journal of Biochemistry* 270:3189–3195.

Bearson, S., Bearson, B., and Foster, J.W. 1997. Acid stress responses in enterobacteria. *FEMS Microbiology Letters* 147:173–180.

Bjornsdottir, K., Breidt, F., Jr., and McFeeters, R.F. 2006. Protective effects of organic acids on survival of *Escherichia coli* O157:H7 in acidic environments. *Applied and Environmental Microbiology* 72:660–664.

Boxman, A.W., Theuvenet, A.P.R., Peters, P.H.J., Dobbelmann, J., and Borts-Heuwels, G.W.F.H. 1985. Regulation of 86RB influx during accumulation of Rb- or K+ in yeast. *Biochimica et Biophysica Acta* 814:50–56.

Brul, S. and Coote, P. 1999. Preservative agents in foods. Mode of action and microbial resistance mechanisms. *International Journal of Food Microbiology* 50:1–17.

Brul, S., Klis, F.M., Oomes, S.J.C.M., et al. 2002. Detailed process design based on genomics of survivors of food preservation processes. *Trends in Food Science and Technology* 13:325–333.

Casey, P.G. and Condon, S. 2002. Sodium chloride decreases the bacteriocidal effect of acid pH on *Escherichia coli* O157:H45. *International Journal of Food Microbiology* 76:199–206.

Chapman, J.S. 1998. Characterizing bacterial resistance to preservatives and disinfectants. *International Biodeterioration and Biodegradation* 41:241–245.

Cheng, H.-Y., Yu, R.-C., and Chou, C.-C. 2003. Increased acid tolerance of *Escherichia coli* O157:H7 as affected by acid adaptation time and conditions of acid challenge. *Food Research International* 36:49–56.

Cheng, L., Moghraby, J., and Piper, P.W. 1999. Weak organic acid treatment causes a trehalose accumulation in low-pH cultures of *Saccharomyces cerevisiae*, not displayed by the more preservative-resistant *Zygosaccharomyces bailii*. *FEMS Microbiology Letters* 170:89–95.

Cloete, T.E. 2003. Resistance mechanisms of bacteria to antimicrobial compounds. *International Biodeterioration and Biodegradation* 51:277–282.

Deak, T. 1991. Food borne yeast. *Advances in Applied Microbiology* 36:179–278.

De Nobel, H., Lawrie, L., Brul, S., et al. 2001. Parallel and comparative analysis of the proteome and transcriptome of sorbic acid stressed *Saccharomyces cerevisiae*. *Yeast* 18:1413–1428.

Dickson, J. and Kunduru, M. 1995. Resistance of acid-adapted Salmonellae to organic acid rinses on beef. *Journal of Food Protection* 58: 973–976.

Doyle, E.M. 1999. *Use of organic acids to control Listeria in meat*. American Meat Institute Foundation. Washington DC.

El-Gedaily, Paesold, G., Chen, C-Y., Guiney, D.G., and Krause, M. 1997. Plasmid virulence gene expression induced by short-chain fatty acids in *Salmonella dublin*: Identification of *rpoS*-dependent and *rpoS*-independent mechanisms. *Journal of Bacteriology* 179:1409–1412.

Fernandes, A.R., Mira, N.P., Vargas, R.C., Canelhas, I., and Sa-Correia, I. 2005. *Saccharomyces cerevisiae* adaptation to weak acids involves the transcription factor Haa1p and Haa1p-regulated genes. *Biochemical and Biophysical Research Communications* 337:95–103.

Ferreira, A., Sue, D., O'Byrne, C.P., and Boor, K.J. 2003. Role of *Listeria monocytogenes* sigma(B) in survival of lethal acidic conditions and in the acquired acid tolerance response. *Applied and Environmental Microbiology* 69:2692–2698.

Ferreira, M.M., Loureiro-Dias, M.C., and Loureiro, V. 1997. Weak acid inhibition of fermentation by *Zygosaccharomyces bailii* and *Saccharomyces cerevisiae*. *International Journal of Food Microbiology* 36:145–153.

Fielding, L.M., Cook, P.E., and Grandison, A.S. 1997. The effect of electron beam irradiation, combined with acetic acid, on the survival and recovery of *Escherichia coli* and *Lactobacillus curvatus*. *International Journal of Food Microbiology* 35:259–265.

Görner, W., Durchschlag, E., Martinez-Pastor, M.T., et al. 1998. Nuclear localization of the C_2H_2 zinc finger protein Msn2p is regulated by stress and protein kinase A activity. *Genes & Development* 12:586–597.

Greenacre, E.J., Brocklehurst, T.F., Waspe, C.R., Wilson, D.R., and Wilson, P.D.G. 2003. Salmonella enterica Serovar Typhimurium and Listeria monocytogenes acid tolerance response induced by organic acids at 20°C: Optimization and modeling. *Applied and Environmental Microbiology* 69:3945–3951.

Greenacre, E.J., Lucchini, S., Hinton, J.C., and Brocklehurst, T.F. 2006. The lactic acid–induced acid tolerance response in *Salmonella enterica* serovar Typhimurium induces sensitivity to hydrogen peroxide. *Applied and Environmental Microbiology* 72:5623–5625.

Guerzoni, M.E., Gardini, F., and Duan, J. 1990. Interaction between inhibition factors on microbial stability of fruit-based systems. *International Journal of Food Microbiology* 10:1–18.

Hazan, R., Levine, A., and Abeliovich, H. 2004. Benzoic acid, a weak organic acid food preservative, exerts specific effects on intracellular membrane trafficking pathways in *Saccharomyces cerevisiae*. *Applied and Environmental Microbiology* 70:4449–4457.

Henriques, M., Quintas, C., and Loureiro-Dias, M.C. 1997. Extrusion of benzoic acid in *Saccharomyces cerevisiae* by an energy-dependent mechanism. *Microbiology* 143:1877–1883.

Herreros, M.A., Sandoval, H., González, L., Castro, J.M., Fresno, J.M., and Tornadijo, M.E. 2005. Antimicrobial activity and antibiotic resistance of lactic acid bacteria isolated from Armada cheese (a Spanish goat's milk cheese). *Food Microbiology* 22:455–459.

Holyoak, C.D., Bracey, D., Piper, P.W., Kuchler, K., and Coote, P.J. 1999. The *Saccharomyces cerevisiae* weak-acid-inducible ABC transporter Pdr12 transports fluorescein and preservative anions from the cytosol by an energy-dependent mechanism. *Journal of Bacteriology* 181:4644–4652.

Holyoak, C.D., Stratford, M., McMullin, Z., Cole, M.B., Crimmins, K., Brown, A.J.P., and Coote, P. 1996. Activity of the plasma-membrane H⁺-ATPase and optimal glycolic flux are required for rapid adaptation and growth in the presence of the weak acid preservative sorbic acid. *Applied and Environmental Microbiology* 62:3158–3164.

Hsiao, C.P. and Siebert, K.J. 1999. Modeling the inhibitory effects of organic acids on bacteria. *International Journal of Food Microbiology* 47:189–201.

Kasemets, K., Kahru, A., Laht, T.M., and Paalme, T. 2006. Study of the toxic effect of short- and medium-chain monocarboxylic acids on the growth of *Saccharomyces cerevisiae* using the CO_2-auxo-accelerostat fermentation system. *International Journal of Food Microbiology* 111:206–215.

Krebs, H.A., Wiggins, D., Sole, S., and Bedoya, F. 1983. Studies on the mechanisms of the antifungal action of benzoate. *Biochemical Journal* 214:657–663.

Kwon, Y.M. and Ricke, S.C. 1998. Induction of acid resistance of *Salmonella typhimurium* by exposure to short-chain fatty acids. *Applied and Environmental Microbiology* 64:3458–3463.

Leyer, G.J. and Johnson, E.A. 1993. Acid adaptation induces cross-protection against environmental stresses in *Salmonella typhimurium*. *Applied and Environmental Microbiology* 59:1842–1847.

Leyva, J.S. and Peinado, J.M. 2005. ATP requirements for benzoic acid tolerance in *Zygosaccharomyces bailii*. *Journal of Applied Microbiology* 98:121–126.

Lillard, H.S., Blankenship, L.C., Dickens, J.A., Craven, S.E., and Schakelford, A.D. 1987. Effect of acetic acid on the microbiological quality of scalded picked and unpicked broiler carcasses. *Journal of Food Protection* 50:112–114.

Macpherson, N., Shabala, L., Rooney, H., Jarman, M.G., and Davies, J.M. 2005. Plasma membrane H+ and K+ transporters are involved in the weak-acid preservative response of disparate food spoilage yeasts. *Microbiology* 151:1995–2003.

Marín, S., Abellana, M., Rubinat, M., Sanchis, V., and Ramos, A.J. 2003. Efficacy of sorbates on the control of *Eurotium* species in bakery products with near neutral pH. *International Journal of Food Microbiology* 87:251–258.

Matsuda, T., Yano, T., Maruyama, A., and Kumagai, H. 1994. Antimicrobial activities of organic acids determined by minimum inhibitory concentrations at different pH ranged from 4.0 to 7.0. *Japan Society for Food Science and Technology* 41:687–701.

Matsushita, K., Inoue, T., Adachi, O., and Toyama, H. 2005. *Acetobacter aceti* possesses a proton motive force-dependent efflux system for acetic acid. *Journal of Bacteriology* 187:4346–4352.

Mellor, S. 2005. Animal feed manufacturers association. http://www.afma.co.za/AFMA_ Template/1,2491,541,00.html. (Accessed November 8, 2005.)

Mollapour, M. and Piper, P.W. 2001. The ZbYME2 gene from the food spoilage yeast *Zygosaccharomyces bailii* confers not only YME2 functions in *Saccharomyces cerevisiae*, but also the capacity for catabolism of sorbate and benzoate, two major weak organic acid preservatives. *Molecular Microbiology* 42:919–930.

Mollapour, M., Fong, D., Balakrishnan, K., et al. 2004. Screening the yeast deletant mutant collection for hypersensitivity and hyper-resistance to sorbate, a weak organic acid food preservative. *Yeast* 21:927–946.

Mollapour, M., Phelan, J.P., Millson, S.H., Piper, P.W., and Cooke, F.T. 2006. Weak acid and alkali stress regulate phosphatidylinositol bisphosphate synthesis in *Saccharomyces cerevisiae*. *Biochemical Journal* 395:73–80.

Nakano, S., Fukaya, M., and Horinouchi, S. 2006. Putative ABC transporter responsible for acetic acid resistance in *Acetobacter aceti*. *Applied and Environmental Microbiology* 72:497–505.

Nielsen, M.K. and Arneborg, N. 2007. The effect of citric acid and pH on growth and metabolism of anaerobic *Saccharomyces cerevisiae* and *Zygosaccharomyces bailii* cultures. *Food Microbiology* 24:101–105.

Nuñez, M., Peña, R.M., Herrero, C., and García, S. 2000. Determination of six metals in Galician wines (in Northwestern Spain) by capillary electrophoresis. *Journal of AOAC International* 83:183–188.

Papadimitriou, M.N., Resende, C., Kuchler, K., and Brul, S. 2007. High Pdr12 levels in spoilage yeast (*Saccharomyces cerevisiae*) correlate directly with sorbic acid levels in the culture medium but are not sufficient to provide cells with acquired resistance to the food preservative. *International Journal of Food Microbiology* 113:173–179.

Pearce, A.K., Booth, I.R., and Brown, A.J.P. 2001. Genetic manipulation of 6-phos-phofructo-1-kinase and fructose 2,6-bisphosphate levels affects the extent to which benzoic acid inhibits the growth of *Saccharomyces cerevisiae*. *Microbiology* 147:403–410.

Piper, P.W. 1999. Yeast superoxide dismutase mutants reveal a pro-oxidant action of weak organic acid food preservatives. *Free Radical Biology & Medicine* 27:1219–1227.

Piper, P., Calderon, C.O., Hatzixanthis, K., and Mollapour, M. 2001. Weak acid adaptation: The stress response that confers yeasts with resistance to organic acid food preservatives. *Microbiology* 147:2635–2642.

Piper, P.W., Mahe, Y., Thompson, S., et al., 1998. The pdr12 ABC transporter is required for the development of weak organic acid resistance in yeast. *The EMBO Journal* 17:4257–4265.

Plumridge, A., Hesse, S.J., Watson, A.J., Lowe, K.C., Stratford, M., and Archer, D.B. 2004. The weak acid preservative sorbic acid inhibits conidial germination and mycelial growth of *Aspergillus niger* through intracellular acidification. *Applied and Environmental Microbiology* 70:3506–3511.

Ponte, J.G., Jr., and Tsen, C.C. 1987. Bakery products. In: L.R. Beuchat (Ed.), *Food and Beverage Mycology*, 2nd ed., pp. 51–100. New York: AVI–Van Nostrand-Reinhold.

Quintavalla, S. and Vicini, L. 2002. Antimicrobial food packaging in meat industry. *Meat Science* 62:373–380.

Ricke, S.C. 2003. Perspectives on the use of organic acids and short chain fatty acids as antimicrobials. *Poultry Science* 82:632–639.

Rico, D., Martin-Diana, A.B., Barat, J.M., and Barry-Ryan, C. 2007. Extending and measuring the quality of fresh-cut fruit and vegetables: A review. *Trends in Food Science and Technology* 18:373–386.

Roering, A.M., Luchansky, J.B., Ihnot, A.M., Ansay, S.E., Kaspar, C.W., and Ingham, S.C. 1999. Comparative survival of *Salmonella typhimurium* DT 104, *Listeria monocytogenes*, and *Escherichia coli* O157:H7 in preservative-free apple cider and simulated gastric fluid. *International Journal of Food Microbiology* 46:263–269.

Ryu, J.-H., Deng, Y., and Beuchat, L.R. 1999. Survival of *Escherichia coli* O157:H7 in dried beef powder as affected by water activity, sodium chloride content and temperature. *Food Microbiology* 16:309–316.

Samelis, J., Sofos, J.N., Kendall, P.A., and Smith, G.C. 2001. Influence of the natural microbial flora on the acid tolerance response of *Listeria monocytogenes* in a model system of fresh meat decontamination fluids. *Applied and Environmental Microbiology* 67:2410–2420.

Samelis, J., Sofos, J.N., Kendall, P.A., and Smith, G.C. 2002. Effect of acid adaptation on survival of *Escherichia coli* O157:H7 in meat decontamination washing fluids and potential effects of organic acid interventions on the microbial ecology of the meat plant environment. *Journal of Food Protection* 65:33–40.

Samson, R.A., Hoekstra, E., Frisvad, J.C., and Filtonborg, O. (Eds.) (1995). *Introduction to Food-Borne Fungi*, Baarn, The Netherlands: CBS.

Schnürer, J. and Magnusson, J. 2005. Antifungal lactic acid bacteria as bio-preser-vatives. *Trends in Food Science and Technology* 16:70–78.

Schüller, C., Mamnun, Y.M., Mollapour, M., et al. 2004. Global phenotypic analysis and transcriptional profiling defines the weak acid stress response regulon in *Saccharomyces cerevisiae*. *Molecular Biology of the Cell* 15:706–720.

Smulders, F.J. and Greer, G.G. 1998. Integrating microbial decontamination with organic acids in HACCP programmes for muscle foods: Prospects and controversies. *International Journal of Food Microbiology* 44:149–169.

Steels, H., James, S.A., Roberts, I.N., and Stratford, M. 2000. Sorbic acid resistance: The inoculum effect. *Yeast* 16:1173–1183.

Steiner, P. and Sauer, U. 2003. Overexpression of the ATP-dependent helicase RecG improves resistance to weak organic acids in *Escherichia coli*. *Applied Microbiology and Biotechnology* 63:293–299.

Stopforth, J.D., Samelis, J., Sofos, J.N., Kendall, P.A., and Smith, G.C. 2003. Influence of organic acid concentration on survival of *Listeria monocytogenes* and *Escherichia coli* O157:H7 in beef carcass wash water and on model equipment surfaces. *Food Microbiology* 20:651–660.

Stratford, M. and Lambert, R.J.W. 1999. Weak-acid preservatives: Mechanisms of adaptation and resistance by yeasts. *Food Australia* 51:26–29.

Suhr, K.I. and Nielsen, P.V. 2004. Effect of weak acid preservatives on growth of bakery product spoilage fungi at different water activities and pH values. *International Journal of Food Microbiology* 95:67–78.

Tamblyn, K.C. and Conner, D.E. 1994. Antibacterial activity of citric, malic and propionic acids against *Salmonella typhimurium* attached to broiler skin. *Poultry Science* 73:23.

Tamblyn, K.C. and Conner, D.E. 1997. Bactericidal activity of organic acids in combination with transdermal compounds against *Salmonella typhimurium* attached to broiler skin. *Food Microbiology* 14:477–484.

Theron, M.M. and Lues, J.F.R. 2007. Organic acids and food preservation: A review. *Food Reviews International* 23:141–158.

Thomas, D.S. and Davenport, R. 1985. *Zygosaccharomyces bailii* a profile of characteristics and spoilage activities. *Food Microbiology* 2:157–169.

Trček, J., Toyama, H., Czuba, J., Misiewicz, A., and Matsushita, K. 2006. Correlation between acetic acid resistance and characteristics of PQQ-dependent ADH in acetic acid bacteria. *Applied Microbiology and Biotechnology* 70:366–373.

Warth, A.D. 1989. Transport of benzoic and propionic acids by *Zygosaccharomyces bailii*. *Journal of General Microbiology* 135:1383–1390.

Warth, A.D. 1991. Effect of benzoic acid on glycolytic metabolite levels and intracellular pH in *Saccharomyces cerevisiae*. *Applied and Environmental Microbiology* 57:3415–3417.

chapter nine

Acid tolerance

9.1 Introduction

Pathogenic organisms are not limited to cardinal ranges of temperature, pH, and water activity, but can adapt to survive outside their known and expected abilities (Hill, O'Driscoll, and Booth, 1995). Some pathogenic organisms have, for example, adapted to survive the various acidic environments in the human body (stomach and small intestines) to be able to cause infections. The acquisition of resistance to inorganic or organic acids may, therefore, pose a serious risk when food is contaminated with such a foodborne pathogen and is critical throughout the entire farm-to-fork process of food production (Merrell and Camilli, 1999; Berry and Cutter, 2000). Acid tolerance response (ATR) is, therefore, a concern to the food processor as well as clinicians, because pathogens may tolerate acidic environments in food processes and also the human body, posing a serious risk in causing disease (Tosun and Aktug Gonul, 2003).

Acid tolerance response is a complex defense system that can minimize the lethal effects of extreme low pH (pH 3) and also the effect of weak organic acids (Baik et al., 1996). Survival of organic and inorganic challenges differs among microorganisms with regard to physiological requirements. One example can be seen in *Listeria monocytogenes* where δ^B may mediate protection against both organic and inorganic acid stresses, but through different mechanisms (Ferreira et al., 2003). Also, *Salmonella enterica* sv. *typhimurium* cells that have been exposed to acid shock, show varying sensitivities to organic acids (Baik et al., 1996; Tetteh and Beuchat, 2003). Two systems exist in *S. typhimurium* that provide redundant protection against inorganic acid stress. These are (1) RpoS dependent, and (2) PhoPQ dependent, where PhoP is an acid-shock protein (Bearson, Wilson, and Foster, 1998).

9.2 Delineating the difference among acid adaptation, acid tolerance, and acid resistance

Acid stress is defined as a combined biological effect of low pH and organic acids present in an environment to which an organism may be exposed (Barua et al., 2002). Various terms have evolved to describe a response following exposure to acid stress: (1) acid adaptation, (2) acid tolerance, (3) inducible acid resistance, and (4) acid habituation (Edelson-Mammel, Porteous, and Buchanan, 2006). When exposed to a moderately acidic environment, microorganisms may develop cells with increased resistance when transferred to more acidic conditions. This is referred to as an acid adaptation response (Cheng, Yu, and Chou, 2003). Acid adaptation generally increases acid tolerance, whereas acid-shocked cells acquire enhanced acid tolerance, not increased tolerance. During fermentation of foods microbial cells may experience acid adaptation, and not acid shock, due to gradual decrease in pH (Ryu and Beuchat, 1999).

The terms "acid resistance" and "acid tolerance" are often used interchangeably and describe growth at moderately acid pH (Russell, 1991; Takahashi et al., 1997), which of course causes much confusion. Sometimes these terms are also used when an organism survives acid shock (Benjamin and Datta, 1995; Diez-Gonzalez and Russell, 1999). Other observations, causing equal confusion, are *Salmonella* that can grow at lower pH values than *Escherichia coli* and *Shigella* but are in fact, more sensitive to acid shock (Diez-Gonzalez and Russell, 1999). The term "extreme acid resistance" is sometimes used to describe viability after acid shock (Lin et al., 1995). Exposure to an environment that induces acid shock may result in a lower production of stress proteins, or different proteins, providing varying protection to stresses (Ryu and Beuchat, 1999).

S. typhimurium is well known for its ability to survive at extreme pH (pH 3.0), but this is dependent on the acid used to acidify the growth medium. The order of development of ATR in *Salmonella* cells by the following acids have been found to increase from ascorbic to citric in the order: ascorbic < HCl ≤ malic < lactic < acetic < citric. Of equal importance is *E. coli* O157:H7, known to be unusually acid tolerant (Álvarez-Ordóñez et al., 2009).

9.3 Role of organic acids in tolerance

When an organism adapts to an acidic environment, this may lead to an acid tolerance response (ATR) which, in turn, may protect against organic acids. ATR is, therefore, a complex defense system generated when acid resistance is induced after adaptation to mildly acidic pH conditions to minimize the lethal effects of extremely low pH and also to protect an organism against organic acids (Baik et al., 1996). In turn, organic acids enhance survivability of acid-sensitive pathogens exposed to low pH by induction of an acid tolerance response. This tolerance is also often linked to increased virulence and concern has risen about implications regarding the use of organic acids (Ricke, 2003). Organic acids, therefore, do not always cause positive results in food safety. Exposure to an organic acid may also induce stress responses in bacteria, rendering them more tolerant of (1) acidity, (2) ethanol, or (3) H_2O_2 (Doyle, 1999). It can be safely assumed that inorganic acid adaptation protects an organism against organic acid stress and vice versa (Baik et al., 1996).

ATR as a function of the acid used may be due to the maintenance of internal pH (pH_i) of the cell (Álvarez-Ordóñez et al., 2009). Some organic acids enter the cell more easily than others, and, as a result, may alter the pH_i of the cell more easily (Greenacre et al., 2003). The pH_i is crucial for maintenance of acid tolerance. The ATR complex phenomenon is strongly affected by environmental factor interaction, prevailing during acid adaptation and acid inactivation (Álvarez-Ordóñez et al., 2009).

An inorganic acid (H^+) can signal a decrease in the proteolysis of δ^s. Inorganic acid stress, therefore leads to an accumulation of δ^s by down-regulating proteolysis. Organic acids increase δ^s stability, and also stimulate translation of the *rpoS* message. Exponential-phase cultures, grown at pH7 and adapted with 100 mM acetate exhibit a 5× increase in *rpoS* translation, but not transcription (Audia, Webb, and Foster, 2001).

Gram-negative bacteria have developed various devices to protect themselves from acid stress (Barua et al., 2002). For example, studies have shown that propionic acid and butyric acid can promote extreme acid resistance of nonpathogenic *E. coli*, as well as the extreme acid resistant *E. coli* O157:H7 (Guilfoyle and Hershfield, 1996), whereas benzoate, on the other hand, is much less active in promoting survival of *E. coli* O157:H7 than acetate (Diez-Gonzalez and Russell, 1999).

The type of organic acid also influences ATR (Álvarez-Ordóñez et al., 2009). For example, acetic is more efficient than lactic acid for induction of ATR in *S. typhimurium*, grown in TSB (Greenacre et al., 2003), whereas citric acid and acetic acid caused the highest ATR in cells grown and challenged in meat extract (Álvarez-Ordóñez et al., 2009). This could have an important impact on food safety, because citric, acetic, and lactic acids are the most commonly used acids in carcass decontamination treatments (Smulders and Greer, 1998) as well as preservatives (Álvarez-Ordóñez et al., 2009). In *S. typhimurium* sudden exposure to reduced pH (acid adaptation) increases resistance to various acids and also enhances survival in fermented and acidic foods. Acid-shocked *E. coli* O157:H7 has also demonstrated increased resistance to lactic acid and enhanced survival in acid foods, such as shredded dry salami and apple juice. Some inorganic acids may, however, induce stronger acid tolerance than organic acids, although they are not commonly used in foods. Acid-adapted *S. typhimurium* cells resulting from exposure to hydrochloric acid have increased resistance to inactivation by organic acids and survived better than unadapted cells during fermentation of milk (Deng, Ryu, and Beuchat, 1999). Acid-adapted *S. typhimurium* strains may also survive during milk fermentation, and have been shown to have increased resistance to inactivation by organic acids (Tosun and Aktug Gonul, 2003). However, resistance to organic acids does not always result from resistance to low pH. Some strains capable of growing at low pH cannot grow when the pH is adjusted with, for example, acetic acid (Nakano, Fukaya, and Horinouchi, 2006).

The acid-tolerance system can protect *Salmonella* against the killing effects of organic acids and, as a result, increase their survival in fermented foods (Leyer and Johnson, 1992; Baik et al., 1996). Inorganic acid resistance of *S. typhimurium* may, however, be increased after exposure to a high concentration of organic acids. This induced resistance is enhanced by (1) acidic pH, (2) anaerobiosis, and (3) prolonged exposure to organic acids (Ricke, 2003). Previous exposure to organic acids also increases *S. typhimurium* survivors against extreme acid (pH 3.0), high osmolarity (2.5 M NaCl), and reactive oxygen (20 mM hydrogen peroxide). This could also cross-protect *Salmonella* against other environmental conditions in animal and processed food products (Kwon et al., 2000).

Another example is *E. coli* O157:H7, where adaptation to acid may negatively influence the effectiveness of acetic acid spray washings on carcasses. It is known that the acid-resistant strain ATCC 43895 could survive at 4 or 10°C for two to seven days in undiluted 2% lactate (pH 2.3–2.5) or 2% acetate (pH 3.0–3.2) spray washings on beef (Samelis et al., 2002b). Cattle are considered the main reservoir of *E. coli* O157:H7. Spraying of carcasses or even primal meat cuts with organic acids may, therefore, result in the development of acid-resistant strains. A serious concern is that the strains may also harbor increased virulence. Another problem may arise when these strains establish niches in plants to cross-contaminate meat (Samelis et al., 2002a). Acid adaptation of *E. coli* O157:H7 has also reduced the effectiveness of various other organic acid sprays to inactivate it on decontaminated beef carcasses (Berry and Cutter, 2000).

Organic-acid-adapted cells of *E. coli* O157:H7 have been found to survive better than unadapted cells during sausage fermentation, indicating enhanced survival in dry salami (pH 5.0) and apple cider (pH 3.4) (Ryu, Deng, and Beuchat, 1999). However, in other organisms such as *L. monocytogenes*, acid shock with lactic or acetic acid produced no enhanced acid tolerance (Ryu and Beuchat, 1999). Glutamic acid, an amino acid, and a component of pickle brines, is known to enhance acid resistance in *E. coli* and also to protect *E. coli* from specific effects of acetic acid (Breidt, Jr., Hayes, and McFeeters, 2004).

9.4   Acid tolerance of gastrointestinal pathogens

The ability of pathogens to cause disease is greatly influenced by acid. Resistance to acid pH may permit pathogens to survive in acid foods and allow them to survive exposure to acid in the human body and then to proceed in causing disease (Roering et al., 1999; Casey and Condon, 2002). The acid tolerance of *E. coli* O157:H7 makes it especially important to elucidate the effect of organic acids on this pathogen (Casey and Condon,

2002). Pathogens associated with oral–fecal routes of transmission are particularly able to survive in extremely acidic environments (Sainz et al., 2005). Acid tolerance of *S. typhimurium*, for example, has been studied extensively, more than most other pathogenic foodborne bacteria, as *S. typhimurium* is frequently exposed to organic acids during its life cycle (Kwon and Ricke, 1998; Ryu and Beuchat, 1999). These pathogenic bacteria, among others, have to withstand the gastric acidity in the stomach (when ingested together with contaminated food) as well as other stresses imposed by the body's defense systems (Kwon and Ricke, 1998; Gauthier, 2005), and also acidification (pH 4–5) of the phagolysosome when phagocytosed by macrophages.

Many foodborne pathogens such as Shiga toxin-producing *E. coli* O157:H7 (STEC) are capable of surviving at pH < 2.5 for more than 2 h in extremely acidic environments, such as the stomach (pH 2.4). As a result of their acid-resistant nature, STEC O157:H7 has a low infectious dose (Barua et al., 2002). Enteric bacteria must, therefore, be able to withstand acid stress under various conditions, pathogenic, but also natural situations. Acid tolerance plays an important role in their survival and growth in fermented foods. This tolerance is enhanced as a result of acid adaptation (Ricke, 2003). Induction of acid tolerance also has an impact on the virulence of a pathogenic invasive organism (Hill et al., 1995). Extended acid stressing may enhance virulence and also cause adaptive mutations of permanent acid resistance (Sheridan and MacDowell, 1998; Stopforth et al., 2003). Virulence of *E. coli* O157:H7 is much dependent on the ability to survive the low pH of the gastric stomach (Benjamin and Datta, 1995). Although *E. coli* is often grown aerobically in the laboratory, the GIT of humans and animals is strictly anaerobic. Anaerobic conditions are known to enhance acid resistance (Diez-Gonzalez and Russell, 1999).

Tolerance by foodborne pathogens sometimes occurs in specific circumstances, such as when exposure to organic acids is less than maximum and the organism then activates specific stress-resistant and virulent mechanisms, which may result in development of resistance to more severe acidic conditions (Ricke, 2003). *Vibrio cholerae* must also survive exposure to inorganic and organic acids in the human stomach and small intestine. The ability of *V. cholerae* to mount an ATR raises questions concerning the role of this system in pathogenicity. An adaptive response to low pH is, therefore, essential for pathogenic organisms. In *S. typhimurium* a regulator known as PhoPQ has been described. This regulator of pathogenicity is absolutely essential for survival when exposed to acid stress. ToxR is another regulatory protein to modulate the expression of virulence factors in response to environmental conditions such as pH (Merrell and Camilli, 1999).

9.5   Cross-resistance to secondary stresses

Cross-protection of foodborne pathogenic bacteria stressed by subsequent exposure to otherwise lethal environmental conditions enhances their survival and growth potential. For example, mild heat treatment of foods containing pathogens may cause heat-shocked cells that will facilitate survival in acidic foods or in the stomach and intestinal tracts (Tetteh and Beuchat, 2003). Cells that have undergone acid adaptation or acid-shock, differ in their resistance to thermal stress. This stress is also different from that of unadapted cells. Heat-shock in *E. coli* O157:H7 also enhances acid tolerance. This increased heat tolerance of acid-adapted cells has serious practical implications for mild thermal processing techniques, as this may jeopardize the safety of fermented foods (Ryu and Beuchat, 1999). Thermal tolerance as a result of bacterial responses to exposure to an acidic environment has been demonstrated in *L. monocytogenes* (Farber and Pagotto, 1992; Lou and Yousef, 1996) and *S. typhimurium* (Leyer and Johnson, 1993). Acid-shocked cells of *L. monocytogenes* also have significantly greater heat tolerance after exposure to HCl, than with acetic acid (Deng, Ryu, and Beuchat, 1999). Furthermore, some *E. coli* O157:H7 strains can be induced to tolerance to NaCl concentrations, with lactic acid as acidulant, whereas exposure to HCl decreases tolerance of *E. coli* O157:H7 to elevated NaCl levels (Ryu, Deng, and Beuchat, 1999; Casey and Condon, 2002).

Acid-adapted *S. typhimurium* also often acquires enhanced osmotic and salt tolerance and cross-protection to an activated lactoperoxidase system. However, it is not the same mechanism associated with induction of acid adaptation or acid-shock that results in cross-protection against dehydration or osmotic stresses (Ryu, Deng, and Beuchat, 1999). The acetic acid ATR provides protection to osmotic stress, and the lactic acid ATR does not provide protection to osmotic stress (Greenacre and Brocklehurst, 2006). Previous exposure to organic acids also protected *S. typhimurium* against reactive oxygen (hydrogen peroxide), the action of surface-active agents, such as crystal violet, and from other environmental conditions in food animal and processed food products (Hill, O'Driscoll, and Booth, 1995; Kwon et al., 2000).

9.6   Mechanisms of acid tolerance development

The mechanisms involved in ATR development in bacteria are not completely understood. However, it is generally recognized that various genes and proteins, including alternative RpoS and shock proteins are involved (Foster, 2000), as well as modifications in membrane fluidity and fatty acid composition (Brown et al., 1997; Álvarez-Ordóñez et al., 2009). There are two systems known to provide extra protection against inorganic acid stress: (1) RpoS dependent, and (2) PhoPQ dependent. PhoP is an acid-shock protein (Bearson, Wilson, and Foster, 1998). Acid tolerance is not dependent on pH, but is dependent on the growth phase of the cells (Deng, Ryu, and Beuchat, 1999). Induction of acid tolerance has been effectively achieved when a fermentable carbohydrate, such as glucose, is included in a growth medium (Buchanan and Edelson, 1996). The ATR system requires protein synthesis and represents a novel genetic response to environmental stress (Tosun and Aktug Gonul, 2003).

Increased acid tolerance correlates strongly with a decrease in proton accumulation in the cytoplasm. This altered proton permeability is again associated with changes in the protein composition of the cell membranes. Acid-adapted *E. coli* changes the lipid composition of its membranes, and elevated levels of cyclopropane fatty acids are often found. This may mean that changes in the protein composition of the cell membrane are a result of changes in the membrane lipid composition. Both lipid and protein alterations may be necessary to protect a bacterial cell in acidic environments (Jordan, Oxford, and O'Byrne, 1999).

Three systems have been identified in *E. coli* to protect cells against pH values of 2–2.5. The first is a glucose-repressed system which is dependent on an alternative sigma factor (δ^s), encoded by the *rpoS* gene and designated as an oxidative system. The second system requires glutamic acid during acid challenge by utilizing an inducible glutamate decarboxylase. The third one requires arginine and an inducible arginine decarboxylase. All three systems are optimally active in stationary-phase cells (Sainz et al., 2005). Inorganic acid stress causes accumulation of δ^s as a result of down-regulating proteolysis, whereas stress imposed by organic acids will also stimulate *rpoS* translation, but not transcription (Audia, Webb, and Foster, 2001).

The ability of *S. typhimurium* to survive at extreme pH (pH 3.0) is dependent on the acid used to acidify the growth medium (Álvarez-Ordóñez et al., 2009). *S. typhimurium* encounters several low pH environments during its life cycle, and the *cadBA* gene has been implicated in a system responsible for pH homeostasis during exposure to acid stress and has been shown to be composed of a complex cascade of proteins. In *S. typhimurium* the ATR is controlled by many regulatory proteins, one of which is RpoS (Merrell and Camilli, 1999). It is not yet certain what the specific mode of action is by which organic acid increases acid resistance in *S. typhimurium*. However, the pH level may be a critical factor in controlling acid resistance induced by organic acids. Several genes have been identified and characterized in *S. typhimurium* that are implicated in expressing ATR phenotypes. These include *rpoS, phoPQ, mviA, atp,* and *atrB* (Kwon and Ricke, 1998). Genes that have been found to increase acid tolerance in *S. typhimurium* without the cells being adapted, are shown in Table 9.1, together with their actual role in the bacterial cell (Audia, Webb, and Foster, 2001). In the case of *V. cholerae,* it mounts an ATR to organic

Table 9.1 Genes Involved in Increased Acid Tolerance in
Salmonella enterica sv. *typhimurium* Cells Not Adapted to
Low pH

Gene	Original function
aceEF	Pyruvate dehydrogenase
atrA	Unknown
clpP	Protease, degrades RpoS
crp	Regulator
icd	Isocitrate dehydrogenase
pgi(atbR)	Phosphoglucoisomerase
rssB(mviA)	Regulator of RpoS degradation by ClpP

Sources: From Foster and Hall *Journal of Bacteriology* 173:5129–5135, 1991; Bearson, S. M. et al., *Journal of Bacteriology* 180:2409–2417, 1996; Bearson, B.L. et. al., *Journal of Bacteriology*, 178:2572–2579, 1998; Audia et al., *International Journal of Medical Microbiology*, 291:97–106, 2001.

acid-shock and the *toxR* gene is involved in mediating ATR to an organic acid challenge, and different regulatory genes mediate ATR to inorganic and organic acids. Both *cadA* and *toxR* are, however, necessary to mount a productive ATR in response to an organic acid challenge (Merrell and Camilli, 1999). In some *Shigella boydii* strains an arginine decarboxylase has been found, which previously was thought to be an acid-resistance system unique to *E. coli* (Chan and Blaschek, 2005).

pH homeostasis is defined as the ability of an organism to maintain its cytoplasmic pH at a value close to neutrality despite changes in the external pH. An example can be found in *E. coli* where the cytoplasmic pH (pH$_i$) will change by less than 0.1 unit per pH unit change in external pH, providing the range of external pH falls within 4.5–7.9. Homeostasis in bacterial cells is achieved by a combination of passive and active mechanisms (Hill, O'Driscoll, and Booth, 1995).

9.6.1   Passive homeostasis

Impermeability of the membrane to protons and other ions plays a major role in preventing severe changes in pH$_i$ in response to varying environmental changes. Another factor is the high buffering capacity of the cell due to the protein, glutamate, and polyamines (Hill, O'Driscoll, and Booth, 1995).

9.6.2   Active pH homeostasis

Active pH homeostasis depends primarily on the potassium ion and proton circuits. The genetic control of ATR in *S. typhimurium* consists of two stages, the preacid-shock phase and the postacid-shock phase. The pre-acid shock is induced at an external pH of ~pH 5.8, and the postacid shock is induced at or below an external pH 4.0. During postacid shock, several acid-shock proteins are also produced that are essential for survival. When a cell is transferred from pH 7.0 to pH 3.3, these acid-shock proteins are not synthesized and result in cell death. The preacid shock at pH 5.8 is, therefore, necessary to enable the cell to maintain pH homeostasis long enough at pH 3.3 to allow for synthesis of acid-shock proteins (Hill, O'Driscoll, and Booth, 1995).

Facultative pathogenic bacteria use many complex systems to enable them to sense and adapt to environmental stresses. These systems then provide signals for the appropriate expression of genes to encode colonization and pathogenicity factors (Merrell and Camilli, 1999).

9.7   Known acid-tolerant organisms

Acid tolerance has been studied for many years, but mainly on *E. coli* and *Salmonella*, with a few studies on *L. monocytogenes* and other Gram-positive bacteria. Similar to observations in other bacteria, it has been found in *Listeria* that at the same external pH_0, weak organic acids are more damaging than strong inorganic acids such as HCl (Phan-Thanh, Mahouin, and Alige, 2000). Acid adaptation mechanisms, such as preexposing cells to a moderately acidic environment, have been observed in a number of foodborne pathogens, including *L. monocytogenes*, EHEC, *S. typhimurium*, *Shigella flexneri*, *Aeromonas hydrophila*, and *Vibrio vulnificus* (Edelson-Mammel, Porteous, and Buchanan, 2006). *E. coli* O157:H7, *Salmonella*, and other enteric bacteria are generally neutrophiles, but capable of adapting to severe acid stress frequently encountered in animal and human bodies (Lin et al., 1995).

E. coli generally does not readily grow at low pH values (<5.5), but if acid resistance genes are induced, the organism is able to survive at pH values as low as 2.0 (Lin et al., 1995; Diez-Gonzalez and Russell, 1999). Different VTEC serotypes will develop diverse stress resistance when challenged by an acidic environment (Molina et al., 2005). However, since the first recognized outbreak in 1982, *E. coli* O157:H7 has emerged as a serious, potentially life-threatening, human foodborne pathogen (Jordan, Oxford, and O'Byrne, 1999). Relative resistance to acidic environments is a characteristic of enterohemorrhagic *E. coli* and provides the organism the temporary ability to endure acidic environments. This ability is enhanced by the presence of several growth-phase-related and inducible acid-resistance systems (Buchanan et al., 2004). The emergence of *E. coli* O157:H7 as a foodborne pathogen is partly due to its increased acid resistance (Leyer et al., 1995; Uyttendaele, Taverniers, and Debevere, 2001). However, survival and growth are influenced by temperature and the type of acidulant (Connor and Kotrola, 1995; Uyttendaele, Taverniers, and Debevere, 2001). *E. coli* O157:H7 can survive in a variety of acidic foods, and as a result of this organism being involved in foodborne illness out-breaks associated with the consumption of acidic foods, much attention is focused on the acid resistance properties of this pathogen. *E. coli* O157:H7 is often shed from cattle, and the acid tolerance status is significant to microbial safety of meat products (Berry and Cutter, 2000). *E. coli* O157:H7 has a low infectious dose (<100 cfu) and as such poses an even greater problem (Phan-Thanh and Gormon, 1997).

Listeria monocytogenes is another major concern to the food industry, due to its inherent resistance to several preservation practices and the high mortality rate of listeriosis in susceptible populations. Of particular concern is its ability to grow at refrigeration temperatures and on dry sur-faces, and to tolerate acidic conditions (Barker and Park, 2001).

A significant ATR has been found in *S. typhimurium*, following culture in acidic environments (pH 3.0). Modifications in the membrane fatty acid composition of *S. typhimurium*, however, have not been found dependent on the pH or the acid used for acidification (Álvarez-Ordóñez et al., 2009). *Shigella* is another organism known to survive for extended periods in unfavorable conditions such as high acid or high temperature (Lin et al., 1995; Tetteh and Beuchat, 2003).

Although the pathogen *Campylobacter jejuni* cannot survive in extremely acidic conditions (below pH 3.0), it is not affected at pH higher than 3.6 and its normal growth pH range is 5.5–5.7. At low pH levels (4.0 and 4.5) HCl has limited inhibitory effect on *Campylobacter*, and no effect at higher pH levels (5.0 and 5.5). At low pH (4.0 and 4.5) organic acids have a strong bactericidal effect. *C. jejuni* and *C. coli* rapidly die. At pH 5.0–5.5, the reduction rate is much higher, compared to HCl, but *Campylobacter* could survive (Chaveerach et al., 2002).

Lactic acid bacteria (LAB) are more resistant to acidic conditions and can tolerate a lower intracellular pH than many other bacteria (Adams and Nicolaides, 1997). Acetic acid bacteria (AAB) are commonly found in nature and frequent food spoilage microorganisms. Due to their high resistance to acidity and the wide range of substances that they are able to use, their presence is mostly related to food modification and human activities involved in food preservation (De Ley, Gillis, and Swings, 1984). AAB are also important oxidative microorganisms and able to survive in high ethanol conditions. This, together with their acid resistance, are rea-sons why AAB are commonly found in wine (Gonzalez et al., 2005).

Heavy losses are experienced in the feed and beverage industry because of certain yeasts being able to grow at low ambient pH in the presence of weak-acid preservatives (Macpherson et al., 2005). The yeast *Z. bailii* is able to grow at pH as low as 2.2, although it is known that these abilities can be affected by the acidification agent and the solute used to reduce the a_w, as well as by other solutes present in food (Thomas and Davenport, 1985; Lenovich et al., 1988).

Yarrowia lipolytica is a yeast species that has been attracting attention due to its numerous biotechnological applications. *Y. lipolytica* is not considered a dangerous food spoilage yeast, but is often reported from food environments to be involved in some damage as a result of food spoilage. The organism is usually isolated from lipid-rich environments such as mayonnaise and salad dressing as a result of their resistance or tolerance toward weak carboxylic acids and other stress environments. The response pattern to an acidic environment expressed by *Y. lipolytica* resembles *Z. bailii* rather than *S. cerevisiae*, and is related to the capacity of the organism to use most of the acids in the presence of glucose, as described for *Z. bailii* (Rodrigues and Pais, 2000).

9.8 Development of acid tolerance

Acid adaptation is a complicated process in which many physiological changes occur, such as production of protective acid stress proteins and alterations in cell membranes (Deng, Ryu, and Beuchat, 1999). Bacteria can survive in acidic environments, depending on their ability to regulate their cytoplasmic pH (pH$_i$) (Hill, O'Driscoll, and Booth, 1995). Some bacteria such as *L. monocytogenes* have a significant acid adaptive response, following 1 h exposure to mild pH (5.5) (Cheng, Yu, and Chou, 2003). Several factors will determine the ability of a microorganism to survive exposure to an acid environment. These include pH, type of acidulant, concentration of acidulant, temperature, water activity, atmosphere, and the presence of other inhibitory substances (Edelson-Mammel, Porteous, and Buchanan, 2006). Acid adaptation in *S. typhimurium* can be described as a two-stage process, beginning with an initial pre-shock exposure to a mild pH in the range of 5.0-6.0, and then followed by acid challenge exposure to a pH below 4.0 (acid shock) (Tosun and Aktug Gonul, 2003). *Salmonella* also encounters acid stress in the form of organic acids produced by commensal microbes present in the human intestinal tract (Audia, Webb, and Foster, 2001).

In *E. coli* O157:H7 acid tolerance is also strongly dependent on the growth phase, with higher levels of tolerance or resistance found in stationary-phase or starved cultures compared to exponential-phase cultures (Jordan, Oxford, and O'Byrne, 1999). The level of acid tolerance is influenced by extracellular components, often proteins, that are produced in response to chemical and physical stresses (Ryu and Beuchat, 1999). Decreased membrane fluidity may also result in increased acid resistance in *E. coli* O157:H7 (Yuk and Marshall, 2005). *E. coli* O157:H7 cultures, grown anaerobically, will more easily be induced to extreme acid resistance than those grown aerobically (Diez-Gonzalez and Russell, 1999). Acid resistance is an important property of *E. coli* to enable the organism to survive gastric acidity and volatile fatty acids, resulting from fermentation in the intestine (Sainz et al., 2005).

Spraying with diluted solutions of organic acids causes a decrease in the pH value of the meat. Such acidification could result in moderate acidic conditions, leading the way toward ATR development (Álvarez-Ordóñez et al., 2009). Acid-containing spray washings often select for specific types of microorganisms, especially acid-tolerant spoilage bacteria or yeasts, able to assimilate a specific organic acid. Survival may even become higher in number and also last longer with consecutive adaptations (Samelis et al., 2002b). Contaminants on freshly slaughtered beef carcasses typically originate from the bovine gastrointestinal tract, arriving in feces or ingesta from intestinal organs. The normal pH of a rumen is neutral, whereas the human stomach is extremely acidic. However, the corn in the feedlot renders the pH of the rumen unnaturally acidic (Robbins, 2005). Cattle have also been fed high-grain, growth-promoting diets for more than 40 years, with the potential to select for acid-resistant forms (Martins, 2000). Normally, these microorganisms would be killed off by the acids in the human stomach, but the digestive tract pH of modern feedlot cattle may result in acid-resistant strains of *E. coli* emerging from such a manmade environment (Robbins, 2005). Acid adaptation in the bovine gastrointestinal tract may also be prevalent due to exposure to low pH and organic acids (Berry and Cutter, 2000).

9.9   Implications of acid tolerance

Foodborne pathogenic bacteria are commonly exposed to weak organic acids, either as by-products of bacterial metabolism in fermentation processes or as additions to foods as preservatives. ATR expressed in response to these conditions is, therefore, of most relevance to food (Greenacre et al., 2003). Acid-adapted cells, as a result of preservation with organic acids could be more resistant to the strong acidic environment found in the GIT, increasing the risk of foodborne infections (Álvarez-Ordóñez et al., 2009). Organic acids have demonstrated the enhancement of survivability of acid-sensitive pathogens exposed to a low pH, by inducing an acid tolerance response with an associated increase in virulence. This has created a situation that may have implications regarding the use of organic acids. However, this may only apply to circumstances where low acid levels have induced these resistance and virulence mechanisms (Ricke, 2003).

Safety concerns have intensified with regard to the potential implications of pathogen behavior and the microbial ecology of carcass decontamination with organic acid spray washes of meat. On beef decontaminated with lactic acid increases of nearly 3 logs within five days of storage at 10°C under air or vacuum have been recorded, whereas increases were only 1 log and virtually nonexistent on untreated beef. Meat-packing plants that use organic acid treatments may harbor reduced numbers of competitive flora, because of bacteria that survive meat decontamination and establish niches in plants where organic acid rinses are mixed with water washes. It is imperative to recognize and take into consideration the potential long-term effects of organic acid interventions on the microbial ecology of meat-packing plants and products when managing potential safety risks (Samelis et al., 2002b).

9.10   Contribution of acidic foodstuffs

The use of a food source for acid challenge can provide protection against low pH. It is, however, not known how food can protect bacteria from extreme acidic conditions (Álvarez-Ordóñez et al., 2009). One explanation is the presence of amino acids that could play a significant protective role (Lin et al., 1995). Juices and other fruit products are preserved by hurdles such as low pH, depression of water activity, chemical preservatives, and thermal treatment (Hathcox and Beuchat, 1996; Stiles, Duffy, and Schaffner, 2002). However, spoilage yeasts *S. cerevisiae*, *Z. bailii*, and *Candida lipolytica* are able to overcome these hurdles (Thomas and Davenport, 1985; Lenovich et al., 1988). Acid-sensitive enteric pathogens often survive longer in extreme pH conditions when inoculated onto foodstuffs such as ground beef or boiled eggs. The protein content may protect these bacteria against the low pH of gastric fluid. Also under laboratory conditions, *S. typhimurium* was protected against acid inactivation when meat extract was used as a challenge medium (Álvarez-Ordóñez et al., 2009).

Regulations on food additives often do not take into account the amount or type of organic acid present in acidified food (Breidt, Jr., Hayes, and McFeeters, 2004). Because acidified food products have been produced safely for many years without heat treatments (Zagory and Garren, 1999), the organic acid(s) present in these products may have contributed to their excellent safety record. However, independent efficacies of organic acids and pH on the inhibition or killing of pathogens in acidified food products have not been quantitatively measured (Breidt, Jr., Hayes, and McFeeters, 2004). Much more information and guidance is still needed in the food safety arena to prevent contamination by hazardous organisms, such as *E. coli* O157:H7, especially low pH foods (Cheng, Yu, and Chou, 2003). This pathogen has even been implicated in dairy products preserved by acid pH, for example, yogurt and fermented cheeses (Lekkas et al., 2006).

9.11   Analytical procedures

Most studies on the ATR are executed using a mineral acid such as HCl as acidulant (Greenacre et al., 2003). Studies also typically measure survival of cells at an arbitrary time as indicators of ATR, which as a result ignores certain factors of inactivation kinetics, for example, shoulder duration and rate of inactivation. There also exists some concern that pathogens, when part of a mixed microbial culture in a food environment, may react differently to decontamination stresses than their pure cultures in a controlled environment (Samelis et al., 2001b; Stopforth et al., 2003). In the determination of survival and growth characteristics of *E. coli* O157:H7 in foods, the type of acid that the cells have been exposed to, as well as the procedure used to achieve exposure, is important in the design of acid challenge studies (Deng, Ryu, and Beuchat, 1999).

Extreme acid resistance of *E. coli* is triggered by acids and low pH (Goodson and Rowbury, 1989), but the relative importance of each factor still needs to be investigated systematically. It is, however, known that undissociated organic acids, more specifically acetic acid and not the dissociated molecule, can induce extreme acid resistance of *E. coli* O157:H7, where the pH effects are mediated via dissociation of acetate (Diez-Gonzalez and Russell, 1999). It should also be possible to obtain more information about induction of acid-tolerance mechanisms of enteropathogenic *E. coli* and *E. coli* serotype O157:H7. Gene arrays have proven potentially useful in identifying connections between regulatory or metabolic pathways not yet known (Sainz et al., 2005).

9.12   Interacting mechanisms

The type and concentration of organic acid and low pH can either independently affect the growth or death of bacterial cells, or they can interact. There must exist an overlap between genetic systems for ATR and organic acid-induced acid resistance (Kwon and Ricke, 1998). Acid stress is a combination of low pH and the concentration of organic acids present in an environment to which microbial cells are exposed (Bearson, Wilson, and Foster, 1998). There exist at least four overlapping acid resistance systems in *E. coli*, which include the glucose-repressed system and the three amino acid decarboxylase systems (Bjornsdottir, Breidt, Jr., and McFeeters, 2006).

9.13   Control strategies

An important control strategy in the combat against extreme acid-tolerant food pathogens such as *E. coli* O157:H7 is found in decontamination of meat carcasses. The ideal may be to control the extent of exposure of *E. coli* O157:H7 to acid during meat processing and to increase the use of nonacid interventions such as hot water spraying and steam pasteurization, and would be a logical approach to sensitize *E. coli* O157:H7 to acid in the absence of acid stress in food environments. Enhanced meat safety may be promoted by more intense use of water-based decontamination

technologies either alone or in combination (or rotation) with acidic interventions, which would lead to modulation of acid resistance of *E. coli*
O157:H7 as well as other pathogens (Samelis et al., 2002a). Predicting survival and growth of an organism in acidic foods will be valuable in better
understanding the factors that influence adaptation of such organisms to
acid stress (Tetteh and Beuchat, 2003; see Chapter 10 this book). Exposure
to NaCl can also induce marked sensitivity to acid challenge in *E. coli*
(Casey and Condon, 2002).

References

Adams, M.R. and Nicolaides, L. 1997. Review of the sensitivity of different food-
 borne pathogens to fermentation. *Food Control* 8:227–239.
Álvarez-Ordóñez, A., Fernandez, A., Bernardo, A., and Lopez, M. 2009. Comparison
 of acids on the induction of an acid tolerance response in *Salmonella typhimu-
 rium*, consequences for food safety. *Meat Science* 81:65–70.
Audia, J.P., Webb, C.C., and Foster, J.W. 2001. Breaking through the acid barrier:
 An orchestrated response to proton stress by enteric bacteria. *International
 Journal of Medical Microbiology* 291:97–106.
Baik, H.S., Bearson, S., Dunbar, S., and Foster, J.W. 1996. The acid tolerance
 response of *Salmonella typhimurium* provides protection against organic
 acids. *Microbiology* 142:3195–3200.
Barker, C. and Park, S.F. 2001. Sensitization of *Listeria monocytogenes* to low pH,
 organic acids, and osmotic stress by ethanol. *Applied and Environmental
 Microbiology* 67:1594–1600.
Barua, S., Yamashino, T., Hasegawa, T., Yokoyama, K., Torii, K., and Ohta, M. 2002.
 Involvement of surface polysaccharides in the organic acid resistance of Shiga
 toxin-producing *Escherichia coli* O157:H7. *Molecular Microbiology* 43:629–640.
Bearson, B.L., Wilson, L., and Foster, J.W. 1998. A low pH-inducible, PhoPQ-
 dependent acid tolerance response protects *Salmonella typhimurium* against
 inorganic acid stress. *Journal of Bacteriology* 180:2409–2417.
Bearson, S.M., Benjamin, W.H., Jr., Swords, W.E., and Foster, J.W. 1996. Acid-shock
 induction of rpoS is mediated by the mouse virulence gene *mviA* of *Salmonella
 typhimurium*. *Journal of Bacteriology* 178:2572–2579.
Benjamin, M.M. and Datta, A.R. 1995. Acid tolerance of enterohemorrhagic
 Escherichia coli. *Applied and Environmental Microbiology* 61:1669–1672.
Berry, E.D. and Cutter, C.N. 2000. Effects of acid adaptation of *Escherichia coli*
 O157:H7 on efficacy of acetic acid spray washes to decontaminate beef carcass tissue. *Applied and Environmental Microbiology* 66:1493–1498.
Bjornsdottir, K., Breidt, F., Jr., and McFeeters, R.F. 2006. Protective effects of organic
 acids on survival of *Escherichia coli* O157:H7 in acidic environments. *Applied
 and Environmental Microbiology* 72:660–664.
Breidt, F., Jr., Hayes, J.S., and McFeeters, R.F. 2004. Independent effects of acetic
 acid and pH on survival of *Escherichia coli* in simulated acidified pickle products. *Journal of Food Protection* 67:12–18.
Brown, J.L., Ross, T., McMeekin, T.A., and Nichols, P.D. 1997. Acid habituation of
 Escherichia coli and the potential role of cyclopropane fatty acids in low pH
 tolerance. *International Journal of Food Microbiology* 37:163–173.

Buchanan, R.L. and Edelson, S.G. 1996. Culturing enterohemorrhagic *E. coli* in the presence and absence of glucose as a simple means of evaluating the acid tolerance of stationary-phase cells. *Applied and Environmental Microbiology* 62:4009–4013.

Buchanan, R.L., Edelson-Mammel, S.G., Boyd, G., and Marmer, B.S. 2004. Influence of acidulant identity on the effects of pH and acid resistance on the radiation resistance of *Escherichia coli* O157:H7. *Food Microbiology* 21:51-57.

Casey, P.G. and Condon, S. 2002. Sodium chloride decreases the bacteriocidal effect of acid pH on *Escherichia coli* O157:H45. *International Journal of Food Microbiology* 76:199-206.

Chan, Y.C. and Blaschek, H.P. 2005. Comparative analysis of *Shigella boydii* 18 food-borne outbreak isolate and related enteric bacteria: Role of *rpoS* and *adiA* in acid stress response. *Journal of Food Protection* 68:521-527.

Chaveerach, P., Keuzenkamp, D.A., Urlings, H.A., Lipman, L.J., and Van, K.F. 2002. In vitro study on the effect of organic acids on *Campylobacter jejuni/coli* populations in mixtures of water and feed. *Poultry Science* 81:621-628.

Cheng, H.Y., Yu, R.C., and Chou, C.C. 2003. Increased acid tolerance of *Escherichia coli* O157:H7 as affected by acid adaptation time and conditions of acid challenge. *Food Research International* 36:49-56.

Connor, D.E. and Kotrola, J.S. 1995. Growth and survival of *Escherichia coli* O157:H7 under acidic conditions. *Applied and Environmental Microbiology* 61:382–385.

De Ley, J., Gillis, M., and Swings, J. 1984. Family VI. *Acetobacteraceae*. In: N.R. Krieg and J.G. Holt (Eds.), *Bergey's Manual of Systematic Bacteriology*, Vol. 1, pp.267–278. Baltimore: Williams and Wilkins.

Deng, Y., Ryu, J.-H., and Beuchat, L.R. 1999. Tolerance of acid-adapted and non-adapted *Escherichia coli* O157:H7 cells to reduced pH as affected by type of acidulant. *Journal of Applied Microbiology* 86:203–210.

Diez-Gonzalez, F. and Russell, J.B. 1999. Factors affecting the extreme acid resistance of *Escherichia coli* O157:H7. *Food Microbiology* 16:367–374.

Doyle, E.M. 1999. *Use of organic acids to control Listeria in meat*. American Meat Institute Foundation. Washington, DC.

Edelson-Mammel, S., Porteous, M.K., and Buchanan, R.L. 2006. Acid resistance of twelve strains of *Enterobacter sakazakii*, and the impact of habituating the cells to an acidic environment. *Journal of Food Science* 71:M201–M207.

Farber, J.M. and Pagotto, F. 1992. The effect of acid shock on the heat resistance of *Listeria monocytogenes*. *Letters of Applied Microbiology* 15:197–201.

Ferreira, A., Sue, D., O'Byrne, C.P., and Boor, K.J. 2003. Role of *Listeria monocytogenes* sigma(B) in survival of lethal acidic conditions and in the acquired acid tolerance response. *Applied and Environmental Microbiology* 69:2692–2698.

Foster, J.W. 2000. Microbial responses to acid stress. In: G. Storz and R. Hengge-Aronis (Eds.), *Bacterial Stress Responses*, pp. 99–115. Washington, DC: ASM Press.

Foster, J.W. and Hall, H.K. 1991. Inducible homeostasis and the acid tolerance response of *Salmonella typhimurium*. *Journal of Bacteriology* 173:5129–5135.

Gauthier, R. 2005. Organic acids and essential oils, a realistic alternative to antibiotic growth promoters in poultry. *I Forum Internacional de avicultura* 17–19 August, 2005, pp. 148–157.

Gonzalez, A., Hierro, N., Poblet, M., Mas, A., and Guillamon, J.M. 2005. Application of molecular methods to demonstrate species and strain evolution of acetic acid bacteria population during wine production. *International Journal of Food Microbiology* 102:295–304.

Goodson, M. and Rowbury, R.J. 1989. Resistance of acid-habituated *Escherichia coli* to organic acids and its medical and applied significance. *Letters in Applied Microbiology* 8:211–214.

Greenacre, E.J. and Brocklehurst, T.F. 2006. The acetic acid tolerance response induces cross-protection to salt stress in *Salmonella typhimurium*. *International Journal of Food Microbiology* 112:62–65.

Greenacre, E.J., Brocklehurst, T.F., Waspe, C.R., Wilson, D.R., and Wilson, P.D.G. 2003. *Salmonella enterica* serovar Typhimurium and *Listeria monocytogenes* acid tolerance response induced by organic acids at 20°C: Optimization and modeling. *Applied and Environmental Microbiology* 69:3945–3951.

Guilfoyle, D.E. and Hershfield, I.N. 1996. The survival benefit of short-chain organic acids and the inducible arginine and lysine decarboxylase genes for *Escherichia coli*. *Letters in Applied Microbiology.*22:393–396.

Hathcox, A.K. and Beuchat, L.R. 1996. Inhibitory effects of sucrose fatty acid esters, alone and in combination with ethylenediaminetetraacetic acid and other organic acids, on viability of *Escherichia coli* O157:H7. *Food Microbiology* 13:213–225.

Hill, C., O'Driscoll, B., and Booth, I. 1995. Acid adaptation and food poisoning microorganisms. *International Journal of Food Microbiology* 28:245–254.

Jordan, K.N., Oxford, L., and O'Byrne, C.P. 1999. Survival of low-pH stress by *Escherichia coli* O157:H7: Correlation between alterations in the cell envelope and increased acid tolerance. *Applied and Environmental Microbiology* 65:3048–3055.

Kwon, Y.M. and Ricke, S.C. 1998. Induction of acid resistance of *Salmonella typhimurium* by exposure to short-chain fatty acids. *Applied and Environmental Microbiology* 64:3458–3463.

Kwon, Y.M., Park, S.Y., Birkhold, S.G., and Ricke, S.C. 2000. Induction of resistance of *Salmonella typhimurium* to environmental stresses by exposure to short-chain fatty acids. *Journal of Food Science* 65:1037–1040.

Lekkas, C., Kakouri, A., Paleologos, E., Voutsinas, L.P., Kontominas, M.G., and Samelis, J. 2006. Survival of *Escherichia coli* O157:H7 in Galotyri cheese stored at 4 and 12 degrees C. *Food Microbiology* 23:268–276.

Lenovich, L.M., Buchanan, R.L., Worley, N.J., and Restaino, L. 1998. Effect of solute type on sorbate resistance in *Zygosaccharomyces rouxii*. *Journal of Food Science* 53:914–916.

Leyer, G.J. and Johnson, E.A. 1992. Acid adaptation promotes survival of *Salmonella* spp. in cheese. *Applied and Environmental Microbiology* 58:2075-2080.

Leyer, G.J. and Johnson, E.A. 1993. Acid adaptation induces cross-protection against environmental stresses in *Salmonella typhimurium*. *Applied and Environmental Microbiology* 59:1842–1847.

Leyer, G.J., Wang, L.L., and Johnson, E.A. 1995. Acid adaptation of *Escherichia coli* O157:H7 increases survival in acidic foods. *Applied and Environmental Microbiology* 61:3752–3755.

Lin, J., Lee, I.S., Frey, J., Slonczewski, J.L., and Foster, J.W. 1995. Comparative analysis of extreme acid survival in *Salmonella typhimurium*, *Shigella flexneri* and *Escherichia coli*. *Journal of Bacteriology* 177:4097–4104.

Lou, Y. and Yousef, A.E. 1996. Resistance of *Listeria monocytogenes* to heat after adaptation to environmental stresses. *Journal of Food Protection* 59:465–471.

Macpherson, N., Shabala, L., Rooney, H., Jarman, M.G., and Davies, J.M. 2005. Plasma membrane H^+ and K^+ transporters are involved in the weak-acid preservative response of disparate food spoilage yeasts. *Microbiology* 151:1995–2003.

Martins, M.-H. 2000. Debunking the Industrial Agriculture Myth that Organic Foods Are More Likely to Be Carriers of Dangerous Bacteria such as e-Coli 0157:H7 or PlantFungus such as Fuminosin. *Organic Consumers Association.* http://www. organicconsumers.org/Organic/ecolimyths.cfm (accessed September 18, 2005.)

Merrell, D.S. and Camilli, A. 1999. The *cadA* gene of *Vibrio cholerae* is induced during infection and plays a role in acid tolerance. *Molecular Microbiology* 34:836–849.

Molina, P.M., Sanz, M.E., Lucchesi, A., Padola, N.L., and Parma, A.E. 2005. Effects of acidic broth and juices on the growth and survival of verotoxin-producing *Escherichia coli* (VTEC). *Food Microbiology* 22:469–473.

Nakano, S., Fukaya, M., and Horinouchi, S. 2006. Putative ABC transporter responsible for acetic acid resistance in *Acetobacter aceti*. *Applied and Environmental Microbiology* 72:497–505.

Phan-Thanh, L. and Gormon, T. 1997. A chemically defined minimal medium for the optimal culture of *Listeria*. *International Journal of Food Microbiology* 35:91–95.

Phan-Thanh, L., Mahouin, F., and Alige, S. 2000. Acid responses of *Listeria monocytogenes*. *International Journal of Food Microbiology* 55:121–126.

Ricke, S.C. 2003. Perspectives on the use of organic acids and short chain fatty acids as antimicrobials. *Poultry Science* 82:632–639.

Robbins, J. 2005. What about grass-fed beef? *The food revolution: How your diet can help save your life and our world.* http://www.foodrevolution.org/grassfed-beef.htm (accessed August 29, 2005).

Rodrigues, G. and Pais, C. 2000. The influence of acetic and other weak carboxylic acids on growth and cellular death of the yeast *Yarrowia lipolytica*. *Food Technology and Biotechnology* 38:27–32.

Roering, A.M., Luchansky, J.B., Ihnot, A.M., Ansay, S.E., Kaspar, C.W., and Ingham, S.C. 1999. Comparative survival of *Salmonella typhimurium* DT 104, *Listeria monocytogenes,* and *Escherichia coli* O157:H7 in preservative-free apple cider and simulated gastric fluid. *International Journal of Food Microbiology* 46:263–269.

Russell, J.B. 1991. Resistance of *Streptococcus bovis* to acetic acid at low pH: Relationship between intracellular pH and anion accumulation. *Applied and Environmental Microbiology* 57:255–259.

Ryu, J.-H. and Beuchat, L.R. 1999. Changes in heat tolerance of *Escherichia coli* O157:H7 after exposure to acidic environments. *Food Microbiology* 16:317–324.

Ryu, J.-H., Deng, Y., and Beuchat, L.R. 1999. Survival of *Escherichia coli* O157:H7 in dried beef powder as affected by water activity, sodium chloride content and temperature. *Food Microbiology* 16:309–316.

Sainz, T., Perez, J., Villaseca, J., et al. 2005. Survival to different acid challenges and outer membrane protein profiles of pathogenic *Escherichia coli* strains isolated from pozol, a Mexican typical maize fermented food. *International Journal of Food Microbiology* 105:357–367.

Samelis, J., Sofos, J.N., Ikeda, J.S., Kendall, P.A., and Smith, G.C. 2002a. Exposure to non-acid fresh meat decontamination washing fluids sensitizes *Escherichia coli* O157:H7 to organic acids. *Letters in Applied Microbiology* 34:7–12.

Samelis, J., Sofos, J.N., Kendall, P.A., and Smith, G.C. 2002b. Effect of acid adaptation on survival of *Escherichia coli* O157:H7 in meat decontamination washing fluids and potential effects of organic acid interventions on the microbial ecology of the meat plant environment. *Journal of Food Protection* 65:33–40.

Sheridan, J.J. and MacDowell, D.A. 1998. Factors affecting the emergence of pathogens on foods. *Meat Science* 49:S151–167.

Smulders, F.J. and Greer, G.G. 1998. Integrating microbial decontamination with organic acids in HACCP programmes for muscle foods: Prospects and controversies. *International Journal of Food Microbiology* 44:149–169.

Stiles, B.A., Duffy, S., and Schaffner, D. 2002. Modelling yeast spoilage on cold filled ready to drink beverages with *Saccharomyces cerevisiae, Zygosaccharomyces bailii* and *Candida lipolytica*. *Applied and Environmental Microbiology* 68:1901–1906.

Stopforth, J.D., Samelis, J., Sofos, J.N., Kendall, P.A., and Smith, G.C. 2003. Influence of organic acid concentration on survival of *Listeria monocytogenes* and *Escherichia coli* O157:H7 in beef carcass wash water and on model equipment surfaces. *Food Microbiology* 20:651–660.

Takahashi, N., Saito, K., Schachtele, C.-F., and Yamada, T. 1997. Acid tolerance and acid-neutralizing activity of *Porphyromonas gingivalis, Prevotella intermedia* and *Fusobacterium nucleatum*. *Oral Microbiology and Immunology* 12:323–328.

Tetteh, G.L. and Beuchat, L.R. 2003. Exposure to *Shigella flexneri* to acid stress and heat shock enhances acid tolerance. *Food Microbiology* 20:179–185.

Thomas, S. and Davenport, R.R. 1985. *Zygosaccharomyces bailii*—A profile of characteristics and spoilage activities. *Food Microbiology* 2:157–169.

Tosun, H. and Aktug Gonul, S. 2003. Acid adaptation protects *Salmonella typhimurium* from environmental stresses. *Turkish Journal of Biology* 27:31–36.

Uyttendaele, M., Taverniers, I., and Debevere, J. 2001. Effect of stress induced by suboptimal growth factors on survival of *Escherichia coli* O157:H7. *International Journal of Food Microbiology* 66:31–37.

Yuk, H.G. and Marshall, D.L. 2005. Influence of acetic, citric and lactic acids on *Escherichia coli* O157:H7 membrane lipid composition, verotoxin secretion, and acid resistance in simulated gastric fluid. *Journal of Food Protection* 68:673–679.

Zagory, D. and Garren, D. 1999. HACCP: What it is and what it isn't. *The Packer.* August 16, 1999. Online journal. http://www.thepacker.com (accessed August 29, 2005).

chapter ten

Modeling organic acid activity

10.1   Introduction

In initial publications on the subject of "predictive microbiology" (McMeekin et al., 1993) the concept is defined as a quantitative science, enabling the objective evaluation of the influence of processing, distribution, and storage procedures on microbiological safety and also the quality of food. Later, in 2003, another book by the same authors used the expression "quantitative microbial ecology of food," providing a more general description (McMeekin, 2003). In a more recent book (McKellar and Lu, 2003) more emphasis is placed on describing microbial responses to the food environments by mathematical models. This shift in emphasis of a definition for predictive modeling, is a clear indication of its evolution into a more specific science (Baranyi and Roberts, 2004).

In order to reliably calculate microbial behavior, predictive microbiology requires a reliable combination of mathematical and statistical considerations (Roberts, 1995). It is, however, often inappropriate to extrapolate mathematical models used in different applications. In Table 10.1 the differences between mathematical characterization of bacterial growth in food microbiology and mathematical modeling techniques used in biotechnology are stipulated.

There is still not enough understanding of the effectiveness of classical preservatives and "natural" preservatives in concurrence with other common components in food preservation systems. This includes an understanding of whether a microorganism dies, survives, adapts, or grows and also the nature of the physiological and molecular mechanisms within cells that would result in these phenotypes, for example, an understanding of the signal transduction systems involved, the stress proteins induced, induction of these systems, amount of cellular energy (ATP) involved in each system, and also the amount of available energy that will determine the various stress response pathways activated (Brul and Coote, 1999).

It is evident that evaluation of the effectiveness of an organic acid for a specific application will require a much better understanding of general as well as specific stress response potentials of foodborne pathogens (Ricke, 2003). Predictive microbiology may be a handy tool in achieving this via mathematical models to describe the behavior of foodborne microorganisms. The concept has developed very rapidly over the past two decades

225

Table 10.1 Differences between Mathematical Characterization of Bacterial Growth and Mathematical Modeling Techniques Used in Biotechnology

Factor	Mathematical characterization	Modeling techniques in biotechnology
Bacterial concentration	Lower concentration needed	Higher concentration needed
Goal	To prevent growth	To optimize growth
Environmental factors	Often heterogenous with limited quantification	Standardization essential
Nutrient limitation	Foods mainly nutritionally rich, no nutrient limitation of microbial growth	Nutrient limitation essential to control microbial growth

and is increasingly found to be of substantial value to the food industry (Ross and McMeekin, 1994). Amid the development of abundant contemporary quantitative microbiology techniques in recent years, traditional approaches are often still followed for microbial enumeration across various stages of food processing and storage, with the main microbial components identified by their phenotypic characteristics. Although this may be a very fascinating field to a dedicated food microbiologist, it takes time, is costly, and no structured database has been accumulated to provide easily and rapidly accessible information (Baranyi and Roberts, 2004).

Molecular nomenclature refers to growth rate and yield as macroscopic bioenergetic parameters, whereas the microscopic bioenergetic parameters consist of substrate utilization, ATP levels, ATP/ADP ratios, and intracellular redox balance. The link between these parameters and molecular reactions in response to antimicrobial stress has only recently been delineated. A large number of signaling systems have been shown to operate simultaneously in response to, among others, acid preservation regimes, and are readily describable by multivariate models such as neural networks. To understand the links between different branches, there is a need to implement more modern biotechnological tools, such as genomics and proteomics as data sources for utilization in multivariate models. With these techniques rapid assessment can be made of the response and extent of the response to environmental stimuli (Brul and Coote, 1999).

In understanding the underlying "signaling cables" of microbes, it is essential to cautiously examine examples of resistance development, the stress response against natural antimicrobial agents, and also responses against physicochemical factors. Operation of these signaling systems depends on substrates available to provide energy that the cells may require for their housekeeping systems. New insights are implemented to devise knowledge-based combinations of preservation regimes, rather than those based on empirical applications. Risk assessment models are also used to link mechanistic insights to risk management in food manufacturing (Brul et al., 2002a).

The combination of genomic transcript profiling and cellular physiology has provided the potential for developing accurate predictive models for microbial growth inhibition when preserving food via hurdle technologies. This has paved the way in assessing cellular signaling systems involved in the transmission of stress and consequently the regulation of defense systems. Not much information is, however, available on the signals that would initiate a cellular response when microorganisms are exposed to organic acids. It is, however, known that the acids lower the internal pH and have an influence on the cell membrane (Aslim et al., 2005). Therefore, it has been proposed that adjustments of the cell envelope ought to take place to prevent the development of a nonessential ATP-consuming metabolic cycle in which anions and protons are extruded from cells. Reinflux then occurs when the proton and anion reassociate at an extracellular pH equal to the pK. In cells responding to sorbic acid, it was found that stress activated their cell integrity pathway signal transduction system (De Nobel et al., 2001). Closer investigation of induced genes and proteins indicated events at the cellular plasma membrane that must be the start of the signaling cascade (Aslim et al., 2005).

Following studies on the effects of single controlling factors such as temperature, pH, or water activity on microbial growth, it was commonly accepted that specific microorganisms would not grow below a certain temperature or below a certain pH or water activity. The most important environmental factor that would determine growth is the temperature, followed by pH and water activity. Other factors include preservatives, antimicrobials, and atmospheric conditions. The only controllable environmental factor during the storage of food is the temperature. However, these other factors can be changed by the growing bacteria, and can, in turn, also interact. This increases the need for dynamic models, where the constant environment would be regarded as a special case (zero variation) of the general scenario. Any food has a controlling storage factor, which may be water activity, a pH value, or storage temperature. If growth response determined by these factors were to be measured and then modeled, the result would give an indication of the amount of growth that could be attributed to these factors. However, if differences between calculated and observed responses were significant, other factors must also be considered (Baranyi and Roberts, 2004).

10.2 Genomics

Although organic acids have been used for centuries to preserve foods, only recently has the mechanism by which bacterial growth inhibition is achieved been investigated. Genomic and proteomic techniques are increasingly being implemented in identifying genes and proteins in microbes in coping with organic acid stress (Hirshfield, Terzulli, and O'Byrne, 2003). Limited information is available on the molecular signaling that immediately follows exposure to weak-acid stress and also about the initial signal that sets off a cellular response (Brul et al., 2002a). However, molecular tools in the study of pathogen behavior in pre- and postharvest food production are providing insight into the specific microbial genetic regulation involved in the response to antimicrobial agents and preservatives. With molecular techniques, it is also possible to develop more specific strategies to control foodborne pathogens (Ricke, 2003). Genomics and the role it plays in food manufacturing are focused on the mechanistic modeling of stress response in microorganisms. Gathered information can then form part of large sets of databases on food ingredients and food processes. It is here that all physical, biochemical, and nutritional factors with regard to the food being produced, are measured in providing essential information in an integrated way on the prediction of the stability and quality of foods (Brul et al., 2002a).

Sequencing of microbial genomes is being performed on a large scale and has opened the way to a full analysis of microbial behavior. Understanding the cellular physiology depends on proper integration of molecular data, which buttons of a microorganism to push, and the substrates available to cells. A major challenge in preservation and fermented foods is the integration of molecular microbiology and classical physiology into the so-called functional genomics, comprising also the necessary bioinformatics with it (Brul et al., 2002a).

Genomics of food-grade microorganisms, food pathogens, and spoilage organisms, and ultimately also the human body are all relevant to the production of safe foods. Microbial diversity is vast, organisms adapt quickly, and horizontal gene transfer is not exceptional, but with the genomics era comes the promise of great expectations for food biotechnology in the food arena, involving the bioconversion of raw materials into edible products (De Vos, 2001).

Several genes have thus far been implicated in the response to organic acids, resulting in adaptation (Schüller et al., 2004). *Saccharomyces cerevisiae* cells are well known to cope with this stress by inducing Pdr12p, which leads to stress adaptation or resistance development (Piper et al., 1998). Genes upregulated by organic acid stress are primarily involved as resistance factors. The availability of systematic functional assays has been a dynamic achievement of the genomic age. Studies on genes involved in the response and adaptation of fungi to the effects of weak organic acids reported on a combination of functional genomics (phenotypic screening or functional assays) with datasets from mRNA profiles. These can successfully locate essential genes operating in complex biological pathways, such as Pdr12p being the main player in organic acid stress response in yeasts, whereas War1p represents the main stress regulator (Schüller et al., 2004). Underlying mechanisms of the antimicrobial effects of organic acids have not been elucidated, but are at least in part attributed to a cytoplasmic acidification that inhibits essential metabolic functions (Krebs et al., 1983; Pearce, Booth, and Brown, 2001).

It is now also known that only a fraction of a specific genomic response is needed to deal with any hazardous environment and that the transcriptional profile for a given stress condition usually includes many more genes than are immediately required to relieve the stress. For example, adjustment of transcriptional programs will enable a yeast cell to rapidly tune the expression patterns in response to environmental challenges. Combining functional genomic analysis and global expression profiling provides an organized way to get better insight into the different stress response pathways (Schüller et al., 2004). Another advantage of genomics in food preservation studies is the potential for modifying the rate of organic acid production, such as lactic acid, by genetic techniques (Davidson et al., 1995). Genomewide analysis of organic acid resistance, using microarrays and proteomics has also proven to be so much faster than "traditional techniques." Investigating further resistance mechanisms developed by yeasts in particular, are also being facilitated by genetic techniques (Brul et al., 2002b).

Application of genomics may also be beneficial in studies on the virulence of *Salmonella* and the effect of organic acids. *Salmonella* is an opportunistic intracellular pathogen and one of the first steps in pathogenesis is penetration of the intestinal epithelium. This is promoted by invasion genes located on a pathogenicity island (SPI-1). The influence of organic acids on invasion of the epithelium can be explained by changes in SPI-1 expression (Van Immerseel et al., 2006).

Studies on gene expression in *Escherichia coli* K12 during induced acid tolerance response identified gene arrays as having the potential to recognize relations between unknown regulatory or metabolic pathways. With this technique more information is obtained about induction mechanisms of acid tolerance mechanisms in enteropathogenic *E. coli* and *E. coli* O157:H7 (Sainz et al., 2005).

10.3   Growth models in defined systems

Many foodstuffs consist of gelled emulsions, due to deliberate addition of gums and thickeners to increase the mass thickness (e.g., sausages) or due to denaturation of proteins to form protein micelles (e.g., cheese). Food emulsions containing water in oil have an internal water phase that is dispersed as droplets within an oil (or lipid) phase. The microorganisms are mostly found in the droplets (Verrips and Zaalberg, 1980;

Charteris, 1995) and classical theories describing microbial growth usually rely on the microarchitecture of these droplets, which limits the availability of water, space, and growth nutrients (Wilson et al., 2002).

A food surface is a simple form of microarchitecture that affects the growth of microorganisms. Foodstuffs may contain numbers of micro-architectures and the behavior of a microorganism will, therefore, be varied. Differences observed by manufacturers can be explained by organisms growing more slowly in structured systems than in broths (Wilson et al., 2002).

Growth models that describe the dependence of primary model parameters on environmental factors (i.e., temperature, water activity, pH, and organic acids) are referred to as secondary growth models. A number of different secondary model types exist in predictive microbiology and are discussed in the following section (Wilson et al., 2002).

10.4 Different predictive models

Primary models describe the growth, inactivation, or survival of microorganisms as a function of time, whereas secondary model types are broadly classified as (Wilson et al., 2002)

1. Arrhenius-type models
2. Bělehrádek-type models
3. Polynomial models
4. Cardinal models
5. Artificial networks

There should always be a distinction made between static models and dynamic models. Static models are only valid under constant environmental conditions, whereas dynamic models are developed to address time-varying environmental factors (Wilson et al., 2002).

Most effective models thus far have focused on modeling the specific growth rate of an organism. Those attempting to model the population lag phase duration were not as successful (Delignette-Muller et al., 1995; Ross and Olley, 1997; Augustin and Carlier, 2000). One example is a CDC model, designed in ModelMaker®, which has the advantage of permitting combinations of differential end explicit equations as well as distinct measures that are required when applying individual cell adaptations (McKellar and Knight, 2002).

10.4.1   Partial least squares regression (PLS)

The PLS model calculates the principal components through independent variables. It also models the dependent variable as a function of the principal components (Beebe, Pell, and Seascholtz, 1998). This is a rigorous technique that can be used against various occurrences that defy assumptions of multiple linear regression. These include high correlation between independent variables and low ratio of sample to measurement. Cross-validation is also employed by PLS, which refers to the interactive recalculation of the model, while each time excluding a different sample point. With this method the sensitivity of the model is tested toward a particular sample. With principal components analysis four fundamental properties can represent the information contained in other acid data used. The model can also be applied to predict the MIC (minimum inhibitory concentration) of an organic acid for a modeled organism, but this may only be more in the nature of an extrapolation and most of the other data parameters needed for the acid must first be obtained (Hsiao and Siebert, 1999).

10.4.2   Stoichiometric models

When a stoichiometric model is used, local changes in weak-acid concentrations as a result of microbial growth can be predicted. Information on the stoichiometry of the conversion of carbon source to metabolic products is essential in order to relate the increase in cell biomass to the changes in the local chemical environment. This information is then used to predict changes in pH. However, this is not in any way straightforward, because of unknown properties of the buffering capacity of the microbial growth medium (Wilson et al., 2002).

Listeria monocytogenes is increasingly becoming a serious problem in the control of food safety. As a result, various models have been developed and specifically adapted for studies on this pathogen. Of these, various models have focused on the influence of p_0 (continuous adaptation phase) on the growth of individual *L. monocytogenes* cells. An older version of such a model was based on the mean individual cell lag times (t_L), where variability of t_L is expressed as the standard deviation, obtained from Bioscreen turbidimetric data (McKellar and Knight, 2000, 2002). This model was, however, limited due to the individual lag times being preset at t_0 which rendered the model inactive in the lag. A revised model was developed that had a continuous adaptation step incorporated prior to the shift from adaptation phase to exponential growth. This model provided improvement over existing dynamic models, which described the adaptation of homogeneous populations of cells (Alavi et al., 1999).

10.5 Predictive indices for organic acids

Minimum inhibitory concentration (MIC) results for microorganisms have been found to be somewhat higher in practice than the predicted values (Nakai and Siebert, 2003). In the development of regression models of organic acid MIC as a function of acid properties, developed by Hsiao and Siebert (1999) for six bacteria, MIC values were successfully predicted for three acids, different from those used to build the models. This was a clear indication that the models have reasonably good predictive ability (Nakai and Siebert, 2003).

In another study, prediction models based on MICs of eight organic acids (acetic, benzoic, butyric, caprylic, citric, lactic, malic, and tartaric acids) were quite successful for each of six different bacteria. R^2 values were found to range from 0.621 to 0.966, and the susceptibility pattern and at least two patterns of acid resistance (E. coli-type and Lactobacillus-type) were apparent (Nakai and Siebert, 2004). These models were also validated by predictions made of the MICs of acids other than those used in the construction of the models. Actual MICs are determined and can then be compared with the predictions (Nakai and Siebert, 2003).

Quantitative determination of organic acids is an essential factor in food control. It is important in the monitoring of bacterial growth and activity, as well as for nutritional reasons (Tormo and Izco, 2004). Mathematical modeling is used to describe growth responses of microorganisms to combinations of preservative factors (McMeekin et al., 1993; McClure et al., 1994). Combinations that are often used include different preservatives (often a combination of organic acids), temperature, and pH (George, Richardson, and Peck, 1996). Other factors that may influence organic acid are (Diez-Gonzalez and Russell, 1997; Stratford and Anslow, 1998; Alakomi et al., 2000; Breidt, Jr., Hayes, and McFeeters, 2004):

1. Specific effects of the acid or acid anion on cellular enzymes or membranes.
2. Internal pH values and buffering capacity of cells.
3. Proton pumping at the expense of cellular atp, and facilitated transport of acid molecules.

Many fields of predictive microbiology have been developed and applied to the food industry over the last decades in controlling food safety and quality, and in understanding the behavior of microorganisms in response to this process (Coroller, Guerrot, and Huchet, 2005). With predictive modeling, information is provided on the interactions between two or more factors in a bacterial growth environment. It also permits interpolation in a region of observations (George, Richardson, and Peck, 1996). Models are used to predict microbial growth in food, and depending on environmental factors, the prediction is made by the complexity of the foodstuff (Coroller, Guerrot, and Huchet, 2005). Such predictions reduce the need for challenge testing in food safety (George, Richardson, and Peck, 1996). The influence of a preservative such as an organic acid or acid mixture has to be included for a model to be complete (Coroller, Guerrot, and Huchet, 2005).

Food preservation has for a long time employed a combination of factors in which acids have played an important role (Knochel and Gould, 1995). It is, therefore, of great interest to effectively apply the combined effects of several factors that inflict obstructions to the growth of microbes (Chirife and Favetto, 1992). Studies on a specific organism, and inclusion of several factors, such as water activity, temperature, pH, and acid concentration, have often been carried out in model systems, and mathematical models have been developed describing the behavior (Hsiao and Siebert, 1999).

Different organisms demonstrate different susceptibilities for the inhibitory effects of organic acids (Matsuda et al., 1994). Substantial variations are found in the inhibition effects of organic acids, because they are not members of a homologous series and vary in numbers of carboxy groups, hydroxy groups, and carbon–carbon double bonds in the molecule. It is, therefore, not possible to predict either the magnitude of change when one organic acid is substituted for another, or even the direction of the change. As a result, it has not been possible to predict what will happen in real food with a mixture of different organic acids. It is, however, possible to model the inhibitory effect of an organic acid as a function of its molecular properties (Hsiao and Siebert, 1999).

In 1999, Hsiao and Siebert applied the principal components analysis model to 11 different properties of 17 organic acids. Results from this study enabled them to construct four fundamental scales (the first four principal components after Varimax rotation). The points of each of the 17 organic acids on each of these four scales represented their relationships. These were referred to as "principle properties" and could then be used to construct multiple regression models relating the MIC of an acid to its principal properties of a set of organisms (Nakai and Siebert, 2004).

The following equation applied to the model:

$$MIC = b_0 + b_1 PC_1 + b_2 PC_2 + b_3 PC_3 + b_4 PC_4$$

(where the bs are fitted regression coefficients, and the PCs are the four principal component scores for an acid).

10.6   Toward improving on existing models

In 1991, Cole commented that the models used at that time were merely descriptive and practical. Nevertheless, premature modeling may actually be the most appropriate means of quantifying interactions as well as synergism between various factors (Cole, 1991). A very important factor in preservation processes is the ability to predict safe limits in the application of organic acids. This has already provided opportunity for the development of empirical models (Stiles Battey, Duffy, and Schaffner, 2002). However, this was followed by a more mechanistic approach to predict the behavior of food spoilage yeasts in the presence of preservatives. Such an approach uses a mathematical model based on physiological observations and analyzes the decrease in growth rate of yeasts, induced by organic acids in terms of changes occurring in yield and consumption rates of the energy source, in a well-characterized medium (Quintas et al., 2005).

Following this model, the need emerged to develop a more structured mathematical model in order to relate the changes in growth rate with increased energy requirements resulting from preservative stress. As a first step to setting up a model that could predict the influence of organic acids on the growth of yeasts, benzoic acid was used and evaluated with regard to its effect on the kinetic and energetic parameters of growth. The model was found to be able to predict growth of spoilage yeasts within the concentration range of the organic acid, but could not predict the ultimate rate of spoilage. This may be due to the fact that spoilage depends not only on the yeast population, but also on their metabolic activity (Quintas et al., 2005).

The Gamma concept of predictive microbiology is described as a cornerstone hypothesis, this being that different antimicrobial effects add together independently (Zwietering et al., 1992), and should, in turn, assist in additional efforts to model the growth of other microorganisms under various conditions (Lambert and Bidlas, 2007). Two strong views have been suggested with respect to the effect of combined primary environmental factors: the first is that "interactions exist,"and with the other "interactions do not exist." Another suggestion has been proposed, which states that interactions already exist between factors at the growth–no growth interface, but in other places they are additive (McMeekin et al., 2000; Le Marc et al., 2002).

Sorbic acid and benzoic acid are often used in combination, due to acclaimed antimicrobial synergy between them (Cole and Keenan, 1986). This synergy has been attributed to their difference in pK_a values (4.67 vs. 4.19, respectively), and lipid solubility. Analysis by the Gamma hypothesis shows a good fit of the reciprocal model with other organic acids, and the parameters obtained are in general agreement with those previously found. However, the question arises whether the Gamma hypothesis may be discarded without a fundamental basis in synergy, especially in light of opposing views as to whether such interactions exist (Lambert and Bidlas, 2007).

In 1983, Eklund developed a mathematical model for the antimicrobial activity of organic acids, which described the antimicrobial action of the dissociated as well as the undissociated organic acid. In contrast to a model assuming the activity of the acid form only, this model provided a good description of the actions of a variety of organic acids (Eklund, 1985). The model was suggested to have practical value, because the determination of MICs of a specific substance at only two different pH levels could be used to predict the MIC of that same substance at other pH levels. However, the model failed to consider the antimicrobial effect of the low pH on its own (Lambert and Bidlas, 2007).

Prediction of the log reduction of an inoculated organism as a function of acid concentration, time, and temperature can also be done by a mathematical model developed for this purpose, using the second-order polynomial equation to fit the data. The following tests justified the reliability of the model: the analysis of variance for the response variable indicated that the model was significant ($P \leq 0.05$ and $R^2 = 0.9493$) and had no significant lack of fit ($P > 0.05$). Assumptions underlying the ANOVA test were also investigated and it was demonstrated that with the normal probability plot of residuals, plot of residuals versus estimated values for the responses, and plot of residuals versus random order of runs, that the residuals satisfied the assumptions of normality, independence, and randomness (Jimenez et al., 2005).

In 2003 a square root-type model for *E. coli* growth in response to temperature, water activity, pH, and lactic acid was developed by Ross et al. (2003). The model was found to predict well for 1025 growth rate estimates reported in the literature, by exclusion of poor quality or unrepresented data ($n = 215$), with a bias factor of 0.92, and an accuracy factor of 1.29. This model performed better than two other predictive models for *E. coli* growth, *Pathogen Modeling Program*, and *Food MicroModel*, and consistently gave better predictions at generation times ≤5 h. The lactic acid term included in the model has been proposed as being responsible for the consistent performance for comparisons to growth in meat, as this parameter is not explicitly included in all other models (Mellefont, McMeekin, and Ross, 2003).

Despite the arrival of rapid genetic as well as immunological techniques for the detection of foodborne pathogens, evaluation of the microbiological quality and safety of foods is usually done in retrospect and is not in reality effective in protecting consumers from foodborne hazards. There still exists a predictive approach. As a result, predictive modeling has developed as an addition to traditional microbiological techniques (Mellefont, McMeekin, and Ross, 2003).

In developing an appropriate predictive model, factors that are measurable and relevant to the food and the conditions that may be encountered during the whole process of manufacturing through to consumption must be included. Temperature, water activity (a_w), and pH, are considered to have the predominant influence on microbial growth in fresh meats (Nottingham, 1982). Successful application of predictive modeling is dependent on the development of appropriate models. Prior to application in industry, predictive models must be evaluated under novel conditions not used to derive the models. There are many methods by which the performance of a model can be assessed, such as the use of subsets of data from which the model is derived. This then serves to generate new data by either growth in liquid growth media or by direct inoculation into a food product. Other methods include comparison to data from the literature and industrial trials (Mellefont, McMeekin, and Ross, 2003).

An example of a response surface model of *L. monocytogenes* growth rate under different temperatures and acid concentrations was able to show that growth inhibition was more effective with citric acid than ascorbic acid because the major dissociation of citric acid occurs inside microbial cells. Different conditions in the model have the potential to allow *L. monocytogenes* response prediction in foods consisting of various chemical and physical factors. There are two crucial steps in evaluating a predictive model (Carrasco et al., 2006):

1. Ensuring that the model accurately describes the data from which it has been generated and represents its biological trend
2. Comparing the prediction with other data

Data should also be obtained under well-controlled experimental conditions (Ross, 1996).

10.7   Significance of modeling

In food preservation it is essential to control microorganisms throughout the food chain. There is, therefore, an inevitable need to predict the behavior of microorganisms and to get an insight to their cellular functioning under conditions generally encountered under relevant food manufacturing and preservation scenarios. Optimal use of this knowledge can only be made if this data is placed in context with single cell versus cell population behavior as well as the context of the physicochemical parameters used to determine food taste, flavor, and nutritional value (Brul et al., 2002a).

Translation of cellular stress models needs to be linked to the cellular physiology that occurs in single cells. This will ensure that the probability of events in terms of modeling will play a prime role. These events include the amount of transcripts and, therefore, molecules of certain transcription factors regulating the onset of stress response routes (Brul et al., 2002a).

As a result of frequent food poisoning outbreaks, often including those attributable to *L. monocytogenes*, the food industry, government authorities, and the public are questioning food preservation methods (Carrasco et al., 2006). The continuing demand for natural and minimally processed food is also causing an increased interest in naturally produced antimicrobial agents (McEntire, Montville, and Chikindas, 2003). The effects of these compounds in relation to other factors, such as temperature, NaCl, water activity, and pH are included in the development of several mathematical models (Carrasco et al., 2006).

It is essential to have a clear understanding of the effect of a popular organic acid such as sorbic acid on the molecular physiology of yeast cells, in order to provide the food industry with the necessary knowledge to develop strategies to continue to make optimal use of the application of sorbic acid as a preservative. In the design of improved preservation strategies, knowledge of how a cell responds and often becomes resistant to preservatives is important (Papadimitriou et al., 2007). Sorbic acid is increasingly being used in intermediate-moisture foods, low pH foods, and also shelf-stable products (Guilbert, Gontard, and Gorris, 1996). It is, therefore, important to be able to predict its migration between phases during

1. Treatment of various foods (e.g., absorption of sorbic acid during processing of dried prunes, or loss of sorbic acid during cooking of prepared foods).
2. Storage of complex foods (e.g., dairy products or cakes that contain acid-treated fruits).
3. Storage of food that is in contact with sorbic acid-containing wrapping materials.
4. Storage of foods coated with an edible layer containing a high level of sorbic acid.

The general understanding of factors limiting microbial growth and their contribution(s) to safe and shelf-stable food products is limited, despite the availability of a wide literature, particularly with regard to conditions in modern food distribution systems (Roberts, 1995).

When different inhibitory effects on microbial activity need to be compared, it is also necessary to determine the parameters that would state the influence of the antimicrobial compound (Ferreira, Loureiro-Dias, and Loureiro, 1997).

References

Alakomi, H.L., Dkytta, E., Saarela, M., Mattila-Sandholm, T., Latva-Kala, K., and Helander, I.M. 2000. Lactic acid permeabilizes Gram-negative bacteria by disrupting the outer membrane. *Applied and Environmental Microbiology* 66:2001–2005.

Alavi, S.H., Puri, V.M., Knabel, S.J., Mohtar, R.H., and Whiting, R.C. 1999. Development and validation of a dynamic growth model for *Listeria monocytogenes* in fluid whole milk. *Journal of Food Protection* 62:170–176.

Aslim, B., Yuksekdag, Z.N., Sarikaya, E., and Beyatli, Y. 2005. Determination of the Bacteriocin-like substances produced by some lactic acid bacetria isolated fromTurkish dairy products. *Swiss Society of Food Science and Technology* 38:691–694.

Augustin, J.C. and Carlier, V. 2000. Mathematical modelling of the growth rate and lag time for *Listeria monocytogenes*. *International Journal of Food Microbiology* 56:29–51.

Baranyi, J. and Roberts, T.A. 2004. Predictive microbiology—Quantitative microbial ecology. *Culture*, 25:14–16.

Beebe, K.R., Pell, R.J., and Seascholtz, M.B. 1998. In: *Chemometrics: A Practical Guide*, p. 348. New York: John Wiley and Sons.

Breidt, F., Jr., Hayes, J.S., and McFeeters, R.F. 2004. Independent effects of acetic acid and pH on survival of *Escherichia coli* in simulated acidified pickle products. *Journal of Food Protection* 67:12–18.

Brul, S. and Coote, P. 1999. Preservative agents in foods. Mode of action and microbial resistance mechanisms. *International Journal of Food Microbiology* 50:1–17.

Brul, S., Coote, P., Oomes, S., Mensonides, F., Hellingwerf, K., and Klis, F. 2002a. Physiological actions of preservative agents: Prospective of use of modern microbiological techniques in assessing microbial behaviour in food preservation. *International Journal of Food Microbiology* 79:55–64.

Brul, S., Klis, F.M., Oomes, S.J.C.M., et al. 2002b. Detailed process design based on genomics of survivors of food preservation processes. *Trends in Food Science and Technology* 13:325–333.

Carrasco, E., Garcia-Gimeno, R., Seselovsky, R., et al. 2006. Predictive model of *listeria monocytogenes'* growth rate under different temperatures and acids. *Food Science and Technology International* 12:47–56.

Charteris, W.P. 1995. Physicochemical aspects of the microbiology of edible table spreads. *Journal of the Society of Dairy Technology* 48:87–96.

Chirife, J. and Favetto, G.J. 1992. Some physico-chemical basis of food preservation by combined methods. *Food Research International* 25:389–396.

Cole, M.B. 1991. Predictive modeling—Yes it is! *Letters in Applied Microbiology* 13:218–219.

Cole, M.B. and Keenan, M.H.J. 1986. Synergistic effects of weak-acid preservatives and pH on the growth of *Zygosaccharomyces bailii*. *Yeast* 2:93–100.

Coroller, L., Guerrot, V., and Huchet, V. 2005. Modelling the influence of single acid and mixture on bacterial growth. *International Journal of Food Microbiology* 100:167–178.

Davidson, B.E., Llanos, R.M., Cancilla, M.R., Redman, N.C., and Hillier, A.J. 1995. Current research on the genetics of lactic acid production in lactic acid bacteria. *International Dairy Journal* 5:763–784.

Delignette-Muller, M.L., Rosso, L., and Flandrois, J.P. 1995. Accuracy of microbial growth predictions with square root and polynomial models. *International Journal of Food Microbiology* 27:139–146.

De Nobel, H., Lawrie, L., Brul, S., et al. 2001. Parallel and comparative analysis of the proteome and transcriptome of weak-acid adapted *Saccharomyces cerevisiae* reveals a crucial function for the small heat-shock protein Hsp26. *Yeast* 18:1413–1428.

De Vos, W.M. 2001. Advances in genomics for microbial food fermentations and safety. *Current Opinion in Biotechnology* 12:493–498.

Diez-Gonzalez, F. and Russell, J.B. 1997. The ability of *Escherichia coli* O157:H7 to decrease its intracellular pH and resist the toxicity of acetic acid. *Microbiology* 143:1175–1180.

Eklund, T. 1985. Inhibition of microbial growth at different pH levels by benzoic and propionic acids and esters of *p*-hydroxybenzoic acid. *International Journal of Food Microbiology* 2:159–167.

Ferreira, M.M., Loureiro-Dias, M.C., and Loureiro, V. 1997. Weak acid inhibition of fermentation by *Zygosaccharomyces bailii* and *Saccharomyces cerevisiae*. *International Journal of Food Microbiology* 36:145–153.

George, S.M., Richardson, L.C., and Peck, M.W. 1996. Predictive models of the effect of temperature, pH and acetic and lactic acids on the growth of *Listeria monocytogenes*. *International Journal of Food Microbiology* 32:73–90.

Guilbert, S., Gontard, N., and Gorris, L.G.M. 1996. Prolongation of the shelf-life of perishable food products using biodegradable films and coatings. *Lebensmittel-Wissenschaft und-Technologie* 29:10–17.

Hirshfield, I.N., Terzulli, S., and O'Byrne, C. 2003. Weak organic acids: A panoply of effects on bacteria. *Science Progress* 86:245–269.

Hsiao, C.P. and Siebert, K.J. 1999. Modeling the inhibitory effects of organic acids on bacteria. *International Journal of Food Microbiology* 47:189–201.

Jimenez, S.M., Destefanis, P., Salsi, M.S., Tiburzi, M.C., and Pirovani, M.E. 2005. Predictive model for reduction of *Escherichia coli* during acetic acid decontamination of chicken skin. *Journal of Applied Microbiology* 99:829–835.

Knochel, S. and Gould, G. 1995. Preservation microbiology and safety: Quo vadis? *Trends in Food Science and Technology* 6:127–131.

Krebs, H.A., Wiggins, D., Stubbs, M., Sols, A., and Bedoya, F. 1983. Studies on the mechanism of the antifungal action of benzoate. *Biochemical Journal* 214:657–663.

Lambert, R.J. and Bidlas, E. 2007. An investigation of the Gamma hypothesis: A predictive modelling study of the effect of combined inhibitors (salt, pH and weak acids) on the growth of *Aeromonas hydrophila*. *International Journal of Food Microbiology* 115:12–28.

Le Marc, Y., Huchet, V., Bourgeois, C.M., Guyonnet, J.P, Mafart, P., and Thuault, D. 2002. Modeling the growth kinetics of *Listeria* as a function of temperature, pH and organic acid concentration. *International Journal of Food Microbiology* 73:219–237.

Matsuda, T., Yano, T., Maruyama, A., and Kumagai, H. 1994. Antimicrobial activities of organic acids determined by minimum inhibitory concentrations at different pH ranges from 4.0 to 7.0. *Journal of the Japanase Society for Food Science and Technology* 41:687–701.

McClure, P.J., Blackburn, C. de W., Cole, M.B., et al. 1994. Modeling the growth, survival and death of microorganisms in foods: The UK Food Micromodel approach. *International Journal of Food Microbiology* 23:265–275.

McEntire, J.C., Montville, T.J., and Chikindas, M.L. 2003. Synergy between nisin and select lactates against *Listeria monocytogenes* is due to the metal cations. *Journal of Food Protection* 66:1631–1636.

McKellar, R.C. and Knight, K.P. 2000. A combined discrete-continuous model describing the lag phase of *Listeria monocytogenes*. *International Journal of Food Microbiology* 54:171–180.

McKellar, R.C., Lu, X., and Knight, K.P. 2002. Proposal of a novel parameter to describe the influence of pH on the lag phase of *Listeria monocytogenes*. *International Journal of Food Microbiology* 73:127–135.

McKellar, R.C. and Lu, X. (Eds.) 2003. *Modelling Microbial Responses in Foods*. Boca Raton, FL: CRC Press.

McKellar, R.C., Lu, X., and Knight, K.P. 2002. Growth pH does not affect the initial physiological state parameter (p_0) of *Listeria monocytogenes*. *International Journal of Food Microbiology* 73:137–144.

McMeekin, T.A. 2003. An essay on the unrealized potential of predictive microbiology. In *Modelling Microbial Responses in Food*, McKellar, R.C. and Lu, X. (Eds.), pp. 231–235. Boca Raton, FL: CRC Press.

McMeekin, T.A., Olley, J.N., Ross, T., and Ratkowsky, D.A. 1993. *Predictive Microbiology: Theory and Application*. Taunton, Somerset, UK: Research Studies Press.

McMeekin, T.A., Presser, K., Ratkowsky, D., Ross, T., Salter, M., and Tienungoon, S. 2000. Quantifying the hurdle concept by modeling the bacterial growth/no growth interface. *International Journal of Food Microbiology* 55:93–98.

Mellefont, L.A., McMeekin, T.A., and Ross, T. 2003. Performance evaluation of a model describing the effects of temperature, water activity, pH and lactic acid concentration on the growth of *Escherichia coli*. *International Journal of Food Microbiology* 82:45–58.

Nakai, S.A. and Siebert, K.J. 2003. Validation of bacterial growth inhibition models based on molecular properties of organic acids. *International Journal of Food Microbiology* 86:249–255.

Nakai, S.A. and Siebert, K.J. 2004. Organic acid inhibition models for *Listeria innocua*, *Listeria ivanovii*, *Pseudomonas aeruginosa* and *Oenococcus oeni*. *Food Microbiology* 21:67–72.

Nottingham, P.M. 1982. Microbiology of carcass meats. In: M.H. Brown, *Meat Microbiology*, pp. 13–66. London: Applied Science.

Papadimitriou, M.N., Resende, C., Kuchler, K., and Brul, S. 2007. High Pdr12 levels in spoilage yeast (*Saccharomyces cerevisiae*) correlate directly with sorbic acid levels in the culture medium but are not sufficient to provide cells with acquired resistance to the food preservative. *International Journal of Food Microbiology* 113:173–179.

Pearce, A.K., Booth, I.R., and Brown, A.J. 2001. Genetic manipulation of 6-phosphofructo-kinase and fructose 2,6-biphosphate levels affects the extent to which benzoic acid inhibits the growth of *Saccharomyces cerevisiae*. *Microbiology* 147:403–410.

Piper, P., Mahé, Y., Thompson, S., et al. 1998. The Pdr12 ABC transporter is required for the development of weak organic acid resistance in yeast. *EMBO Journal* 17:4257–4265.

Quintas, C., Leyva, J.S., Sotoca, R., Loureiro-Dias, M.C., and Peinado, J.M. 2005. A model of the specific growth rate inhibition by weak acids in yeasts based on energy requirements. *International Journal of Food Microbiology* 100:125–130.

Ricke, S.C. 2003. Perspectives on the use of organic acids and short chain fatty acids as antimicrobials. *Poultry Science* 82:632–639.

Roberts, T.A. 1995. Microbial growth and survival: Developments in predictive modelling. *International Biodeterioration and Biodegradation* 36:297–309.

Ross, T. 1996. Indices for performance evaluation of predictive models in food microbiology. *Journal of Applied Bacteriology* 81:501–508.

Ross, T. and McMeekin, T.A. 1994. Predictive microbiology. *International Journal of Food Microbiology* 23:241–264.

Ross, T. and Olley, J. 1997. Problems and solutions in the application of predictive microbiology. In: F. Shahidi, Y. Jones, and D.D. Kitt (Eds.), *Seafood Safety, Processing and Biotechnology*, pp. 101–118. Lancaster, PA: Technomic.

Ross, T., Ratkowski, D.A., Mellefont, L.A., and McMeekin, T.A. 2003. Modelling the effect of temperature, water activity, pH and lactic acid concentration on the growth rate of *Escherichia coli*. *International Journal of Food Microbiology* 82:33–43.

Sainz, T., Perez, J., Villaseca, J., et al. 2005. Survival to different acid challenges and outer membrane protein profiles of pathogenic *Escherichia coli* strains isolated from pozol, a Mexican typical maize fermented food. *International Journal of Food Microbiology* 105:357–367.

Schüller, C., Mamnun, Y.M., Mollapour, M., et al. 2004. Global phenotypic analysis and transcriptional profiling defines the weak acid stress response regulon in *Saccharomyces cerevisiae*. *Molecular Biology of the Cell* 15:706–720.

Stiles Battey, A., Duffy, S., and Schaffner, D.W. 2002. Modeling yeast spoilage in cold-filled ready-to-drink beverages with *Saccharomyces cerevisiae*, *Zygosaccharomyces bailii*, and *Candida lipolytica*. *Applied and Environmental Microbiology* 68:1901–1906.

Stratford, M. and Anslow, P.A. 1998. Evidence that sorbic acid does not inhibit yeast as a classic 'weak acid preservative'. *Letters in Applied Microbiology* 27:203–206.

Tormo, M. and Izco, J.M. 2004. Alternative reversed-phase high-performance liquid chromatography method to analyse organic acids in dairy products. *Journal of Chromatography A* 1033:305–310.

Van Immerseel, F., Russell, J.B., Flythe, M.D., et al. 2006. The use of organic acids to combat *Salmonella* in poultry: A mechanistic explanation of the efficacy. *Avian Pathology* 35:182–188.

Verrips, C.T. and Zaalberg, J. 1980. The intrinsic microbial stability of water-in-oil emulsions: I. Theory. *European Journal of Applied Microbiology and Biotechnology* 10:187–196.

Wilson, P.D.G., Brocklehurst, T.F., Arino, S., et al. 2002. Modelling microbial growth in structured foods: Towards a unified approach. *International Journal of Food Microbiology* 73:275–289.

Zwietering, M.H., Wijtzes, T., De Wit, J.C., and Van't Riet, K. 1992. A decision support system for prediction of the microbial spoilage in foods. *Journal of Food Protection* 55:973–979.

chapter eleven

Legislative aspects

11.1 Introduction

Newly recognized types of foodborne illnesses are emerging globally. These include verocytotoxin-producing *Escherichia coli* (VTEC) infections, *E. coli* O157:H7, listeriosis, and other bacterial infections, and pose serious challenges to public health authorities. Food contamination, particularly with pathogenic organisms plays an increasingly important role in the economy of countries all over the world. It may even have devastating effects on their foreign trade, as was the case in Peru during the early stages of the cholera epidemic in the beginning of the 1990s (Petrera and Montoya, 1992). Microbial contamination of foods may even result in serious economic losses to industries, even if the hazard were identified before consumers were affected (Molins, Motarjemi, and Käferstein, 2001). The high cost of foodborne illness to society has been shown in numerous studies in different countries, including the United States (Buzby et al., 1996) and Canada (Dean, 1990; Unnevehr and Jensen, 1999). Pathogenic bacteria, such as *Salmonella* and *Campylobacter* are commonly found in poultry and this problem is worsened by modern mass rearing practices (Molins, Motarjemi, and Käferstein, 2001). Various foodstuffs may be implicated in foodborne illness outbreaks (Smith deWaal, 2003) and the most notorious are summarized in Table 11.1.

Throughout the white papers on food safety the guiding principle is that food safety policies should be based on an all-inclusive, integrated approach throughout the whole food chain, across all food sectors, among the member states, outside and inside the European Union, in international and E.U. decision-making forums, and at all stages of the policy-making cycle (Domenech, Escriche, and Martorell, 2007). Today's E.U. policy is to implement improved food quality and safety. The European Union's first priority is quality and not quantity. The White Paper on Food Safety, therefore, proposes a series of measures for the production of safe foods (Drosinos, Mataragas, and Metaxopoulos, 2005).

11.2 Differences in regulatory authorities

European Union: Strictly hygienic processing is proposed to adequately ensure product safety. As a result, the European Union regards

Table 11.1 Foodstuff Commonly Associated with Foodborne Illness Outbreaks

Food linked to outbreak	Additional information
Seafood	Several have been linked to shellfish, such as oysters, clams, mussels
Eggs and egg dishes	
Beef	Very often ground beef is implicated
Produce	Cantaloupe, tomatoes, strawberries, watermelon, potatoes, scallions, lettuce, raspberries, sprouts, basil, parsley
Poultry	Although outbreaks linked to poultry are less common than those linked to beef, more individual cases of foodborne diarrhea linked to poultry than other food
Dairy products	Cheese, pasteurized milk, raw milk, ice cream
Pork	Ham, pork sausage
Game	Venison, bear meat, cougar meat
Juices	Apple cider, apple juice, orange juice
Lunch meats	Hot dogs, bologna
Foods with multiple ingredients	

Source: Data from Smith DeWaal, *Food Control* 14:75–79, 2003.

decontamination strategies as a means of concealing poor hygiene (Smulders and Greer, 1998).

U.S.D.A.: Propose that even with the best current hygienic practices contamination of the carcass is foreseeable. The U.S.D.A. regards interventions as an addition to the Hazard Analysis and Critical Control Point (HACCP) in meeting performance standards in an effort to reduce pathogens (Smulders and Greer, 1998). HACCP is a production control system that has been established for the food industry, focused on food safety prevention systems (Cross, 1994). Decontamination technologies of animal carcasses are increasingly used commercially in North America to control microbial surface contamination of meat, specifically aimed at *E. coli* O157:H7 (Smulders and Greer, 1998; Sofos and Smith, 1998; Samelis et al., 2002a). Organic acids are widely used in the USA for this purpose, but are not permitted under E.U. regulations (Smulders and Greer, 1998; Bolton, Doherty, and Sheridan, 2001).

The use of organic acids is not practiced widely for meat decontamination. Within the European Union any other method of product decontamination other than washing with potable water is not allowed for meat hygiene regulations. Legislators are averse to grant permission for accepting this technology, because it is perceived to be a means of concealing or compensating for poor hygienic practices in the slaughterhouse (Smulders and Greer, 1998).

11.3 Application guidelines for organic acid preservation

Substances recommended for reduction of pH mainly include the organic acids (Barbosa-Cánovas et al., 2003). Maximum concentrations of organic acid permitted in different foods are regulated by legislation (EC, 1995; State Bureau for Quality Supervision, Inspection, and Quarantine, 1996; Wen, Wang, and Feng, 2007). United States FDA regulations do not allow the use of preservatives as a main hurdle to microbial growth in acidified foods. In the control of the growth and toxin production by *Clostridium botulinum*, the Code of Federal Regulations (21 CFR part 114) states that only acid or acid ingredients should be added to maintain the pH at or below 4.6, and if necessary, a heat treatment must be included in the process. However, regulations do not take into account the amount or type of organic acid present in acidified food. Regulations (21 CFR part 114) were established in 1979 for the safe production of acidified foods in the United States. It was at that time, however, not known that any vegetative pathogenic organisms were able to survive at or below pH 4.6. Acidified food products have increased continuously since promulgation of this regulation. Currently, various acidified foods are produced and there may be a risk of their being contaminated with acid-resistant pathogens (Breidt, Jr., Hayes, and McFeeters, 2004).

Processing and sanitizing practices and also HACCP programs in the meat industry are not always adequate in preventing *Listeria monocytogenes* growth. Postpackaging and postprocessing hurdle technologies are, therefore, needed for its control (Samelis et al., 2005). This is particularly important in food consumed without further cooking (Calicioglu, Sofos, and Kendall, 2003). In the application of organic acids in retarding growth of *Zygosaccharomyces bailii* and *Saccharomyces cerevisiae* very high concentrations are required, which may be nearer to or greater than the legal limits (Hazan, Levine, and Abeliovich, 2004).

The presence of fumaric acid in fruit juices can be the result of added synthetic malic acid and it is then considered as an index of adulteration. This is confirmed by analysis of D-malic acid, which is not normally present in malic acid from natural sources. Guidelines of the Association of the Industry of Juices and Nectars from Fruits and Vegetables of the European Economic Community (AIJN) considers 5 mg/kg as the limiting concentration for fumaric acid content in apple juices (Trifiro, Saccani, and Gherardi, 1997).

Table 11.2 summarizes various organic acids that are included in the FDA's list of food additives that are generally regarded as safe (GRAS).

Table 11.2 Organic Acids Included in the FDA's List of GRAS Food Additives

Organic acid	Type of additive	Salt	Type of additive	Ester	Type of additive
Acetic acid	Various	Calcium acetate Sodium acetate	Sequestrant Various	Ethyl acetate Geranyl acetate Linalyl acetate	Flavoring Flavoring Flavoring
Ascorbic acid	Preservative Nutrient/dietary suppl	Calcium ascorbate Sodium ascorbate	Preservative Preservative		
Benzoic acid	Preservative	Sodium benzoate	Preservative		
Butyric acid	Flavoring			Ethyl butyrate	
Caprylic acid	Preservative				
Citric acid	Sequestrant Various	Calcium citrate Calcium diacetate Manganese citrate Potassium citrate Sodium citrate	Nutrient/dietary suppl Sequestrant Various Sequestrant Nutrient/dietary suppl Sequestrant Various Sequestrant Various	Isopropyl citrate Monoisopropyl citrate	Sequestrant Sequestrant
Formic acid	Various			Ethyl formate	Various
Lactic acid	Flavoring	Calcium lactate	Various		
Malic acid	Various				

Propionic acid	Preservative	Calcium propionate	Preservative
		Sodium propionate	Preservative
Sorbic acid	Preservative	Calcium sorbate	Preservative
		Potassium sorbate	Preservative
		Sodium sorbate	Preservative
Succinic acid	Various		
Tartaric acid	Sequestrant Various	Sodium tartrate	Sequestrant

Sources: Data from Alakomi et al., *Applied and Environmental Microbiology,* 66:2001–2005, 2000; Banerjee and Sarkar, *Food Microbiology,* 21:335–342, 2004; De la Rosa et al., *Journal of Food Protection,* 68:2465–2469, 2005; Huang et al. *Journal of Food Science,* 70:M382–M387, 2005; Valli et al., *Applied and Environmental Microbiology,* 72:5492–5499, 2006.

11.4 The role of general food safety regulations

Except for the food grown in your garden, all food products contain preservatives, as every manufacturer adds a preservative or preservatives
to the food during processing (*Food Additives World*, 2008). In the whole
process of food safety, the key issue is to support all links of the complex
food production and distribution process until the consumer is reached,
ranging from (1) harvesting, (2) storage, (3) processing, (4) packaging, (5)
sale, and (6) consumption. This network of food control was traditionally aimed at the intermediate stages of the food chain, where the food
processing takes place, not considering the ends of the chain (Domenech,
Escriche, and Martorell, 2007). The introduction of pathogen reduction
steps in food processing has been implemented in the United States by
regulation (Smith et al., 2002) and a consistent decline in the incidence of
food-transmitted salmonellosis (Olsen et al., 2001), was seen to be resulting
from new developments. Protection of the consumer population against
food-transmitted viruses has also been improved in a similar manner
(Daniels et al., 2000; Koopmans et al., 2000; Lopman et al., 2003; Struijk,
Mossel, and Moreno Garcia, 2003).

All leading food industries have to perform thorough risk assessments
in the preservation of food products (Van Gerwen et al., 2000; Hoornstra
and Notermans, 2001). The crucial point then is analysis of the production
process as a whole, the so-called from farm to fork and consists of four
subprocesses that are executed systematically (1) hazard identification, (2)
hazard characterization, (3) exposure assessment, and (4) risk characterization (Brul et al., 2002).

The definition of food quality and safety requirements involves consumers, primary production, plus the agro-food industry and the administration agents. Government and its role in food control is one of the
main factors in enforcing minimum requirements as well as implementing quality systems that would ensure food safety (Domenech, Escriche,
and Martorell, 2007). Consumers are increasingly questioning the ability
of modern food systems to provide safe food, which causes frustration
by authorities and industry with the public's attitudes toward food risks
(Macfarlane, 2002). They believe that the public worries about the wrong
things, which then twists the outline and results in the misallocation of
efforts (Dunlap and Beus, 1992; Grobe and Douthitt, 1995).

Most food producers run a quality management system, according
to DIN EN ISO 9000ff in which quality is defined as the "degree to which
a set of inherent (existing) characteristics fulfills requirements." This
definition entails a key issue—customer satisfaction—in the revision of
DIN EN ISO 9000ff that was published in the year 2000. Food producers
are required to monitor information relating to customer perception to
determine whether the organization has fulfilled customer requirements.

This is one of the measurements of the performance of the quality management system. Food producers are, therefore, consequently involved in communicating food safety (Rohr et al., 2005).

The so-called QS label for meat and meat products has recently been launched in Germany. The aim of this QS is to establish a system for quality management and control to cover all stages from (1) birth to slaughtering, (2) cutting and processing, (3) transportation and storage of products, and (4) regaining consumers' confidence over the long term (Rohr et al., 2005). Governments and the meat industry have been forced, as a result of the meat crisis and consequent decline in beef consumption, particularly in Europe, to react and work toward restoring consumer confidence in meat (Verbeke et al., 2007).

Huge economic losses are caused by visible growth of spoilage molds on food such as bread, but even greater concerns are the health hazards associated with the presence of mycotoxins (Arroyo, Aldred, and Magan, 2005). The most hazardous mycotoxins facing the food industry are the aflatoxins, due to their potential of being highly carcinogenic. Aflatoxins are rated as Class I human carcinogens by the International Agency for Research on Cancer (IARC). Approximately 100 countries have implemented specific limits for mycotoxins in food and feedstock, with the populations of these countries representing 87% of the world's inhabitants. All these countries have regulatory limits for at least aflatoxin B1 or the sum of aflatoxins B1, B2, G1, and G2 in foods or feeds (Binder, 2007).

As mentioned before, E.U. legislators propose that strict hygiene measures should be sufficient in combating the risk of contaminating *E. coli* O157:H7, *Salmonella*, and other pathogens (Bolton, Doherty, and Sheridan, 2001). Use of organic acids for meat decontamination is, therefore, not practiced widely. In the European Union, regulations in meat handling do not allow any method of product decontamination except for washing with potable water. This is mainly due to reluctance of legislators to grant permission to implement this technology, as it is perceived to be a way of covering up or compensating for poor hygienic practices in the abattoir. It may be absolutely possible that American abattoirs were the first to incorporate organic acids sprays as a component of carcass dressing processes (Smulders and Greer, 1998).

Food-processing and antimicrobial treatments are only confirmed to be adequate if effective against postprocessing contamination, as cross-contamination of food with pathogenic bacteria often occurs after processing. Dried products are regarded to be one of the safest food groups due to several hurdles to microbial survival or growth, which include: (1) drying temperature, (2) low water activity (<0.85), and (3) preservatives (may vary, depending on the composition of the marinating mixture) (Calicioglu, Sofos, and Kendall, 2003).

The FDA and USDA provide widespread regulatory oversight in ensuring food safety (Donoghue, 2003). In the United States, the foods regulated by the FDA have been found to cause four out of every five food poisoning outbreaks. However, the budget of the FDA for regulating foods is only one third of the USDA's food inspection budget (Office of Budget and Program Analysis, 2001), which implies that the FDA regulates more food with less money. Moreover, the seafood industry in the United States has a very poor record of compliance with the FDA's HACCP regulation. There is also no government testing that would enable monitoring of success. The farm is another area where insufficient government food safety oversight is evident. This poses a serious problem, as bacteria, parasites, and viruses frequently enter the food supply on the farm, mainly due to manure contamination. It is essential for animal producers to control manure in such a manner that it does not contaminate water or crops (Smith DeWaal, 2003).

In the private markets adequate food safety is either not always provided, or the safety is not readily apparent to consumers, due to the cost involved with safety testing. Producers or retailers may also not be able to determine or certify the safety of foods, considering the wide range of microbial agents and the hazards associated with them (Unnevehr and Jensen, 1999). Enumeration of *E. coli* biotype 1 has been established as a requirement by the U.S. Department of Agricultural Food Safety and Inspection Service as a control in the slaughtering process. As such the interest in meat decontamination technologies has reemerged. These technologies include the spraying of carcasses with either (1) organic acid solutions, (2) hot water, (3) steam, or (4) nonacid chemical solutions (Smulders and Greer, 1998; Sofos and Smith, 1998; Samelis et al., 2002b).

Listeria monocytogenes is a serious cause of foodborne disease and government agencies and the food industry have implemented regulations to control contamination of food by this pathogen. The food is monitored regularly by the FDA and U.S. Dept of Agriculture (USDA) (Carroll et al., 2007). When a processed food product is contaminated, this monitoring, as well as plant inspections, is intensified. Should it be necessary, the food implicated is recalled (CDC, 2003). The use of organic acid salts provides establishments two alternatives against *L. monocytogenes* contamination. These are: alternative one, to use a postlethality treatment and growth inhibitor for *L. monocytogenes* on RTE (ready-to-eat) products, or alternative two, to use either a postlethality treatment or a growth inhibitor for *L. monocytogenes* on RTE products. When an establishment uses one of these alternatives, it should receive fewer verification activities requests from the USDA/FSIS (Carroll et al., 2007).

11.5   Codex Alimentarius Commission

The Codex Alimentarius Commission (Codex) was established in the 1960s by the Food and Agriculture Organization (FAO) and the World

Health Organization (WHO), two United Nations organizations. Codex is an international organization established for encouraging fair international trade in food and to protect the health and economic interests of consumers. The Codex Alimentarius Commission is responsible for establishing a system of guidelines, standards, and recommendations that guides the direction of the global food supply. It strives to ensure that the world's food supply is sound, through adoption of food standards, codes of practice, and other guidelines developed by its committees. Codex also promotes adoption and implementation of these guidelines by governments. The Codex Alimentarius Commission implements the Joint FAO/WHO Food Standards Program, the purpose of which is to protect the health of consumers and to ensure fair practices in the food trade.

Under the World Trade Organization, Codex decisions override national and local decisions. Codex, therefore, sets the standards, and the World Trade Organization creates and implements policy. Codex has nothing to do with trade policy; it applies to the manufacture and labeling of products before they are traded. It has over 170 member countries within the framework of the Joint FAO/WHO Food Standards Programme. There exists, however, concern that some of the Codex decisions may adversely affect consumers' ability to manage their desired and healthy lifestyle (FAO/WHO, 2009).

11.6   Proposed amendments

Public health officials are predicting the number of underlying factors that increasingly cause foodborne diseases to become even more of a problem. These may include: (1) emerging pathogens, (2) improper food storage as well as food preparation practices among consumers, (3) insufficient training of food handlers, (4) increase in global food supply, and (5) increase in immunocompromised people (Medeiros et al., 2001).

Throughout the White Paper on Food Safety the guiding principle states that food safety policies should be comprehensive and integrated (EPC, 2004). This would entail integration (1) throughout the whole food chain, (2) across all food sectors, (3) between the member states, (4) outside and inside the European Union, (5) in international and E.U. decision-making forums, and (6) at all stages of the policy-making cycle (Domenech, Escriche, and Martorell, 2007).

Public authorities are pushing the food industry as well as the feed industry to develop all-inclusive quality management systems and to improve safety. This is essential in attempting to regain consumers' trust in food (Rohr et al., 2005). Organic agriculture and food processes are widely practiced in developed countries, seeking the development of a sustainable food production system, on social, ecological, and economical levels (Rekha, Naik, and Prasad, 2006).

Various different sets of criteria are used to interpret the quality of a food product. The term "acceptability" is a practical approach when comparing it to the "quality limit". Below this limit a product is then rejected. The acceptance limit is mainly defined by economic and physiological factors, whereas the quality of a product is basically defined by its fundamental characteristics. For food products such as fruits and vegetables, properties such as color, firmness, and taste will change over time. Shelf life is calculated as the time before the product qualities drop below the acceptance limit under standardized storage conditions (Tijskens, 2000; Rico et al., 2007).

In the USDA-FSIS compliance guidelines (USDA-FSIS, 1999, 2001) for chilling of thermally processed meat and poultry products it is stated that the products should be:

OPTION 1: Chilled from 54.4 to 26.7°C within 1½ h, and further to 4.4°C within an additional 5 h

OPTION 2: Chilled from 48 to 12.7°C in 6 h with continued cooling to 4.4°C, with the stipulation that the product not be shipped until the temperature reaches 4.4°C (Juneja and Thippareddi, 2004)

Depending on treatment conditions, organic acids can produce irreversible sensory changes when applied directly to meat cuts. These adverse effects occur frequently (Smulders and Greer, 1998). In 1987 an integrated approach was considered in addressing the problem of assessing the efficacies of various production practices (Smulders, 1987). A consensus was reached on strict adherence to good financial practices (GFP) and good manufacturing practices (GMP) measures during production and processing and that it should be the basis of muscle food safety strategies. Decontamination steps of whatever nature should then also serve only as supplementary ways of ensuring safety (Smulders and Greer, 1998). The issue of decontamination has since been extensively studied and in 1997 (James et al., 1997), an overview was constructed in the United Kingdom of the various methods of meat decontamination. The use of organic acids was recognized and proposed as one of the most practical options (Smulders and Greer, 1998). Organic acids as a decontaminant have been recommended in North American abattoirs (USDA, 1996) and in 1996, the European Union's Advisory Scientific Veterinary Committee issued a report evaluating the desirability of decontaminating with irradiation, organic acids, alkaline compounds (trisodium phosphate), hyperchlorinated water, steam, or hot water. However, this treatment was considered for poultry carcasses only, due to the view that in processing of the other major meat species, adherence to strict hygiene measures should be sufficient in avoiding major problems (Smulders and Greer, 1998).

Clostridium perfringens is an important foodborne pathogen, estimated to cause 248,000 cases annually of foodborne illnesses in the United States. Several processed meat products have been implicated in *C. perfringens* outbreaks. Raw meat and poultry are often contaminated with spores of *C. perfringens* that are widely found in soil and water. Outbreaks occur primarily because of consumed food that was improperly handled after cooking. However, spores present in raw materials can survive the traditional heat processing schedules and these heat-activated surviving spores then have the potential to germinate, outgrow, and multiply during subsequent chilling. Guidelines do not take into consideration the antimicrobial efficacies of the organic acid salts used in meat processing, as either flavor enhancers or preservatives. Also in the USDA-FSIS performance standard for germination and outgrowth of spore-forming bacteria it is specified that germination and outgrowth should be <1.0 $\log_{10}$ CFU/g, and *C. perfringens* can be used in microbial challenge studies to demonstrate that the cooling performance standard is met for both *C. perfringens* and *C. botulinum* (Juneja and Thippareddi, 2004).

11.7 Role of government and parastatals

Foodborne illness is a major cause of economic burden and death in the United States and it is estimated that each year more than one quarter of Americans become infected with some form of foodborne disease. The estimated death toll as a result of foodborne diseases in the United States is 5000 per year (Medeiros et al., 2001). New organisms are constantly added to the list of potential pathogens, although those not previously considered as food pathogens are gaining most of the attention (Cabe-Sellers and Beattie, 2004). Foods most often implicated in food poisoning outbreaks are seafood, eggs, beef, and produce, of which three of these top four high-risk foods are regulated by the U.S. Food and Drug Administration (FDA) (Smith DeWaal 2003).

Contamination of ready-to-eat meats occurs mainly at postprocessing, and many of these products are consumed without further heating. Although RTE meats generally contain salts such as sodium chloride, nitrite, and nitrate, known to possess antimicrobial activities, they are not effective in inhibiting the growth of *L. monocytogenes* during storage at low temperatures. RTE meats are frequently contaminated with *L. monocytogenes* and *Salmonella* spp., posing a serious hazard at refrigeration temperatures or when temperature abused. *Salmonella enterica* sv. Enteritidis is one of the serovars most frequently implicated and identified in foodborne salmonellosis outbreaks in the United States and eggs, poultry, meat, and meat products are the most common vehicles of salmonellosis (Mbandi and Shelef, 2002).

In the U.S. Code of Federal Regulations acid and acidified foods are defined as foods having a pH of 4.6 or lower. Acid foods naturally have a pH below 4.6, whereas in acidified foods, acid or acid food ingredients are added to reach a final pH of 4.6 or lower. In acidified foods, treatment must be applied if needed to destroy microbial pathogens (21 CFR part 114). Current FDA regulations for acidified foods do not take into account the amount or type of organic acid needed to lower pH. Outbreaks of *E. coli* O157:H7 have raised concern about the safety of acidified food in general. Although acidified foods have an outstanding safety record, it is essential to have a better understanding of the microbial response to organic acids in foods (Bjornsdottir, Breidt, Jr., and McFeeters, 2006).

The first recognized outbreak of *E. coli* O157:H7 was in 1982 and the organism has since emerged as a serious, potentially life-threatening, human foodborne pathogen. Outbreaks involving acidic foods have especially drawn attention to the acid tolerance of this organism. Various acidic foods have been implicated in outbreaks of food poisoning attributed to *E. coli* O157:H7. These include apple cider, dry-fermented sausage, mayonnaise, and yogurt (Jordan, Oxford, and O'Byrne, 1999).

E. coli O157:H7 also survives well in reduced-a_w meat products and has been associated with eating contaminated undercooked beef (Ryu, Deng, and Beuchat, 1999; Ryu and Beuchat, 1999). The most probable mode of transmission to food is through bovine fecal contamination (Armstrong, Hollingsworth, and Morris, 1996; Samelis et al., 2002a). At the WHO food strategic planning meeting (WHO, 2002) the microbiological problems associated with food safety were seriously addressed. A major concern was the emergence of new pathogens. Also apparent was the problem with microorganisms being able to adapt and change with the changing methods of food production, preservation, and packaging (Galvez et al., 2007).

According to the CDC *Campylobacter* is the most commonly identified bacterial cause of diarrheal illness in the world. The natural habitat of these bacteria is the intestines of healthy birds and most raw poultry meat is infested with *Campylobacter* (Anonymous, 2005).

11.8   Feed preservation

Contaminated feed poses a serious source of infection for livestock. Animals rarely show symptoms of salmonellosis and there is always the risk of undetected carriers entering the food production chain. Safe feed is the first step in a farm-to-fork food safety concept to guarantee safe food. All raw feed components should be considered as a potential source of *Salmonella*, and, therefore, decontamination steps are of utmost importance to avoid contaminated feed spreading to herds (Sauli, Danuser, and Geeraerd, 2005).

Large quantities of feed are produced, transported, and stored daily for use in industry and even minor *Salmonella* contamination may affect many herds. Bigger producers, therefore, add organic acids or heat-treat their feed, or both. Inclusion of organic acids as a preservative or dietary additive has positive effects on the growth rate and also the effectiveness of feed utilization (Giesting and Easter, 1991; Mroz et al., 2000). The heat treatment of the feed increases the digestibility of nitrogen and amino acids. Smaller producers, however, do not treat their feed, which may have a large impact on the infection of the animals (Sauli, Danuser, and Geeraerd, 2005).

Feed additives authorized by E.U. legislation include organic acids as a preservative, and are considered safe substances because of abnormal residues detected in meat. However, the high cost of organic acids poses a major problem (Castillo et al., 2004). Lactic acids have traditionally been used as natural biopreservatives of feed (Schnürer and Magnusson, 2005). Originally organic acids were added to animal feeds as fungistats (Paster, 1979; Dixon and Hamilton, 1981; Ricke, 2003).

11.9 Commercial trials

Commercial trials are necessary to develop and evaluate organic acid decontamination systems to improve meat safety without giving up desirable sensory properties. Such information is essential to convince regulatory agencies and meat processors to redirect their decontamination strategies to postcarcass processing and storage (Smulders and Greer, 1998). Limited studies have been conducted on combinations of antimicrobials for antilisterial activity, and no studies have appeared on the sequential application of dipping or spraying antimicrobial treatments for the control of *L. monocytogenes* in RTE meat products (Geornaras et al., 2005).

A food treatment and conservation subprogram has played an important role in research projects on intermediate-moisture foods and also the development of technologies for preservation of fruits by combined methods. This subprogram now promotes the amalgamation of working groups on postharvest technologies in fruits and vegetables, physical characteristics of foods, food engineering, and special regime foods as well as food packaging (Parada-Arias, 1995). More research is needed in the development of fruit preservation technology, to (1) reduce postharvest losses, (2) increase added value of raw materials, and (3) to secure the use of full capacity of processing industries in preserving fruits in big containers, without the use of refrigeration (Parada-Arias, 1994).

References

Alakomi, H.L., Skyttä, E., Saarela, M., Mattila-Sandholm, T., Latva-Kala, K., and Helander, I.M. 2000. Lactic acid permeabilizes gram-negative bacteria by disrupting the outer membrane. *Applied and Environmental Microbiology* 66:2001–2005.

Anonymous. 2005. Preventing foodborne illness. *National Provisioner* 219:98–102.

Armstrong, G.L., Hollingsworth, J., and Morris, J.G. 1996. Emerging foodborne pathogens: *Escherichia coli* O157:H7 as a model of entry of a new pathogen into the food supply of the developed world. *Epidemiological Reviews* 18:29–51.

Arroyo, M., Aldred, D., and Magan, N. 2005. Environmental factors and weak organic acid interactions have differential effects on control of growth and ochratoxin A production by *Penicillium verrucosum* isolates in bread. *International Journal of Food Microbiology* 98:223–231.

Banerjee, M. and Sarkar, P.K. 2004. Antibiotic resistance and susceptibility to some food preservative measures of spoilage and pathogenic micro-organisms from spices. *Food Microbiology* 21:335–342.

Barbosa-Cánovas, G.V., Fernández-Molina, J.J., Alzamora, S.M., Tapia, M.S., López-Malo, A., and Chanes, J.W. 2003. General considerations for preservation of fruits and vegetables. In: *Handling and Preservation of Fruits and Vegetables by Combined Methods for Rural Areas*. Rome: Food and Argriculture Organization of the United Nations.

Binder, E.M. 2007. Managing the risk of mycotoxins in modern feed production. *Animal Feed Science and Technology* 133:149–166.

Bjornsdottir, K., Breidt, F., Jr., and McFeeters, R.F. 2006. Protective effects of organic acids on survival of *Escherichia coli* O157:H7 in acidic environments. *Applied and Environmental Microbiology* 72:660–664.

Bolton, D.J., Doherty, A.M., and Sheridan, J.J. 2001. Beef HACCP: Intervention and non-intervention systems. *International Journal of Food Microbiology* 66:119–129.

Breidt, F., Jr., Hayes, J.S., and McFeeters, R.F. 2004. Independent effects of acetic acid and pH on survival of *Escherichia coli* in simulated acidified pickle products. *Journal of Food Protection* 67:12–18.

Brul, S., Klis, F.M., Oomes, S.J.C.M., et al., 2002. Detailed process design based on genomics of survivors of food preservation processes. *Trends in Food Science and Technology* 13:325–333.

Buckley, D.J., Morrissey, P.A., and Gray, J.I. 1995. Influence of dietary vitamin E on the oxidative stability and quality of pig meat. *Journal of Animal Science* 73:3122–3130.

Buzby, J.C., Roberts, T., Lin, C.-T.J., and MacDonald, J.M. 1996. Bacterial foodborne disease: Medical costs and productivity losses. Agricultural Economic Report No. 741. Economic Research Service, US Department of Agriculture.

Cabe-Sellers, B.J. and Beattie, S.E. 2004. Food safety: Emerging trends in foodborne illness surveillance and prevention. *Journal of the American Dietetic Association* 104:1708–1717.

Calicioglu, M., Sofos, J.N., and Kendall, P.A. 2003. Fate of acid-adapted and non-adapted *Escherichia coli* O157:H7 inoculated post-drying on beef jerky treated with marinades before drying. *Food Microbiology* 20:169–177.

Carroll, C.D., Alvarado, C.Z., Brashears, M.M., Thompson, L.D., and Boyce, J. 2007. Marination of turkey breast fillets to control the growth of *Listeria monocytogenes* and improve meat quality in deli loaves. *Poultry Science* 86:150–155.

Castillo, C., Benedito, J.L., Mendez, J., et al. 2004. Organic acids as a substitute for monensin in diets for beef cattle. *Animal Feed Science and Technology* 115:101–116.

Centers for Disease Control. 2003. Subject: Listeriosis. http://www.cdc.gov/ncidod/dbmd/diseaseinfo/ listeriosis_g.htm (accessed Febuary 27, 2006).

Cross, R. 1994. What HACCP really means. *Webb Agricultural Publications.* http://www.haccpalliance.org/alliance/haccp.htm (accessed April 12, 2010).

Daniels, N.A., Bergmire-Sweat, D.A., Schwab, K.J., et al. 2000. A foodborne outbreak of gastroenteritis associated with Norwalk-like viruses: First molecular traceback to deli sandwiches contaminated during preparation. *Journal of Infectious Diseases* 181:1467–1470.

Dean, K.H. 1990. HACCP and food safety in Canada. *Food Technology* 44:172–178.

De la Rosa, P., Cordoba, G., Martín, A., Jordano, R., and Medina, L.M. 2005. Influence of a test preservative on sponge cakes under different storage conditions. *Journal of Food Protection* 68:2465–2469.

Dixon, R.C. and Hamilton, P.B. 1981. Effect of feed ingredients on the antifungal activity of propionic acid. *Poultry Science* 60:2407–2411.

Domenech, E., Escriche, I., and Martorell, S. 2007. Quantification of risks to consumers' health and to company's incomes due to failures in food safety. *Food Control* 18:1419–1427.

Donoghue, D.J. 2003. Antibiotic residues in poultry tissues and eggs: Human health concerns? *Poultry Science* 82:618–621.

Drosinos, E.H., Mataragas, M., and Metaxopoulos, J. 2005. Biopreservation: A new direction towards food safety. In: Riley, A.P. (Ed.), *New Developments in Food Policy Control and Research,* pp. 31–64. New York: Nova Science.

Dunlap, R.E. and Beus, C.E. 1992. Understanding public concerns about pesticides: An empirical examination. *Journal of Consumer Affairs* 26:418–438.

EC. 1995. European Parliament and Council Directive No. 95/2/EC. European Communities, Brussels.

EPC. 2004. Regulation No. 852/2004 of the European Parliament and of the Council of 29 April 2004 on the hygiene of foodstuffs. *Official Journal* L 139:1–54. http://www.food.gov.uk/multimedia/pdfs/h3ojregulation.pdf> (accessed August 29, 2005).

FAO/WHO Food Standards. 2009. http://www.codexalimentarius.net (accessed November 11, 2009).

Food Additives World. 2008. Preservatives. *http://www.foodadditivesworld.com/* (accessed May 12, 2009).

Galvez, A., Abriouel, H., Lopez, R.L., and Ben, O.N. 2007. Bacteriocin-based strategies for food biopreservation. *International Journal of Food Microbiology* 120:51–70.

Geornaras, I., Belk, K.E., Scanga, J.A., Kendall, P.A., Smith, G.C., and Sofos, J.N. 2005. Postprocessing antimicrobial treatments to control *Listeria monocytogenes* in commercial vacuum-packaged bologna and ham stored at 10 degrees C. *Journal of Food Protection* 68:991–998.

Giesting, D.W. and Easter, R.A. 1991. Effect of protein source and fumaric acid supplementation on apparent ileal digestability of nutrients by young pigs. *Journal of Animal Science* 69:2497–2503.

Grobe, D. and Douthitt, R. 1995. Consumer acceptance of recombinant bovine growth hormone—Interplay between beliefs and perceived risks. *Journal of Consumer Affairs* 29:128–143.

Hazan, R., Levine, A., and Abeliovich, H. 2004. Benzoic acid, a weak organic acid food preservative, exerts specific effects on intracellular membrane trafficking pathways in *Saccharomyces cerevisiae. Applied and Environmental Microbiology* 70:4449–4457.

Hoornstra, E. and Notermans, S. 2001. Quantitative microbiological risk assessment. *International Journal of Food Microbiology* 66:21–29.

Huang, N.-H., Ho, C.-P., and McMillan, K.W. 2005. Retail shelf-life of pork dipped in organic acids before modified atmosphere or vacuum packaging. *Journal of Food Science* 70:M382–M387.

James, C., Goksoy, E., and James, S. 1997. *Past, present and future methods of meat decontamination.* University of Bristol/MAFF Fellowship in Food Process Engineering, Langford.

Jordan, K.N., Oxford, L., and O'Byrne, C.P. 1999. Survival of low-pH stress by *Escherichia coli* O157:H7: Correlation between alterations in the cell envelope and increased acid tolerance. *Applied and Environmental Microbiology* 65:3048–3055.

Juneja, V.K. and Thippareddi, H. 2004. Inhibitory effects of organic acid salts on growth of *Clostridium perfringens* from spore inocula during chilling of marinated ground turkey breast. *International Journal of Food Microbiology* 93:155–163.

Koopmans, M.P.G., Vinjé, J., De Wit, M., Leenen, I., van der Poel, W., and Duynhoven, Y. 2000. Molecular epidemiology of human enteric caliciviruses in the Netherlands. *Journal of Infectious Diseases* 181(Suppl. 2):S262–S269.

Lopman, B.A., Reacher, M.H., van Duijnhoven, Y., Hanon, F.-X., Brown, D., and Koopmans, M. 2003. Viral gastroenteritis outbreaks in Europe, 1995–2000. *Emerging Infectious Diseases* 9:90–96.

Macfarlane, R. 2002. Integrating the consumer interest in food safety: The role of science and other factors. *Food Policy* 27:65–80.

Mbandi, E. and Shelef, L.A. 2002. Enhanced antimicrobial effects of combination of lactate and diacetate on *Listeria monocytogenes* and *Salmonella* spp. in beef bologna. *International Journal of Food Microbiology* 76:191–198.

Medeiros, L.C., Kendall, P., Hillers, V., Chen, G., and DiMascola, S. 2001. Identification and classification of consumer food-handling behaviors for food safety education. *Journal of the American Dietetic Association* 101:1326–1339.

Molins, R.A., Motarjemi, Y., and Käferstein, F.K. 2001. Irradiation: A critical control point in ensuring the microbiological safety of raw foods. *Food Control* 12:347–356.

Mroz, Z., Jongbloed, A., Partanen, K.H., Vreman, K., Kemme, P.A., and Kogut, J. 2000. The effects of calcium benzoate in diets with or without organic acids on dietary buffering capacity, apparent digestibility, retention of nutrients, and manure characterization in swine. *Journal of Animal Science* 78:2622–2632.

Office of Budget and Program Analysis. 2001. U.S. Department of Agriculture 2001 Budget Summary; Food and Drug Administration. FY 2001. *Congressional Budget Request Table of Contents.*

Olsen, S.J., Bishop, R., Brenner, F.W., et al. 2001. The changing epidemiology of *salmonella:* Trends in serotypes isolated from humans in the United States, 1987–1997. *Journal of Infectious Diseases* 183:753–761.

Parada-Arias, E. 1994. IMF: An Iberoamerican cooperative project. *Journal of Food Engineering* 22:445–452.

Parada-Arias, E. 1995. CYTED—The Iberoamerican cooperative research program in food technology. *Food Research International* 28:343–346.

Paster, N. 1979. A commercial scale study of the efficiency of propionic acid and calcium propionate as fungistats in poultry feed. *Poultry Science* 58:578–576.

Petrera, M. and Montoya, M. 1992. The economic impact of the cholera epidemic, Peru, 1991. *Epidemiological Bulliten,* PAHO 13:9–12.

Rekha, B., Naik, S.N., and Prasad, R. 2006. Pesticide residue in organic and conventional food-risk analysis. *Journal of Chemical Health and Safety* 13:12–19.

Ricke, S.C. 2003. Perspectives on the use of organic acids and short chain fatty acids as antimicrobials. *Poultry Science* 82:632–639.

Rico, D., Martin-Diana, A.B., Barat, J.M., and Barry-Ryan, C. 2007. Extending and measuring the quality of fresh-cut fruit and vegetables: A review. *Trends in Food Science and Technology* 18:373–386.

Rohr, A., Luddecke, K., Drusch, S., Muller, M.J., and Alvensleben, R. 2005. Food quality and safety—consumer perception and public health concern. *Food Control* 16:649–655.

Ryu, J.-H. and Beuchat, L.R. 1999. Changes in heat tolerance of *Escherichia coli* O157:H7 after exposure to acidic environments. *Food Microbiology* 16:317–324.

Ryu, J.-H., Deng, Y., and Beuchat, L.R. 1999. Survival of *Escherichia coli* O157:H7 in dried beef powder as affected by water activity, sodium chloride content and temperature. *Food Microbiology* 16:309–316.

Samelis, J., Bedie, G.K., Sofos, J.N., Belk, K.E., Scanga, J.A., and Smith, G.C. 2005. Combinations of nisin with organic acids or salts to control *Listeria monocytogenes* on sliced pork bologna stored at 4°C in vacuum packages. *Lebensmittel-Wissenschaft und-Technologie* 38:21–28.

Samelis, J., Sofos, J.N., Ikeda, J.S., Kendall, P.A., and Smith, G.C. 2002a. Exposure to non-acid fresh meat decontamination washing fluids sensitizes *Escherichia coli* O157:H7 to organic acids. *Letters in Applied Microbiology* 34:7–12.

Samelis, J., Sofos, J.N., Kendall, P.A., and Smith, G.C. 2002b. Effect of acid adaptation on survival of *Escherichia coli* O157:H7 in meat decontamination washing fluids and potential effects of organic acid interventions on the microbial ecology of the meat plant environment. *Journal of Food Protection* 65:33–40.

Sauli, I., Danuser, J., and Geeraerd, A.H. 2005. Estimating the probability and level of contamination with *Salmonella* of feed for finishing pigs produced in Switzerland—the impact of the production pathway. *International Journal of Food Microbiology* 100:289–310.

Schnürer, J. and Magnusson, J. 2005. Antifungal lactic acid bacteria as bio-preservatives. *Trends in Food Science and Technology* 16:70–78.

Smith, S.E., Orta-Ramirez, A., Ofoli, R.Y., Ryser, E.T., and Smith, D.M. 2002. R-phycoerythrin as a time-temperature integrator to verify the thermal processing adequacy of beef patties. *Journal of Food Protection* 65:814–819.

Smith DeWaal, C. 2003. Safe food from a consumer perspective. *Food Control* 14:75–79.

Smulders, F. 1987. Prospectives for microbiological decontamination of meat and poultry by organic acids with special reference to lactic acid. In: F.J.M. Smulders (Ed.), *Elimination of Pathogenic Organisms from Meat and Poultry,* pp. 319–344. London: Elsevier Science.

Smulders, F.J. and Greer, G.G. 1998. Integrating microbial decontamination with organic acids in HACCP programmes for muscle foods: Prospects and controversies. *International Journal of Food Microbiology* 44:149–169.

Sofos, J.N. and Smith, G.C. 1998. Nonacid meat decontamination technologies: Model studies and commercial applications. *International Journal of Food Microbiology* 44:171–188.

State Bureau for Quality Supervision, Inspection and Quarantine. 1996. Hygienic standards for the use of food additives. National Standards of the People's Republic of China GB 2760. *Standards Press of China*, p. 290. Beijing.

Struijk, C.B., Mossel, D.A.A., and Moreno Garcia, B. 2003. Improved protection of the consumer community against food-transmitted diseases with a microbial aetiology: A pivotal food safety issue calling for a precautionary approach. *Food Control* 14:501–506.

Tijskens, L.M.M. 2000. Acceptibility. In: L.R. Shewfelt and B. Bruckner (Eds.), *Fruit and Vegetable Quality: An Integrated View*, pp. 125–143. New York: CRC Press.

Trifiro, A., Saccani, G., and Gherardi, S. 1997. Use of ion chromatography for monitoring microbial spoilage in the fruit juice industry. *Journal of Chromatography A* 770:243–252.

Unnevehr, L.J. and Jensen, H.H. 1999. The economic implications of using HACCP as a food safety regulatory standard. *Food Policy* 24:625–635.

US Department of Agriculture, Food Safety and Inspection Service. 1996. Food labeling: Health claims and labeling statements; Folate and Neural Tube Defects; (Rules and Regulations). *Federal Register* 61: 8752–8781.

US Department of Agriculture, Food Safety and Inspection Service. 1999. Performance standards for the production of certain meat and poultry products. *Federal Register* 64:732–749.

US Department of Agriculture, Food Safety and Inspection Service. 2000. Food additives for use in meat and poultry products: Sodium diacetate, sodium acetate, sodium lactate and potassium lactate: Direct final rule. *Federal Register* 65:3121–3123.

US Department of Agriculture, Food Safety and Inspection Service. 2001. Performance standards for the production of meat and poultry products: Proposed rule. *Federal Register* 66:12589–12636.

Valli, M., Sauer, M., Branduardi, P., Borth, N., Porro, D. and Mattanovich, D. 2006. Improvement of lactic acid production in *Saccharomyces cerevisiae* by cell sorting for high intracellular pH. *Applied and Environmental Microbiology* 72:5492–5499.

Van Gerwen, S.J., te Giffel, M.C., van't Riet, K., Beumer, R.R., and Zwietering, M.H. 2000. Stepwise quantitative risk assessement as a tool for characterization of microbiological food safety. *Journal of Applied Microbiology* 88:938–951.

Verbeke, W., Frewer, L.J., Scholderer, J., and De Brabander, H.F. 2007. Why consumers behave as they do with respect to food safety and risk information. *Analytica Chimica Acta* 586:2–7.

Wen, Y., Wang, Y., and Feng, Y.Q. 2007. A simple and rapid method for simultaneous determination of benzoic and sorbic acids in food using in-tube solid-phase microextraction coupled with high-performance liquid chromatography. *Analytical and Bioanalytical Chemistry* 388:1779–1787.

WHO. 2002. Food safety strategic planning meeting: Report of a WHO strategic planning meeting, HO headquarters, Geneva, Switzerlang, 20–22 February 2001. World Health Organization, Geneva. http://whqlibdoc.who.int/hq/2001/WHO_SDE_PHE_FOS_01.2 .pdf.

chapter twelve

Incidental and natural organic acid occurence

12.1 Introduction

In the ongoing search for natural and effective preservatives, interest has shifted to the natural contents of some foodstuffs that have in the past received little or no attention, apart from their function as complementary food or beverages. In this chapter some of these comestibles are mentioned and described, specifically with regard to their organic acid content.

12.2 Honey

Not much work has been published on the varying composition of organic acids in different types of honey. Evaluation of their organic acid profile and pattern may, however, provide additional information on honey samples from various sources (Anklam, 1998). Nineteen organic acids have thus far been identified in honey and this may also be used in characterizing the different types of honey. Organic acids are also components playing a role in the flavor of the honey. Although organic acids comprise only a small portion of honey (0.5%), acidity of honey as well as the antimicrobial activity of organic acids contributes in the preservation process. The total acidity can be used as an indicator of deterioration due to storage, aging, or even to measure the purity and authenticity (Suarez-Luque et al., 2002).

12.3 Sourdough

Sourdough is also classified as a natural preservative and has long been known to improve the shelf lives of bread and bakery products. Phenyllactic acid (PLA) has been found in cultures of *Lactococcus plantarum* with antifungal activity in sourdough breads. Relevant amounts of PLA naturally occur in several honeys from different geographical areas. It is nontoxic for human and animal cells and may also be safely used in food. However, information is needed on its effect on rheology and flavor, although PLA solutions are relatively odorless (Lavermicocca, Valerio, and Visconti, 2003).

12.4 Berries

Natural preservatives that can be derived from plants or microbes have the potential to reduce microbial growth in various foodstuffs. However, limited information is available on the efficiency of plant-derived antimicrobials on food, as opposed to laboratory testing done on agar-based tests (Penney et al., 2004). Chopped cranberries as a source of antimicrobial compounds have been tested, and cranberry juice was found to exhibit antimicrobial properties. Several different phenolic acids and flavonoids, in free form as well as glycosides and esters, have been found in cranberries (Chen, Zuo, and Deng, 2001). Benzoic acid is the most common phenolic in cranberries and the main source of antifungal activity (Penney et al., 2004). Uncharacterized polymeric compounds have also been detected in cranberries and adversely affect the adherence of pathogenic bacteria, such as *Escherichia coli*. This may most likely contribute to the effectiveness of cranberry juice, when ingested, on a decreased risk of infections (Avorn et al., 1994).

Even in foodstuffs such as yogurt, cranberries were found to suppress microbial growth. This suppression was most evident within 24 h after addition of fruit (Penney et al., 2004). Benzoic acid occurs naturally in lingonberries, cranberries, cloudberries, and cinnamon (Davidson, 1997; Heimhuber et al., 1990; Visti, Viljakainen, and Laakso, 2003). Benzoic acid concentration is especially high in lingonberries (0.6–1.3 g/l free benzoic acid), whereas the pH is low (pH 2.6–2.9), and is increased if the juice is stored under warm conditions. Not many microorganisms can ferment lingonberries and this fruit can normally be conserved without preservatives (Be-Onishi et al., 2004).

12.5 Wine

Ethanol and organic acids are major constituents of wine (Fernandes et al., 2007). It has been found that people who drink wine during a meal are less susceptible to foodborne toxin infections (Correia et al., 2003). The exact mechanisms involved in the antimicrobial activity of wine are not fully understood (Fernandes et al., 2007). Wine is a beverage possessing some unique properties. It contains relatively high ethanol content as well as other antimicrobial agents, such as organic acids, and polyphenol compounds (Just and Daeschel, 2003). Most studies on the antimicrobial activity of wine employed experimental conditions much different from *in vivo* conditions, using mostly a mixture of wine and a low pH solution to simulate the inactivation. This combination of acids plus ethanol shows higher bactericidal effect than a mixture of acids and ethanol separately, an effect similar to that obtained with wine (Fernandes et al., 2007).

The antimicrobial effect of the organic acids fraction in wine is mainly due to malic and lactic acids, and other constituents influencing the bactericidal effect of wine are phenolic compounds (Rhodes et al., 2006). When comparing the antimicrobial effect of white wine with red wine, it was found that red wine generally contains high concentrations of tartrate and lactate. Tartrate is, however, also the predominant anion in white wines, but lactate concentration is much less than in red wines (Arellano et al., 1997). In studies white wine showed a lighter effect in the beginning of inactivation (10 min), whereas at the end of experimental time (60 min), lower survivor concentrations were found than when using red wine. One possible explanation may be due to the fact that the white wine may not have undergone malolactic fermentation, and possesses higher concentrations of malic acid than the red wine. Chardonnay has been found more lethal to *Salmonella* populations than Pinot Noir. In 2004 it was established that red wine was more potent than white wine in killing *Listeria monocytogenes*. Red wine had a higher ethanol content and titrable acidity than white wine. Inactivating bacteria in the wine is strongly dependent on specific wine composition (Fernandes et al., 2007).

Organic acid content in wine has an important impact on organoleptic characteristics of the wine. α-Hydroxy acids (tartaric, malic, lactic, and citric acids) are mainly responsible for these characteristics. Other acids in wines include acetic, ascorbic, gluconic, and sorbic acids, as well as sulfite, sulphate, phosphate, and malonate (Masar et al., 2001).

Organic acids are also prominent in wine-derived products; some are originally present in the grape, whereas others are produced by fermentation. Organic acid concentration in wine-derived products may provide information relating to (1) the origin of the raw material, (2) microbial growth, and also (3) techniques used in processing. Acetic acid is the main organic acid in vinegars, tartaric acid the major acid in wine, and malic acid is prevalent in low amounts in wine vinegars, although very inconsistently, depending on the origin of the wine and also the winemaking processes. Malic acid is the most representative organic acid in cider vinegars and is converted to lactic acid during malolactic fermentation after alcoholic fermentation. Tartaric acid originally occurred in the grape and subsequently in wine and wine vinegars (Morales, Gonzalez, and Troncoso, 1998).

Only a few yeasts and bacteria are able to spoil wine, because of the high sugar content, low pH, high alcohol concentrations, and anaerobic fermentation conditions (Du Toit, Pretorius, and Lonvaud-Funel, 2005).

12.6   Coffee

Organic acids contribute to both the taste and flavor of coffee, specifically the acidic taste sensation, because of their volatility. In roasted coffee organic acids are classified into four groups: aliphatic, chlorogenic, alicyclic, and phenolic (Galli and Barbas, 2004).

12.7   Vinegar

Tartaric acid is well represented in grape-derived products and is also present in considerable amounts in wine vinegar, including balsamic vinegar, but in low concentrations in cider vinegars. Malic acid is present in significant amounts in cider vinegars and sherry vinegars with the highest concentration found in balsamic vinegar. High malic acid content of apples is reduced by malolactic fermentation. Citric acid is only found in small amounts in vinegars, more specifically cider vinegars. Acetic acid levels are high in sherry vinegars. The high amount of lactic acid is partially oxidized by acetic bacteria in cider vinegars (Morales, Gonzalez, and Troncoso, 1998).

12.8   Acid foods

Acidified food products have for many years been produced without heat treatments, greatly due to the organic acids present in these products (Zagory and Garren, 1999; Breidt, Jr., Hayes, and McFeeters, 2004). The effects of pH and organic acids on the survival of pathogenic bacteria have been investigated in a variety of acid and acidified food products, including apple cider (Uljas and Ingham, 1998; Hsin-Yi and Chou, 2001), mayonnaise, dressings, condiments (Raghubeer et al., 1995; Tsai and Ingham, 1997; Smittle et al., 2000; Mayerhauser, 2001), and fermented meats (Glass et al., 1992; Duffy and Vanderlinde, 2000; Pond et al., 2001). However, this is complicated by the concentration of the undissociated (protonated) form of organic acid and the pH being interdependent variables, linked by the Henderson–Hasselbalch equation (Breidt, Jr., Hayes, and McFeeters, 2004). Citric acid is present in most juices, malic acid is also present in high concentrations, and minute concentrations of lactate can be found (Arellano et al., 1997).

12.9   Kombucha

Kombucha is a health drink made by adding kombucha culture to tea and allowing it to ferment. It originated in the Far East and has been consumed for at least 2,000 years. Although many health claims are made for kombucha, not much research has been done on the benefits of this drink. It has, however, been shown to have antibiotic, antiviral, and anti-fungal properties in laboratory tests. The following organic acids have been found in kombucha: glucuronic acid, lactic acid, acetic acid, usnic acid, oxalic acid, malic acid, gluconic acid, and butyric acid (Blanc, 1996; Lynch et al., 2006).

12.10   Edible films

According to an estimate made by the WHO, without effective packag-ing, developing countries would lose up to 50% of their produce before it reached the consumer. However, after serving its purpose, packaging becomes a burdensome waste. Studies on edible films are continuously designed to promote food safety. Such films are based on food ingredi-ents, for example, pureed spinach and apples infused with oils such as carvacrol, a compound found in oregano (Friedman et al., 2009).

12.11   Summary

In Table 12.1 a summary of complementary foodstuffs is presented. In a quest for alternative preservation methods and compounds, interest in the antimicrobial contents, and specifically the organic acid contents of such products, has increased.

Table 12.1 Organic Acids Prevalent in Various Other Frequently Consumed Complementary Foodstuffs Considered as Potential Alternative Preservatives

Organ acid or derivative	Berries			Coffee	Fruit juices	Honey	Kombucha	Sourdough	Vinegar	Wine
	Cloudberry	Cranberry	Lingonberry (Cowberry)							
Acetic acid				+			+	+	+	+
Acetate						+				
Ascorbic acid	+	+	+							
Ascorbate							+			+
Benzoic acid	+	+	+			+				
Butyric acid				+			+			
Cinnamic acid		+				+				
Citric acid	+	+	+	+	+	+			+	+
Formic acid				+						
Fumaric acid				+		+				
Gluconate						+	+			
Gluconic acid						+	+			
Glucuronic							+			
Lactic acid			+	+						
Lactate								+	+	+
Malic acid	+	+	+	+	+	+	+		+	+
Maleic acid				+		+				
Malonate						+				
Oxalic					+					

	1	2	3	4	5	6	7	8
Phenyllactic acid						+		
Propionic acid			+		+	+		
Pyruvic acid			+		+	+		
Quinic acid				+				
Salicylic acid					+	+	+	+
Sorbate								
Sorbic acid			+		+	+		
Succinic acid							+	
Tartaric acid					+	+	+	+
Usnic acid						+	+	+

References

Anklam, E. 1998. A review of the analytical methods to determine the geographical and botanical origin of honey. *Food Chemistry* 63:549–562.

Arellano, M., Andrianary, J., Dedieu, F., Couderc, F., and Puig, P. 1997. Method development and validation for the simultaneous determination of organic and inorganic acids by capillary zone electrophoresis. *Journal of Chromatography A* 765:321–328.

Avorn, J., Monane, M., Gurwit, J., Glynn, R., Choodnovskiy, I., and Lipsitz, L. 1994. Reduction of bacteriuria and pyuria after ingestion of cranberry juice. *Journal of the Americal Medical Association* 271:751–754.

Be-Onishi, Y., Yomota, C., Sugimoto, N., Kubota, H., and Tanamoto, K. 2004. Determination of benzoyl peroxide and benzoic acid in wheat flour by high-performance liquid chromatography and its identification by high-performance liquid chromatography-mass spectrometry. *Journal of Chromatography A* 1040:209–214.

Blanc, P.J. 1996. Characterization of the tea fungus metabolites. *Biotechnology Letters* 18:139–142.

Breidt, F., Jr., Hayes, J.S., and McFeeters, R.F. 2004. Independent effects of acetic acid and pH on survival of *Escherichia coli* in simulated acidified pickle products. *Journal of Food Protection* 67:12–18.

Chen, H., Zuo, Y., and Deng, Y. 2001. Separation and determination of flavonoids and other compounds in cranberry juice by high-performance liquid chromatography. *Journal of Chromatography A* 913:387–395.

Correia, A., Gomes, A., Oliveira, B., Gonçalves, G., Miranda, M., and Almeida, O. 2003. The protective effect of alcoholic beverages in foodborne outbreaks of *Salmonella enteritidis* PT1 in northern Portugal. *Eurosurveillance Weekly* 7,13, Article 3.

Davidson, P.M. 1997. Chemical preservatives and natural antimicrobial compounds. In: M.P. Doyle, L.R Beuchat, and T.J. Montville (Eds.), *Food Microbiology—Fundamentals and Frontiers*, pp. 520–556. Washington DC: ASM Press.

Duffy, L.L., Frau, F.H., and Vanderlinde, P.B. 2000. Acid resistance of enterohemorrhagic and generic *Escherichia coli* associated with foodborne disease and meat. *International Journal of Food Microbiology* 60:183–189.

Du Toit, W.J., Pretorius, I.S., and Lonvaud-Funel, A. 2005. The effect of sulphur dioxide and oxygen on the viability and culturability of a strain of *Acetobacter pasteurianus* and a strain of *Brettanomyces bruxellensis* isolated from wine. *Journal of Applied Microbiology* 98:862–871.

Fernandes, J., Gomes, F., Couto, J.A., and Hogg, T. 2007. The antimicrobial effect of wine on *Listeria innocua* in a model stomach system. *Food Control* 18:1477–1483.

Friedman, M., Zhu, L., Feinstein, Y., and Ravishankar, S. 2009. Carvacrol facilitates heat induced inactivation of *Escherichia coli* O157:H7 and inhibits formation of heterocylic amines in grilled ground beef patties. *Journal of Agricultural and Food Chemistry*. 57:1848–1853.

Galli, V. and Barbas, C. 2004. Capillary electrophoresis for the analysis of short-chain organic acids in coffee. *Journal of Chromatography A* 1032:299–304.

Glass, K.A., Loeffelholz, J.M., Ford, J.P., and Doyle, M.P. 1992. Fate of *Escherichia coli* O157:H7 as affected by pH or sodium chloride and in feremented, dry sausage. *Applied and Environmental Microbiology* 58:2513–2516.

Heimhuber, B., Wray, V., Galensa, R., and Herrmann, K. 1990. Benzoylglucoses from two *Vaccinum* species. *Phytochemistry* 29:2726–2727.

Hsin-Yu, C. and Chou, C.-C. 2001. Acid adaptation and temperature effect on the survival of *E. coli* O157:H7 in acidic fruit juice and lactic fermented milk product. *International Journal of Food Microbiology* 70:189–195.

Just, J.R. and Daeschel, M.A. 2003. Antimicrobial effects of wine on *Escherichia coli* O157:H7 and *Salmonella typhimurium* in a stomach model system. *Journal of Food Science* 68:285–290.

Lavermicocca, P., Valerio, F., and Visconti, A. 2003. Antifungal activity of phenyllactic acid against molds isolated from bakery products. *Applied and Environmental Microbiology* 69:634–640.

Lynch, M., Painter, J., Woodruff, R., and Braden, C. 2006. Surveillance for foodborne-disease outbreaks—United States, 1998–2002. *Morbidity and Mortality Weekly Report* 55:1–42.

Masar, M., Kaniansky, D., Bodor, R., Johnck, M., and Stanislawski, B. 2001. Determination of organic acids and inorganic anions in wine by isotachophoresis on a planar chip. *Journal of Chromatography A* 916:167–174.

Mayerhauser, C.M. 2001. Survival of enterohemorrhagic *Escherichia coli* O157:H7 in retail mustard. *Journal of Food Protection* 64:783–787.

Morales, M.L., Gonzalez, A.G., and Troncoso, A.M. 1998. Ion-exclusion chromatographic determination of organic acids in vinegars. *Journal of Chromatography A* 822:45–51.

Penney, V., Henderson, G., Blum, C., and Johnson-Green, P. 2004. The potential of phytopreservatives and nisin to control microbial spoilage of minimally processed fruit yogurts. *Innovative Food Science and Emerging Technologies* 5:369–375.

Pond, T.J., Wood, D.S., Mumin, I.M., Barbut, S., and Griffiths, M.W. 2001. Modeling the survival of *Escherichia coli* O157:H7 in uncooked, semidry, fermented sausage. *Journal of Food Protection* 64:759–766.

Raghubeer, E.V., Ke, J.S., Campbell, M.L., and Meyer, R.S. 1995. Fate of *Escherichia coli* O157:H7 and other coliforms in commercial mayonnaise and refrigerated salad dressings. *Journal of Food Protection* 58:13–18.

Rhodes, P.L., Mitchell, J.W., Wilson, M.W., and Melton, L.D. 2006. Antilisterial activity of grape juice and grape extracts derived from *Vitis vinifera* variety Ribier. *International Journal of Food Microbiology* 107:281–286.

Smittle, R.B. 2000. Microbiological safety of mayonnaise, salad dressings, and sauces produced in the United States: A review. *Journal of Food Protection* 63:1144–1153.

Suarez-Luque, S., Mato, I., Huidobro, J.F., Simal-Lozano, J., and Sancho, M.T. 2002. Rapid determination of minority organic acids in honey by high-performance liquid chromatography. *Journal of Chromatography A* 955:207–214.

Tsai, Y.-W. and Ingham, S.C. 1997. Survival of *Escherichia coli* O157:H7 and *Salmonella* spp. in acidic condiments. *Journal of Food Protection* 60:751–755.

Uljas, H.E. and Ingham, S.C. 1998. Survival of *Escherichia coli* O157:H7 in synthetic gastric fluid after cold and acid habituation in apple juice or trypticase soy broth acidified with hydrochloric acid or organic acids. *Journal of Food Protection* 61:939–947.

Visti, A., Viljakainen, S., and Laakso, S. 2003. Preparation of fermentable lingon-
berry juice through removal of benzoic acid by *saccharomyces cerevisiae* yeast.
Food Research International 36:597–602.
Zagory, D. and Garren, D. 1999. HACCP: What It Is and What It Isn't. *The Packer.*
August 16, 1999. Online journal. http://www.thepacker.com (accessed
August 29, 2005).

chapter thirteen

Biopreservation

13.1   Introduction

Biopreservation refers to a combination of fermentation and preservation processes and entails the extension of shelf life and improving the safety of food by employing microorganisms or their metabolites. The main goal of biopreservation is, therefore, the enhancement of safety using bacteria with antimicrobial capabilities or their antimicrobial substances (Ross, Morgan, and Hill, 2002). The production of large quantities of food and the ability to store food, even in such large quantities, is one of the most important developments permitting the formation of civilization. There exist microorganisms that are nontoxic and have been found even to be beneficial, directly or indirectly, to human health. The whole process of biopreservation is based on the idea of nonpathogenic microorganisms or their products being selected to replace chemicals added to inhibit pathogenic or spoilage microorganisms in food products (Stiles, 1996) (Figure 13.1). Empirical use of microorganisms or their products for food preservation has for many years in the history of man been common practice and evidence of food preservation has come from as early as the postglacial era, from 15,000 to 10,000 BC. Fermentation as a procedure for food preservation has also been traced back thousands of years, where the first use was recorded from 6000 to 1000 BC, when fermentation was used to produce beer and wine, vinegar, bread, yogurt, cheese, and butter (Soomro, Masud, and Anwaar, 2002).

Microorganisms used in fermentation provide a desired flavor and texture, while also producing metabolites that inhibit the growth of other microorganisms. However, this process can also be used in nonfermented foods such as milk, meat and meat products, fruits, and vegetables (Stiles, 1996). These protective cultures are specially selected based on their ability to control the growth of unwanted microorganisms in fermented food. This inhibition is accomplished as a result of competition for nutrients and also by the production of antimicrobial substances (Maragkoudakis and Tsakalidou, 2007). Typical products of the metabolism of these microbial cultures are the organic acids (Steiner and Sauer, 2003).

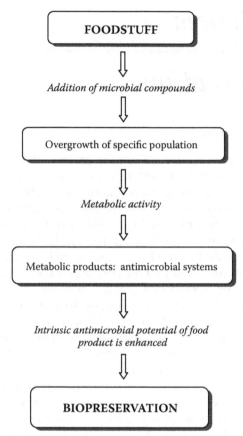

Figure 13.1 Schematic representation of the processes involved in biopreservation (Data from Riley, A.P. (Ed.) *New Developments in Food Policy Control and Research*, pp. 31–64. New York: Nova Science. 2005. Settani, L. and Corsetti, A. *International Journal of Food Biology* 121:123–138. 2008.

13.2   *LAB and biopreservation*

Residues of cheese found in an Egyptian pot that dates from 2300 BC (McGee, 1984) as well as passages from the Bible indicate that lactic acid bacteria have been used for at least 4,000 to 5,000 years for the fermentation and preservation of food (Davidson et al., 1995). Lactic acid bacteria (LAB) are naturally found in a wide range of foodstuffs and may under certain circumstances cause spoilage. However, they are often deliberately introduced in the production of a wide range of fermented foods, as it was found that certain treatments under certain storage conditions caused LAB outgrowth in raw materials, which produced desirable sensory characteristics, and also improved the shelf life of the products. Various LAB currently play an important role in the production of safe fermented foods that are also shelf stable. LAB have been used as natural biopreservatives of food and feed and are considered harmless or even to improve human and animal health (probiotics). Their primary inhibitory effect is speculated to be some metabolic feature that is present in all of them, which refers to their fermentative pathways. They use these fermentative pathways to generate cellular energy and to produce organic acids wherever they grow (mainly lactic acid). This also results in a decrease in pH of the surrounding medium (Adams and Nicolaides, 1997).

Lactic acid bacteria and bifidobacteria are preferred as protective and probiotic cultures, and have been used since the beginning of history as starter cultures. They have a long history of being safely used and consumed. LAB are widely used for fermentation of milk, meat, and vegetable foods. In fermentation of dairy products, lactose is metabolized to lactic acid. Other metabolic products, hydrogen peroxide, diacetyl, and bacteriocins may also play inhibitory roles and contribute to improving the organoleptic attributes of these foods, as well as their preservation (Messens and De Vuyst, 2002).

Various compounds produced by starter bacteria can be involved in preventing the growth of unwanted bacteria, but the production of sufficient amounts of lactic acid is fundamental to the success of a fermentation process (Davidson et al., 1995). Although lactic acid is the major LAB metabolite, heterofermentative LAB produce relatively high amounts of acetic acid and also trace amounts of propionic acid (Schnürer and Magnussen, 2005). LAB also produce bacteriocins, which provide them with desirable properties and render them suitable for food preservation. Organic acids enhance the activity of bacteriocins, and acidification improves the antibacterial activity of both organic acids and bacteriocins (Jack, Tagg, and Ray, 1995; Stiles, 1996; Galvez et al., 2007). The types of organic acids produced during the fermentation process, and also the concentration produced, depend on (1) the LAB species or strains, (2) culture composition, and (3) growth conditions (Lindgren and Dobrogosz, 1990; Ammor et al., 2006). Most manufacturing processes of vegetable-based foods involve lactic acid fermentation, which produces biological stability and improved organoleptic properties of the food products (Rubia-Soria et al., 2006). Homo- and heterofermentative LAB are also widely used as starter cultures in sourdough and to improve the sensory and technological properties of bread (Gianotti et al., 1997; Meignen et al., 2001; Şimşek, Çon, and Tulumoğlu, 2006). The most important antimicrobial compounds produced by LAB are presented in Table 13.1, and the antimicrobial activity associated with the presence of LAB in foodstuffs are summarized in Table 13.2.

LAB include the genera *Lactococcus, Streptococcus, Pediococcus, Leuconostoc, Enterococcus, Carnobacterium,* and *Propionibacterium,* and a wide variety of strains are routinely used as starter cultures in manufacturing dairy, meat, and vegetable products (O'Sullivan, Ross, and Hill,

Table 13.1 Factors Involved with Antimicrobial Activity of LAB in Biopreservation

Products of LAB metabolism	Function
Organic acids (lactic, acetic)	Increase acidity, antimicrobial compounds
Bacteriocins	Nisin; only bacteriocin permitted as food preservative, disrupts cytoplasmic membrane
CO_2	Reduces membrane permeability
Hydrogen peroxide	Oxidizes proteins
Diacetyl	Interacts with arginine-binding proteins
Reuterin	Not confirmed, may interact with thiol group in proteins that may lead to oxidative stress (Whitehead et al., 2008)
Ethanol	

Source: Adams, M.R. *Microbiology of Fermented Food*, pp. 1–44. London: Blackie Academic and Professional, 1998.

Table 13.2 Antimicrobial Action Performed by the Presence of LAB

Consequence of presence of LAB	Antimicrobial action
Low redox potential	pH reduction of surrounding medium (pH 3.5–4.5)
Nutrient depletion	Growth decreases carbohydrate food content
Crowding	Competitive growth inhibits spoilage or pathogenic organisms

2002). *Lactococcus lactis* is the best characterized species among the lactic acid bacteria (Kim, Jeong, and Lee, 2007).

13.3 Other organisms implicated in biopreservation

Some known commercial products including Microguard® and Bioprofit® are available. In these products the use of a *Propionibacterium freudenreichii* strain, plus *Lactobacillus rhamnosus* is employed to increase the inhibitory activity against fungi and some Gram-positive bacteria (Caplice and Fitzgerald, 1999). In such a mixture propionic acid may be produced from sugars through lactate as an intermediate. This is an alternative from using only propionibacteria in producing acid from sugar (Tyree, Clausen, and Gaddy, 1991).

Acetic acid bacteria (AAB) are known for oxidizing various ethanol-containing substrates into a variety of vinegars, but are also used for the production of some biotechnologically important products, such as sorbose and gluconic acids. However, the presence of AAB is not always welcomed as certain species will spoil wine, beer, and juices (Trček, 2005).

Vinegar has been known and used by most ancient civilizations. Its use as a preservative agent, as well as seasoning additive, is as ancient as the use of wine (Tesfaye et al., 2002). Vinegar is defined as "a liquid fit for human consumption, produced from a suitable raw material of agricultural origin, containing starch, sugars, or starch and sugars by the process of double fermentation, alcoholic and acetous, and contains a specified amount of acetic acid" (FAO/WHO, 1987). It is produced by a two-stage fermentation process, comprising (1) the conversion of fermentable sugars to ethanol by yeasts, usually *Saccharomyces* spp., and (2) the oxidation of ethanol by bacteria, mainly *Acetobacter* spp. (Adams, 1998).

Aspergillus niger is primarily used in industry for the production of enzymes and organic acid and has been used for many years in the food industry. No adverse effects on human health have yet been reported. This yeast has been used for the production of organic acids such as citric acid and gluconic acid since the beginning of the previous century. These organic acids remain the only ones that are used in significant quantities as food additives and are still produced by mycological processes (Bigelis and Lasure, 1987). However, citric acid production has not received much attention concerning modern methods of molecular biology. It may be presumed that this area is considered an established area (Abarca et al., 2004).

Interest in the yeast *Yarrowia lipolytica* has also developed over the past few years, due to potentially numerous biotechnological applications. Although this organism is not considered to be a dangerous food spoilage yeast, various cases have been reported from food environments, some associated with severe damage (Rodrigues and Pais, 2000).

13.4 New technologies and applications

In 2002, a review was published on the position of genomics of food-related bacteria, and at that stage 29 ongoing genome-sequencing projects were described, comprising LAB (24), bifidobacteria (3), propionic acid bacteria (1), and *Brevibacterium* (1) (Nes and Johnsborg, 2004). It is evident that awareness of microorganisms retaining essential activities after being consumed, is becoming more and more obvious. Because these organisms can then interact within the consumer's intestine, genomics of food-grade microorganisms and food pathogens as well as spoilage organisms are, therefore, an essential part of safe food production, in relation to human genomics (De Vos, 2001).

A culture cocktail has been developed, consisting of a lactic acid producing bacterial culture condensate mixture (LCCM) to be applied as an antimicrobial. This mixture has a low pH and also contains bacteriocin-like substances. Ingestion of LCCM after a meal is proposed to help in preventing *Salmonella enterica* sv. *Enteritidis* infection (Park, Seok, and Cho, 2005). In another project LAB was proposed as being effective for the fermentation of raw fish material, in addition to its use as a bioprotector. This is done in attempting to change its sensory qualities and to attain fish products that organoleptically resemble meat products. A starter culture of *Lactobacillus mesenteroides* is considered a possible candidate for achieving this. It would be of great significance to obtain fish with desirable meat-like properties to meet consumer concern for safety and nutritive seafoods of expected quality (Gelman, Drabkin, and Glatman, 2000).

13.5   Consumer acceptance

Current research on original and new food technologies is mainly focused on addressing consumer concerns and consumer acceptance. Much research has already been dedicated to the evaluation of apparent risks and concerns of consumers toward various food-related issues. These commonly include irradiated foods, bioengineered foods, pesticides, foods processed with laser light sources, and microbially contaminated foods (Schutz, Brühn, and Diaz-Knauf, 1989; Bord and Conner, 1990; Dunlap and Beus, 1992; Frewer, Shepherd, and Sparks, 1994; Cardello, Schutz, and Lesher, 2007). It has become obvious that optimal sensory quality will not on its own guarantee success and that foods produced or processed by the various available technologies pose challenging problems for researchers interested in factors implicated in consumer choice, consumer acceptance, and consumer purchase behavior (Cardello, Schutz, and Lesher, 2007).

It has been decided that the global response of consumers to functional food-grade bacteria may actually assist in demonstrating the mechanism by which these bacteria affect human health. It may provide an understanding and allow optimization of the effect of probiotic bacteria, and also LAB. However, legislative bodies will require genomewide profiling, especially when dealing with novel foodstuffs produced by biopreservation techniques (De Vos, 2001).

13.6   Organic acids and probiotics

Probiotics are defined as viable microorganisms that upon ingestion will provide health benefits that are in addition to the inherent basic nutrition. According to the Food and Agriculture Organization (FAO), a "probiotic" is a live microorganism that, when consumed in sufficient amounts, produces a health benefit to the human host (FAO, 2001; Vankerckhoven et al., 2004). LAB are increasingly implemented as probiotics to assist in stimulating the immune response against infection by enteropathogenic bacteria. They are also often used in the treatment and prevention of diarrhea

(Reid, 1999; Mohd Adnan and Tan, 2007). LAB produce organic acids that decrease the pH of the environment. This low pH then becomes selective against sensitive microorganisms. LAB are known to be resistant to a wide range of antibiotics (Herreros et al., 2005) and treatment with such antibiotic-resistant LAB strains can help uphold the normal bacterial population in the intestines, or help in restoring it after antibiotic treatment, as the normal balance of host normal flora are often disrupted by the application of antibiotics, causing intestinal disorders (Gotcheva et al., 2002). Compounds responsible for inhibiting gastrointestinal pathogens are produced by probiotic or potentially probiotic LAB, but the chemical composition is not yet completely described (De Vuyst, Makras, and Holo, 2004).

In 2004 the PROSAFE collection of probiotic lactic acid bacteria was established. This is an E.U.-funded project with the objective to investigate the biosafety of LAB intended for human consumption. The collection comprised 907 LAB strains of nutritional or human origin and included the genera *Lactobacillus*, *Bifidobacterium*, and *Enterococcus*. Most probiotic strains were received as lactobacilli or bifidobacteria (Vankerckhoven et al., 2004).

Probiotics also play a role in dairy fermentations to assist in the preservation of milk as a result of the production of lactic acid and other antimicrobial compounds. They have other functions as well, which are not involved with preservation or organic acids (Parvez et al., 2006).

References

Abarca, M.L., Accensi, F., Cano, J., and Cabanes, F.J. 2004. Taxonomy and significance of black aspergilli. *Antonie Van Leeuwenhoek* 86:33–49.

Adams, M.R. 1998. Vinegar. In: J.B. Wood (Ed.) *Microbiology of Fermented Food*, pp. 1–44. London: Blackie Academic and Professional.

Adams, M.R. and Nicolaides, L. 1997. Review of the sensitivity of different food-borne pathogens to fermentation. *Food Control* 8:227–239.

Ammor, S., Tauveron, G., Dufour, E., and Chevallier, I. 2006. Antibacterial activity of lactic acid bacteria against spoilage and pathogenic bacteria isolated from the same meat small-scale facility: 1—Screening and characterization of the antibacterial compounds. *Food Control* 17:454–461.

Bigelis, R. and Lasure, L.L. 1987. Fungal enzymes and primary metabolites used in food processing. In: L.R. Beuchat (Ed.), *Food and Beverage Mycology*, pp. 473–516. New York: Van Nostrand Reinhold.

Bord, R.J. and Conner, R.E. 1990. Risk communication, knowledge and attitudes: Explaining reactions to a technology perceived as risky. *Risk Analysis* 10:499–506.

Caplice, E. and Fitzgerald, G.F. 1999. Food fermentations: Role of microorganisms in food production and preservation. *International Journal of Food Microbiology* 50:131–149.

Cardello, A.V., Schutz, H.G., and Lesher, L.L. 2007. Consumer perceptions of foods processed by innovative and emerging technologies: A conjoint analytic study. *Innovative Food Science and Emerging Technologies* 8:73–83.

Davidson, B.E., Llanos, R.M., Cancilla, M.R., Redman, N.C., and Hillier, A.J. 1995. Current research on the genetics of lactic acid production in lactic acid bacteria. *International Dairy Journal* 5:763–784.

De Vos, W.M. 2001. Advances in genomics for microbial food fermentations and safety. *Current Opinion in Biotechnology* 12:493–498.

De Vuyst, L., Makras, L.A., and Holo, H. 2004. Antimicrobial potential of probiotic or potentially probiotic lactic acid bacteria, the first results of the International European Research Project PROPATH of the PROEUHEALTH cluster. *Microbial Ecolocy in Health and Disease* 16:125–130.

Dunlap, R.E. and Beus, C.E. 1992. Understanding public concerns about pesticides: An empirical examination. *Journal of Consumer Affairs* 26:418–438.

FAO. 2001. Joint FAO/WHO Expert consultation on evaluation of health and nutritional properties of probiotics in food including powder milk with live lactic acid bacteria. Amerian Córdoba Park Hotel, Córdoba, Argentina.

FAO/WHO Food Standards Programme. 1987. Codex standards for sugars, cocoa products and chocolate and miscellaneous. Codex standard for vinegar. In *Codex alimentarius*. Regional European standard, Codex Stan 162. Ginebra.

Frewer, L.J., Shepherd, R., and Sparks, P. 1994. The interrelationship between perceived knowledge, control and risk associated with a range of food related hazards targeted at the individual, other people and society. *Journal of Food Safety* 14:19–40.

Galvez, A., Abriouel, H., Lopez, R.L., and Ben, O.N. 2007. Bacteriocin-based strategies for food biopreservation. *International Journal of Food Microbiology* 120:51–70.

Gelman, A., Drabkin, V., and Glatman, L. 2000. Evaluation of lactic acid bacteria, isolated from lightly preserved fish products, as starter cultures for new fish-based food products. *Innovative Food Science and Emerging Technologies* 1:219–226.

Gianotti, A., Vannini, L., Gobbetti, M., Corsetti, A., Gardini, F., and Guerzoni, M. 1997. Modeling of the activity of selected starters during sourdough fermentation. *Food Microbiology* 14:327–337.

Gotcheva, V., Hristozova, E., Hristozova, T., Guo, M., Roshkova, Z., and Angelov, A. 2002. Assessment of potential probiotic properties of lactic acid bacteria and yeast strains. *Food Biotechnology* 16:211–225.

Herreros, M.A., Sandoval, H., González, L., Castro, J.M., Fresno, J.M., and Tornadijo, M.E., 2005. Antimicrobial activity and antibiotic resistance of lactic acid bacteria isolated from Armada cheese (a Spanish goat's milk cheese). *Food Microbiology* 22: 455–459.

Jack, R.W., Tagg, J.R., and Ray, B. 1995. Bacteriocins of Gram positive bacteria. *Microbiological Reviews* 59:171–200.

Kim, J.E., Jeong, D.W., and Lee, H.J. 2007. Expression, purification, and characterization of arginine deiminase from *Lactococcus lactis* ssp. lactis ATCC 7962 in *Escherichia coli* BL21. *Protein Expression and Purification* 53:9–15.

Lindgren, S.E. and Dobrogosz, W.J. 1990. Antagonistic activities of lactic acid bacteria in food and feed fermentation. *FEMS Microbiology Reviews* 7:149–163.

Maragkoudakis, P. and Tsakalidou, E. 2007. *PathogenCombat probiotic & protective cultures: Properties & potential applications.* Laboratory of Dairy Research, Agricultural University of Athens.

McGee, H. 1984. *On Food and Cooking*, p. 36. London: Unwin Hyman.

Meignen, B., Onno, B., Gelinas, P., Infontes, M., Guilois, S., and Cahagnier, B. 2001. Optimization of sourdough fermentation with *Lactobacillus brevis* and baker's yeast. *Food Microbiology* 18:239–245.

Messens, W. and De Vuyst, L. 2002. Inhibitory substances produced by *Lactobacilli* isolated from sourdoughs—A review. *International Journal of Food Microbiology* 72:31–43.

Mohd Adnan, A.F. and Tan, I.K.P. 2007. Isolation of lactic acid bacteria from Malaysian foods and assessment of the isolates for industrial potential. *Bioresource Technology* 98:1380–1385.

Nes, I.F. and Johnsborg, O. 2004. Exploration of antimicrobial potential in LAB by genomics. *Current Opinion in Biotechnology* 15:100–104.

O'Sullivan, L., Ross, R.P., and Hill, C. 2002. Potential of bacteriocin-producing lactic acid bacteria for improvements in food safety and quality. *Biochimie* 84:593–604.

Park, J.H., Seok, S.H., and Cho, S.A. 2005. Antimicrobial effect of lactic acid producing bacteria culture condensate mixture (LCCM) against *Salmonella enteritidis*. *International Journal of Food Microbiology* 101:111–117.

Parvez, S., Malik, K.A., Ah, K.S., and Kim, H.Y. 2006. Probiotics and their fermented food products are beneficial for health. *Journal of Applied Microbiology* 100:1171–1185.

Reid, G. 1999. The scientific basis for probiotic strains of *Lactobacillus*. *Applied and Environmental Microbiology* 65:3763–3766.

Riley, A.P. (Ed.). 2005. In: *New Developments in Food Policy Control and Research*, pp. 31–64. New York: Nova Science.

Rodrigues, G. and Pais, C. 2000. The influence of acetic and other weak carboxylic acids on growth and cellular death of the yeast *Yarrowia lipolytica*. *Food Technology and Biotechnology* 38:27–32.

Ross, R.P., Morgan, S., and Hill, C. 2002. Preservation and fermentation: Past, present and future. *International Journal of Food Microbiology* 79:3–16.

Rubia-Soria, A., Abriouel, H., Lucas, R., Ben, O.N., Martinez-Canamero, M., and Galvez, A. 2006. Production of antimicrobial substances by bacteria isolated from fermented table olives. *World Journal of Microbiology and Biotechnology* 22:765–768.

Schnürer, J. and Magnusson, J. 2005. Antifungal lactic acid bacteria as bio-preservatives. *Trends in Food Science and Technology* 16:70–78.

Schutz, H.G., Brühn, C.M., and Diaz-Knauf, K.V. 1989. Consumer attitude toward irradiated foods: Effects of labeling and benefit information. *Food Technology* 43:80–86.

Settani, L. and Corsetti, A. 2008. Application of bacteriocins in vegetable food preservation. *International Journal of Food Microbiology* 121:123–138.

Şimşek, O., Çon, A.H., and Tulumoğlu, S. 2006. Isolating lactic starter cultures with antimicrobial activity for sourdough processes. *Food Control* 17:263–270.

Soomro, A.H., Masud, T., and Anwaar, K. 2002. Role of lactic acid bacteria (LAB) in food preservation and human health—A review. *Pakistan Journal of Nutrition* 1:20–24.

Steiner, P. and Sauer, U. 2003. Overexpression of the ATP-dependent helicase RecG improves resistance to weak organic acids in *Escherichia coli*. *Applied Microbiology and Biotechnology* 63:293–299.

Stiles, M.E. 1996. Biopreservation by lactic acid bacteria. *Antonie van Leeuwenhoek* 70:331–345.

Tesfaye, W., Morales, M.L., Garcia-Parrilla, M.C., and Troncoso, A.M. 2002. Wine vinegar: Technology, authenticity and quality evaluation. *Trends in Food Science and Technology* 13:12–21.

Trček, J. 2005. Quick identification of acetic acid bacteria based on nucleotide sequences of the 16S–23S rDNA internal transcribed spacer region and of the PQQ-dependent alcohol dehydrogenase gene. *Systematic Application of Microbiology* 28:735–745.

Tyree, R.W., Clausen, E.C., and Gaddy, J.L. 1991. The production of propionic acid from sugars by fermentation through lactic acid as an intermediate. *Journal of Chemical Technology and Biotechnology* 50:157–166.

Vankerckhoven, V.V., Van Autgaerden, T., Huys, G., Vancanneyt, M., Swings, J., and Goossens, H. 2004. Establishment of the PROSAFE collection of probiotic and human lactic acid bacteria. *Microbial Ecology in Health and Disease* 16:131–136.

Whitehead, K.J., Versalovic, J., Roos, S., and Britton, R.A. 2008. Genomic and genetic characterization of the bile stress response of probiotic *Lactobacillus reuteri* ATCC 55730. *Applied and Environmental Microbiology* 74:1812–1819.

chapter fourteen

Novel applications for organic acids

14.1   Emerging challenges

Although various criteria are used in interpreting the quality and acceptance of a food product, the acceptance limit is still mainly defined by economic and physiological factors (Rico et al., 2007). Research has been intensified, shifting the focus on the implications of decontamination on pathogen behavior, in association with the overall microbial ecology of food products (Samelis et al., 2002).

Activity of organic acids is mainly inhibitory and not microbicidal, and their even distribution is, therefore, essential to effectively combat contamination. However, suboptimal doses pose a problem by allowing spoilage, very often mold spoilage. In addition to economic losses caused by visible growth of spoilage molds, health hazards associated with mycotoxins are also causing concern (Arroyo, Aldred, and Magan, 2005).

14.2   Consumer satisfaction

Consumers are becoming more and more critical about the use of synthetic additives (Bruhn, 2000). However, they tend to judge the quality of fresh-cut fruit and vegetables on their appearance and freshness at time of purchase, whereas subsequent purchases depend upon the consumer's satisfaction in terms of most other characteristics of the food (Rico et al., 2007).

In recent years, consumer pressure has led to a reduction in the use of preservatives. However, the use of a suboptimal concentration of preservatives may actually stimulate the growth of some spoilage fungi such as *Hyphopichia burtonii* and *Candida guillermondii*, *Eurotium* spp., and *Aspergillus flavus*, which may cause mycotoxin production to be stimulated (Magan and Lacey, 1986; Mutasa and Magan, 1990; Arroyo, Aldred, and Magan, 2005). Public authorities are, therefore, pushing industry in developing comprehensive quality management systems that would enhance availability of information to the consumer to regain consumers' trust in food. Most food producers run a quality management system, according to DIN EN ISO 9000ff, which defines quality as the "degree to which a set of inherent (existing) characteristics fulfills requirements." This definition implies a key issue of customer satisfaction in a revision of DIN EN ISO 9000ff which was published in the year 2000. Food producers are hereby required to monitor information with regard to customer perception, and if the organization has fulfilled customer requirements (Rohr et al., 2005).

14.3 Optimizing organic acid application in animal feed

It has been proposed that the ruminal microbial ecosystem be manipulated to improve the efficiency of converting feeds to products consumable by humans. In so doing, the organic acids may replace antibiotics and be added to feed for ruminants. Fortunately, application in ruminants is less extensive than in other farm animals and organic acids such as aspartate, citric acid, succinic acid, or pyruvic acid have been investigated and proposed as potential agents for use in ruminants. The advantages of these substances in the rumen are described as (Castillo et al., 2004):

- Stimulation of lactate uptake by *Selenomonas ruminatum* to prevent or correct a decrease in ruminal acidosis
- Reduction of methanogenesis, by reducing energy losses associated with CH_4 production in the rumen
- Buffering of ruminal pH

Although limited information is available from research conducted on the ideal dosage and feeding techniques, it is recommended that powder or granular form be used to avoid prolonged residence in the rumen or an increase in the yield of acetate and propionate. The optimal amount of organic acid needed also varies with the final form of the feed, and inclusion of forage in the diet. A proper dosage has been found to be 1–5% by body fat. If the proportion is less than 0.1%, the effect will, therefore, not be sufficient, and a proportion higher than 10% will affect the palatability of the diet. As a result, many more *in vivo* trials are to be conducted to determine the effects of these organic acids on beef cattle performance (Castillo et al., 2004).

Edema disease in piglets, caused by ETEC, normally occurs after weaning. Acidic conditions in the GIT are known to have bactericidal effects on potentially harmful bacteria, as it is a more favorable environment for lactobacilli, which may inhibit the colonization and proliferation of *Escherichia coli*. Lactobacilli also produce metabolites against Gram-negative bacteria. Organic acids may also be applied to reduce coliform along the GIT, and various application methods are being investigated (Tsiloyiannis et al., 2001).

Organic acids have also been proposed to be used in rearing water systems for broilers to reduce cross-infection of *Campylobacter* spp. on a farm. However, further research is necessary to specify the exact effect of acidified water on the transmission of *Campylobacter* among chickens (Chaveerach et al., 2002).

At a pH below 3–3.5, organic acids are very efficacious in controlling bacterial growth, but this does not reflect the situation in the GIT of poultry and pigs. It may, therefore, be a logical idea to add organic acids to feeds and to protect them against dissociation in the crop and in the intestine as a result of the higher pH. This is necessary for the organic acids to reach far into the GIT, where the majority of microbial populations are found (Gauthier, 2005).

14.4 Preservative combinations

Combinations of antimicrobials in formulations significantly cause greater inhibition of growth than antimicrobials used singly. Although combinations of antimicrobials do not decrease, and may sometimes even increase water activity of various food products, they have been reported to effectively inhibit *Listeria monocytogenes*. In addition, the antilisterial effect of combination treatments in antilisterial formulations appears to increase when the product is dipped into solutions of organic acid (Barmpalia et al., 2004).

Lactic acid and propylene glycol have been found to be effective when added to bakery products such as sponge cakes at different doses in attempting to prolong their normally short shelf life. Propylene glycol acts as an emulsifier and stabilizer for cake products and its use is permitted in various preparation processes as a sustaining solvent of preservatives or as a moistening agent (De la Rosa et al., 2005).

14.5 Antimicrobial packaging

According to the WHO, in developing countries effective packaging plays a very essential part in the quality of produce and estimates that without it, 50% of produce may be lost before reaching the consumer. But after performing its function, packaging is just burdensome waste. Edible films are based on food ingredients, and measure only 5/1000 of an inch thick. Currently various films are being studied, including those made of pureed spinach and apples, infused with oils. Carvacrol is a popular choice and is found in oregano (Friedman et al., 2009).

An economical method for manufacturing antimicrobial packaging is achieved by direct incorporation of the preservative into film. Such an example is the incorporation of sorbic and benzoic acid in methylcellulose, hydroxylpropylcellulose and chitosan films (Begin and Van Calsteren, 1999). Antimicrobial packaging is a form of active food packaging, in particular for meat products, showing great promise and attention, as microbial contamination of many foods occurs primarily at the surface, mainly as a result of postprocess handling. "Active packaging" is a type of packaging developed to extend the shelf life, improve the safety, or improve sensory properties while at the same time also preserving the quality of the food (Quintavalla and Vicini, 2002).

The agent is allowed to slowly migrate from the packaging material to the surface of the product. This is an exceptionally challenging technology, but may have a significant impact on extending shelf life and safety, especially of meat products. Direct surface application of antibacterial substances onto the surface of foods has limited benefits, as these substances are rapidly neutralized and also diffuse from the surface into the food. Moreover, if antimicrobial agents are incorporated into meat formulations, this may again result in the active substances being inactivated by the contents of the foodstuff (Quintavalla and Vicini, 2002).

In synthetic polymers, as well as edible films, various classes of antimicrobial compounds have been evaluated. These include organic acids, enzymes, bacteriocins, and other compounds which may consist of triclosans, silver zeolites, and fungicides. Several approaches are followed when incorporating an antimicrobial substance into packaging material (Quintavalla and Vicini, 2002):

1. To put the antimicrobial into a film, at the time when the film is produced. However, this is not cost effective and the antimicrobial material, which is then not exposed to the surface of the film is not effectively active.
2. To apply an antimicrobial additive in a controlled matter where the material is needed and not lost.

In multilayered packaging material, antimicrobials may be incorporated into the food contact layer.

14.5.1   Factors for the design of antimicrobial film or packaging

1. Inherent antimicrobial activity and chemical character of films
2. Storage temperature
3. Mass transfer coefficients
4. Physical properties of the packaging material

The organic acids, sorbate, propionate, and benzoate, as well as their respective anhydrates, have been proposed and tested for antimicrobial activity (Quintavalla and Vicini, 2002).

Organic acids combined with dynamic gas exchange modified atmosphere packaging are also known to improve some shelf life properties of retail pork chops.

14.6   Optimizing commercial trials

Commercial trials are essential in evaluating organic acid systems and also in developing new systems suitable for the commercial environment to improve meat safety without sacrificing sought-after sensory characteristics. With such information it is possible to convince regulatory agencies and food processors to redirect their decontamination efforts in improving food quality as well as food safety (Smulders and Greer, 1998).

The American Heart Association (AHA) promotes the consumption of fatty fish, at least twice a week, in obtaining the health benefits of omega-3 fatty acids (Sallam, 2007). However, fish contain relatively large quantities of free amino acids and volatile nitrogen bases and are more perishable than other foods (Ashie, Smith, and Simpson, 1996). More research and commercial testing of organic acid application have become important, and sodium salts of low molecular weight organic acids (acetic, lactic, and citric acids) are continuously being investigated. Effectiveness of, in particular, sodium lactate on microbial growth in fish products has shown discrepancies and is dependent on several factors, including: (1) the concentration of sodium lactate used, (2) dipping period, (3) fish species, (4) type of fish product, (5) degree of microbial contamination, and (6) storage condition. Many more commercial trials are needed to address these issues (Sallam, 2007).

Because of a concern that continuous application of organic acids is causing microbial resistance, research is needed on alternative application protocols and to expand the diversity of potential antimicrobials with practical application to food production. To successfully achieve this, it is essential to explore the interaction between the food matrix and food-borne pathogens (Ricke et al., 2005).

14.7   New possibilities in minimally
processed foods

Minimal processing techniques have been developed in response to the challenges of preservation while retaining nutritional and sensory quality (Rico et al., 2007). Current interests include the combination of mild treatments to improve food safety and to extend shelf life. Incorporation of organic acids such as ascorbic acid into ground beef and also spices into edible coating film have been proposed to stabilize lipid oxidation and production of –SH radicals in ground beef during storage after irradiation storage (Ouattara et al., 2002).

14.8   Alternatives to washing techniques

The majority of manufacturers of fresh minimally processed produce use chlorine-based washing procedures, which have led to controversies about the formation of carcinogenic compounds in water (chloramines and trihalomethanes) and have questioned the use of chlorine (Wei et al., 1999; Rico et al., 2007). Preservation of fresh-cut fruits and vegetables has mainly consisted of washing treatments, but alternative methods or modified methods have been proposed and are being implemented. These include (1) antioxidants, (2) irradiation, (3) ozone, (4) modified atmosphere packaging, (5) whey permeate, and (6) organic acids. Organic acids have also proved to be effective antimicrobial agents against mesophilic as well as psychrophilic microorganisms in fresh-cut vegetables (Uyttendaele et al., 2004; Bari et al., 2005). However, none have yet gained widespread acceptance by the industry and minimal processing of fresh-cut fruits and vegetables renders the products highly perishable with chilled storage still required to ensure reasonable shelf life (Garcìa and Barret, 2002; Rico et al., 2007). The proposed washing techniques are summarized and presented in Table 14.1 (Rico et al., 2007).

14.9   Alternative application regimes

Studies have been done on testing antimicrobial compounds for antilisterial activity, but none of these reported on the sequential application of dipping or spraying for the control of *L. monocytogenes* in RTE (ready-to-eat) meat products (Geornaras et al., 2005). Studies were performed fairly recently to show the strong antimicrobial action of sodium citrate and sodium lactate in controlling *Clostridium perfringens* during cooling procedures (Sabah, Juneja, and Fung, 2004).

Fumigation of fruit and vegetable crops is not widely used for controlling postharvest decay. However, fumigants do have important properties, which renders them very practical in the application against decay caused by fungi. Fumigants can penetrate into areas that are not easily accessible to liquid pesticide sprays. They then drench and exert their effect during the exposure period but afterwards diffuse away, leaving little or no residue. However, there is always the risk of damaging the fruit in commercial packinghouses. For application to large volumes of fruit, care should be taken to monitor concentration and to maintain a safe level. In addition, acetic acid vapor is very corrosive to equipment and repeated exposure should be avoided. Acetic acid vapor is proposed

Table 14.1 Washing of Minimally Processed Fruits and Vegetables with Chemical-Based Treatments

Chemical	Treatment	Produce treated	Advantages/ future trends
Calcium-based solutions	Calcium lactate Calcium chloride Calcium propionate	Fruits and vegetables, as firming agent (strawberries) or to maintain firmness Delicate fruits Fruit with high senescence index (grapefruit)	Found that final product can increase calcium content Calcium lactate good alternative to calcium chloride, less bitterness or off-colors
Chlorine	Liquid chlorine Hypochlorite	Traditionally vegetables	Effectiveness questioned Possible future regulations
Chlorine dioxide (ClO_2)	ClO_2 combined with ozone, thyme essential oil	Fruit and vegetables	High oxidation capacity Does not form dangerous chloramine compounds
Electrolyzed water	Generated by electrolysis of aqueous NaCl to produce electrolyzed basic solution at cathode and electrolyzed acidic solution at anode	Whole lettuce Fresh-cut vegetables	More effective than chlorine due to high reduction potential Higher effectiveness than ozone in reducing viable aerobes on whole lettuce Neutralizes harmful substances, such as cyanides, ammonium, etc. Few studies on fresh-cut vegetables
Hydrogen peroxide	Dipping in H_2O_2 solution	Fresh-cut peppers, cucumber, zucchini, cantaloupe, honeydew melon	Sporocidal: effective on food contact surfaces, and packaging material

(continued)

Table 14.1 Washing of Minimally Processed Fruits and Vegetables with
Chemical-Based Treatments (Continued)

Chemical	Treatment	Produce treated	Advantages/future trends
Organic acids	Lactic acid Acetic acid Ascorbic acid Citric acid Tartaric acid	Fresh-cut fruits and vegetables	Strong antimicrobial agent against psychrophilic and mesophilic microorganisms Natural preservatives, effective in retaining quality of minimally processed products: (1) antimicrobial activity, (2) inhibition of spoilage, avoiding oxidative processes
Ozone	Ozonated water	Fresh-cut vegetables	Beneficial effect on storage life of fresh noncut commodities: broccoli, cucumbers, apples, grapes, oranges, pears, raspberries, strawberries Studies needed on quality Higher corrosiveness than chlorine use

Source: Data from Rico et al., *Trends in Food Science & Technology*, 18:373–386, 2007.

as an alternative postharvest fungicide as opposed to thiabendazole, but should be applied on small volumes of fruit that could be fumigated in a small room or tent (Sholberg et al., 2004).

14.10   Recognizing the need in RTE foods

Ready-to-eat meat or poultry products are often reported as sources of *L. monocytogenes* infections in humans (Barmpalia et al., 2004). Deli meats and nonreheated frankfurters are the highest risk products for listeriosis with regard to preservation, because they may be consumed without cooking or reheating. There is, therefore, a need for additional precautions to ensure the safety of the product until it is consumed. Application of organic acids to the surface of RTE meats should be most effective, because *L. monocytogenes* contamination is typically concentrated on the surface (Lu et al., 2005). Factors that influence antilisterial activity of organic acids on frankfurters usually act in an interactive manner and combined treatments usually have a greater effect than individual treatments (Harmayani, Sofos, and Schmidt, 1993; Mossel et al., 1995; Davidson, 2002).

Inclusion of antimicrobials in the formulation of RTE meat products is prescribed in effective control. Many U.S. meat processors are, therefore, currently adding preservatives, such as sodium or potassium lactate (up to 2%) in combination with sodium diacetate (0.05–0.15%) to product formulations (Thompkin, 2002). Organic acids (or their salts) plus bacteriocins, applied as immersion or spraying solutions alone or in combination, can also control *L. monocytogenes* contamination on RTE meat products during storage (Geornaras et al., 2006).

References

Arroyo, M., Aldred, D., and Magan, N. 2005. Environmental factors and weak organic acid interactions have differential effects on control of growth and ochratoxin A production by *Penicillium verrucosum* isolates in bread. *International Journal of Food Microbiology* 98:223–231.

Ashie, I.N.A., Smith, J.O., and Simpson, B.K. 1996. Spoilage and shelf life extension of fresh fish/shellfish. *Critical Reviews in Food Science and Nutrition* 36:87–122.

Bari, M.L., Ukuku, D.O., Kawasaki, T., Inatsu, Y., Isshiki, K., and Kawamoto, S. 2005. Combined efficacy of nisin and pediocin with sodium lactate, citric acid, phytic acid, and potassium sorbate and EDTA in reducing the *Listeria monocytogenes* population of inoculated fresh-cut produce. *Journal of Food Protection* 68:1381–1387.

Barmpalia, I.M., Geornaras, I., Belk, K.E., Scanga, J.A., Kendall, P.A., Smith, G.C., and Sofos, J.N. 2004. Control of *Listeria monocytogenes* on frankfurters with antimicrobials in the formulation and by dipping in organic acid solutions. *Journal of Food Protection.* 67:2456–2464.

Begin, A. and Van Calsteren, M.R. 1999. Antimicrobial films produced from chitosan. *International Journal Biological Macromolecules.* 26:63–67.

Bruhn, C. 2000. Food labelling: Consumer needs. In: J.R. Blanchfield (Ed.), *Food Labelling*. Cambridge, UK: Woodhead.

Castillo, C., Benedito, J.L., Mendez, J., Pereira, V., Lopez-Alonso, M., Miranda, M., and Hernandez, J. 2004. Organic acids as a substitute for monensin in diets for beef cattle. *Animal Feed Science and Technology* 115:101–116.

Chaveerach, P., Keuzenkamp, D.A., Urlings, H.A., Lipman, L.J., and Van, K.F. 2002. In vitro study on the effect of organic acids on *Campylobacter jejuni/coli* populations in mixtures of water and feed. *Poultry Science* 81:621–628.

Davidson, P.M. 2002. Control of microorganisms with chemicals. In: V.K. Juneja and J.N. Sofos (Eds.), *Control of Food-Borne Microorganisms*, pp.165–190. New York: Marcel Dekker.

De la Rosa, P., Cordoba, G., Martin, A., Jordano, R., and Medina, L.M. 2005. Influence of a test preservative on sponge cakes under different storage conditions. *Journal of Food Protection* 68:2465–2469.

Friedman, M., Zhu, L., Feinstein, Y., and Ravishankar, S. 2009. Carvacrol facilitates heat induced inactivation of *Escherichia coli* O157:H7 and inhibits formation of heterocylic amines in grilled ground beef patties. *Journal of Agricultural and Food Chemistry*. 57, pp. 1848–1853.

Garcìa, E. and Barrett, D.M. 2002. Preservative treatments for fresh-cut fruits and vegetables. In: O. Lamikanra (Ed.), *Fresh-Cut Fruits and Vegetables. Science, Technology and Market*. Boca Raton, FL: CRC Press.

Gauthier, R. 2005. Organic acids and essential oils, a realistic alternative to antibiotic growth promoters in poultry. I Forum Internacional de avicultura 17–19 August 2005, pp. 148–157.

Geornaras, I., Belk, K.E., Scanga, J.A., Kendall, P.A., Smith, G.C., and Sofos, J.N. 2005. Postprocessing antimicrobial treatments to control *Listeria monocytogenes* in commercial vacuum-packaged bologna and ham stored at 10 degrees C. *Journal of Food Protection* 68:991–998.

Geornaras, I., Skandamis, P.N., Belk, K.E., et al. 2006. Post-processing application of chemical solutions for control of *Listeria monocytogenes*, cultured under different conditions, on commercial smoked sausage formulated with and without potassium actate-sodium diacetate. *Food Microbiology* 23:762–771.

Harmayani, E., Sofos, J.N., and Schmidt, G.R. 1993. Fate of *Listeria monocytogenes* in raw and cooked ground beef with meat processing additives. *International Journal of Food Microbiology* 18:223–232.

Lu, Z., Sebranek, J.G., Dickson, J.S., Mendonca, A.F., and Baily, T.B. 2005. Inhibitory effects of organic acid salts for control of *Listeria monocytogenes* on frankfurters. *Journal of Food Protection* 68:499–506.

Magan, N. and Lacey, J. 1986. Water relations and metabolism of propionate in two yeasts from hay. *Journal of Applied Bacteriology* 60:169–173.

Mossel, D.A.A., Cory, J.E.L., Struijk, C.B., and Baird, R.M. 1995. *Essentials of the Microbiology of Foods. A Textbook for Advanced Studies*. New York: John Wiley & Sons.

Mutasa, E.S. and Magan, N. 1990. Utilization of potassium sorbate by tobacco spoilage fungi. *Mycological Research* 94:965–970.

Ouattara, B., Giroux, M., Yefsah, R., Smoragiewicz, W., Saucier, L., Borsa, J., and Lacroix, M. 2002. Microbiological and biochemical characteristics of ground beef as affected by gamma irradiation, food additives and edible coating film. *Radiation Physics and Chemistry* 63:299–304.

Quintavalla, S. and Vicini, L. 2002. Antimicrobial food packaging in meat industry. *Meat Science* 62:373–380.

Ricke, S.C., Kundinger, M.M., Miller, D.R., and Keeton, J.T. 2005. Alternatives to antibiotics: Chemical and physical antimicrobial interventions and foodborne pathogen response. *Poultry Science* 84:667–675.

Rico, D., Martin-Diana, A.B., Barat, J.M., and Barry-Ryan, C. 2007. Extending and measuring the quality of fresh-cut fruit and vegetables: A review. *Trends in Food Science & Technology* 18:373–386.

Rohr, A., Luddecke, K., Drusch, S., Muller, M.J., and Alvensleben, R. 2005. Food quality and safety—Consumer perception and public health concern. *Food Control* 16:649–655.

Sabah, J.R., Juneja, V.K., and Fung, D.Y. 2004. Effect of spices and organic acids on the growth of *Clostridium perfringens* during cooling of cooked ground beef. *Journal of Food Protection* 67:1840–1847.

Sallam, K.I. 2007. Antimicrobial and antioxidant effects of sodium acetate, sodium lactate, and sodium citrate in refrigerated sliced salmon. *Food Control* 18:566–575.

Samelis, J., Sofos, J.N., Kendall, P.A., and Smith, G.C. 2002. Effect of acid adaptation on survival of *Escherichia coli* O157:H7 in meat decontamination washing fluids and potential effects of organic acid interventions on the microbial ecology of the meat plant environment. *Journal of Food Protection* 65:33–40.

Sholberg, P.L., Shephard, T., Randall, P., and Moyls, L. 2004. Use of measured concentrations of acetic acid vapour to control postharvest decay in d'Anjou pears. *Postharvest Biology and Technology* 32:89–98.

Smulders, F.J. and Greer, G.G. 1998. Integrating microbial decontamination with organic acids in HACCP programmes for muscle foods: Prospects and controversies. *International Journal of Food Microbiology* 44:149–169.

Thompkin, R.B. 2002. Control of *Listeria monocytogenes* in the food processing environment. *Journal of Food Protection* 65:709–725.

Tsiloyiannis, V.K., Kyriakis, S.C., Vlemmas, J., and Sarris, K. 2001. The effect of organic acids on the control of post-weaning oedema disease of piglets. *Research in Veterinary Science* 70:281–285.

Uyttendaele, M., Neyts, K., Vanderswalmen, H., Notebaert, E., and Debevere, J. 2004. Control of *Aeromonas* on minimally processed vegetables by decontamination with lactic acid, chlorinated water, or thyme essential oil solution. *International Journal of Food Microbiology* 90:263–271.

Wei, C.I., Huang, T.S., Kim, J.M., Lin, W.F., Tamplin, M.L., and Bartz, J.A. 1999. Growth and survival of *salmonella montevideo* on tomatoes and disinfection with chlorinated water. *Journal of Food Protection.* 58:829–836.

chapter fifteen

Detection of organic acids

15.1 Introduction

Timely and effective quantitative analysis of organic acids is essential as, together with inorganic anions, amino acids, and carbohydrates, these important compounds are important in many fields not only limited to agriculture and food science. Such areas include chemistry, biochemistry, and pharmaceuticals (Soga and Ross, 1999). Organic acids play pivotal roles in the metabolism of living beings, they occur naturally in a number of foods, and their analysis is of importance in food technology and physiology (Chen et al., 1999). A wide range of analytical methods has been developed for determining preservatives and organic acids, whereas individual determination is usually done by spectroscopic methods (Gonzalez, Gallego, and Valcarcel, 1998). Dissociation constants (pK_a) are fundamental in chromatographic procedures typically used for isolation (Gomis, 1992).

In addition, it is often necessary to determine organic acid composition in food to indicate contamination. Short-chain organic acids are known to significantly affect the flavor and the quality of food (Alur et al., 1995). These acids mainly include acetic acid and less often propionic and butyric acid, and may originate from raw materials or they may be produced by fermentation during processing or storage. Microbial contamination during storage also results in the formation of volatile acids and impairs the quality of products (Yang and Choong, 2001).

15.2 Traditional detection methods

Organic acids have been separated and determined by chromatographic techniques (Tang and Wu, 2007) such as HPLC (Zhao et al., 2001), GC (Deng, 1997), TLC (Thompson and Hedin, 1966), GLC (Sarkar and Malhotra, 1979), and IC (Morales, Gonzalez, and Troncoso, 1998). Most successful techniques for simultaneous determination of several organic acid constituents in samples include (1) GC, (2) IC, or (3) HPLC (Wu et al., 1995). However, GC methods with derivatization steps or HPLC have been adopted for measurement of organic acids, inasmuch as complex extraction steps are necessary and these methods were not specific for organic acids. Moreover, reagents used could potentially interact with various compounds that contained functional groups and active hydrogen. The

GC and HPLC methods, however, lack specificity (Arellano et al., 1997). The GC method is very tedious and unsuitable for studies on large numbers of samples, whereas the HPLC method was found to be much simpler, as it employed an ion-exchange column and a flow gradient with UV detection at 210 nm (Galli and Barbas, 2004). Although GC and HPLC are still commonly used, alternative approaches have been found to give faster analyses, preferably without any tiresome derivatization process needed (Chen et al., 1999). GC with split injection is also not used for analysis of free C1–C12 short-chain organic acids. Their high polarity retards the separation from food and the response of the flame detector to them is, therefore, quite low (Larsson and Roos, 1983; Yang and Choong, 2001).

The most common analytical method for detection of benzoic acid or sorbic acid has been reversed-phase HPLC (Saad et al., 2005). Organic acids in wine vinegars have also been detected by reversed-phase HPLC (RP-HPLC) using two C_{18} columns, UV detection at 210 nm, and sample filtration through Sep-Pack C_{18}. However, complete separation of organic acids has not been achieved with this procedure (Morales, Gonzalez, and Troncoso, 1998).

GC is employed for selective determination of food preservatives. GC and MS are both useful to determine preservatives with limits of detection of 100–500 pg, but these methods involve sample pretreatments similar to those prescribed by the Association of Official Analytical Chemists (AOAC) and are time-consuming (Gonzalez, Gallego, and Valcarcel, 1998).

Quantitation of citric, malic, and tartaric acids that are typically present in fruit juices has been carried out routinely by liquid chromatography, by using either refractive index or low wavelength UV detection. However, HPLC methods are very sensitive to matrix interferences and may often exclude the reliable detection or accurate quantitation of acids at levels below 50 ppm. Although HPLC methodology is known for its shortcomings, GC determination of fruit juice acids has not been widely employed because of the difficulty in isolation and derivatization of the acids (Barden et al., 1997).

The very commonly used sorbic acid has been detected with low recoveries in raw beef (52–84%) by photometry, after distillation and extraction. Better recoveries (~100%) and less manipulation have been achieved by the fluorescence technique in determination of 4-hydroxybenzoic methyl ester in nonalcoholic beverages. Additives can be present in combinations, and chromatographic methods are often used for selective individual or joint determination. Sorbic and benzoic acids have also been determined by thin-layer chromatography in beverages. This method involves minimal sample manipulation. HPLC is often the preferred method for determining additives present in mixtures, which are usually only just volatile. Simultaneous determination of additives, such as sweeteners, preservatives, and colorings in soft drinks is usually done by HPLC with UV detection, but concurrent separation is hindered by polarity differences (Gonzalez, Gallego, and Valcarcel, 1998).

In Table 15.1 the traditional detection methods are more concisely discussed by highlighting the pluses and minuses of application.

15.3   Contemporary methods

The use of capillary electrophoresis (CE) to determine organic acids in food samples was first demonstrated with a commercial reagent kit. A mixture of seven standards (i.e., citric, tartaric, malic, succinic, acetic, lactic, and butyric acids) have been separated in less than 15 min and detected by indirect UV absorbance at the 50 µg/ml level. Not all analytes exhibit high UV absorbtivity, therefore, this indirect absorbance method is welcomed as a universal method for detection of nonabsorbing analytes. Moreover, the indirect absorbance method affords good sensitivity. For many analytes, CE, coupled with indirect UV absorbance detection, has gained increasing popularity. CE methods are successfully used for the detection of various organic acids in sugar refinery juices, fruit juices, and various food samples. For slower moving organic acids, it has been found that a CE, using chromate as BGE, would result in poorer resolution as these acids would appear as trailing peaks (Wu et al., 1995).

In response to the need for rapid and simple methods for the detection of short-chain organic acids in complex matrices, a technique was developed that very well suited the specific purpose of CE. This method employed indirect detection and a surfactant to reverse electro-osmotic flow. Direct UV detection provides higher sensitivity and precision and the ability to detect UV-absorbing organic acids, such as oxalic, fumaric, or pyroglutamic acids, which cannot be detected with the indirect system (Galli and Barbas, 2004).

Many CE methods have been developed for the analysis of inorganic anions, organic acids, amino acids, or carbohydrate analysis. A CZE method has been proposed where indirect UV detection is done using 2,6-pyridine dicarboxylic acid (PDC) and cetyltrimethylammonium hydroxide (CTAH) at pH 12.1. Most inorganic and organic anions, amino acids, and carbohydrates have little or no UV absorbance and require indirect UV detection. Advantages of this method include (Soga and Ross, 1999)

1. Most anionic compounds can be analyzed simultaneously without derivatization.
2. A distinct electrophoretogram is produced without other matrix interferences.
3. Minimum sample preparation.
4. Excellent reproducibility, good linearity, and sensitivity.

Table 15.1 Advantages and Disadvantages of Various Traditional Detection Methods for Organic Acids

Detection method	Specific/ proposed uses	Advantage	Limitations
High performance liquid chromatography (HPLC)	Determination of additives in mixtures	Simultaneous detection of several organic acids Minimal sample preparation Quantitative determination of organic acids in short time	Lacks specificity Sensitive for matrix interferences Longer detection time
Gas chromatography (GC)	Not widely used for detection of organic acids Determination of acids in fruit juices not widely employed Dairy products more often analyzed Selective determination of preservatives Determination of preservatives with detection limits 100–500 pg	Simultaneous detection of several organic acids High sensitivity Good precision High selectivity	Lacks specificity Unsuitable for large numbers of samples Acids difficult to isolate Poor detector response Before analysis, necessary to prepare more volatile, nonreactive derivatives Requires at least two sample preparation steps
Thin-layer chromatography (TLC)	Separation and quantification of organic acids in grape juices and wines	Wide range of sample types Minimal sample manipulation High sensitivity High specificity High sensibility Rapid separation	Cannot be used with volatile compounds

Table 15.1 Advantages and Disadvantages of Various Traditional Detection
Methods for Organic Acids (Continued)

Detection method	Specific/ proposed uses	Advantage	Limitations
Ion chromatography (IC)	Separation of organic and inorganic weak acids	Simultaneous detection of several organic acids	In complex matrices, more extensive sample preparation required, to prevent destruction of column
	Separation of neutral/or basic substance from ionic compounds	Very little sample required when in homogeneous ionic form	
	Specific examples of food analyzed: coffee, milk, wine		
GC coupled with mass spectrometry (MS)	Extensively used for analysis of aromatic compounds in food	Identification based on both retention time and mass spectrum	High price and running cost
	Measuring contaminants from spoilage or adulteration	Allows quantitative detection of analytes	False positive/ negative results possible
	Preferred method for volatile or thermally stable compounds	Enhanced specificity	MS performance dependent on effective GC separation
		Identification of trace components in complex mixtures	
		Two techniques combined provide more definitive substance identification	

Sources: Data from Wu et al., *Journal of Chromatography A* 716:291–301, 1995; Barden et al., *Journal of Chromatography A* 785:251–261, 1997; Dashek and Micales, *Methods in Plant Biochemsitry and Molecular Biology*, pp. 107–113. Boca Raton, FL: CRC Press , 1997; González, Gallego, and Valcarcel, *Journal of Chromatography A* 823:321–329, 1998; Xiong and Li, *Journal of Chromatography A* 822:125-136, 1998; Soga and Ross, *Journal of Chromatography A* 837:231–239, 1999; Yang and Choong, *Food Chemistry* 75:101–108, 2001; Mato, Suarez-Luque, and Huidobro, *Food Research International* 38:1175–1188, 2005.

Problems encountered with CE include poor solubility of fatty acids in aqueous solutions and the unavailability of sensitive modes for aliphatic acids. This is often counteracted by nonaqueous capillary electrophoresis (NACE) with indirect detection. NACE-IA (indirect absorbance) is a simple, sensitive, and effective technique for the determination of fatty acids using *i*-PrOH and 40% acetonitrile (as organic modifier) and HIQSA (as chromophores). NACE-IA or NACE-ILIF (indirect laser-induced fluo-rescence) methods are also used in the separation of acids at low cost, providing fast analysis and good sensitivity, apart from the problem with instability of the baseline (Chen et al., 1999).

Compared to HPLC, GC, TLC, GLC, and IC, CE offers (1) high resolu-tion efficiency, (2) rapid separation speed, (3) simple sample pretreatment, (4) small consuming sample, and (5) short analysis time. No derivatiza-tion is needed as in GC or a sample extraction as in HPLC or IC. Addition of SDS (anionic surfactant) causes an increase in migration times of all organic acids. (Tang and Wu, 2007).

Capillary zone electrophoresis (CZE), with direct or indirect photom-etry and conductivity has become popular in wine analysis. Very little, or sometimes no sample preparation is needed and short analysis times are also apparent advantages of CE and CZE in the analysis of wine. Capillary isotachophoresis (ITP), with conductivity, thermometric, and UV absorp-tion detection, is suitable for the separation of various anionic constitu-ents (organic acids and inorganic anions), currently occurring in wines (Masar et al., 2001).

Ion-exclusion chromatography is a useful technique for separation of organic and inorganic weak-acids (Chinnici et al., 2005; Morales, Gonzalez, and Troncoso, 1998). It is especially valuable for separating neutral and weakly acidic or basic substances from ionic compounds based on an ion-exclusion mechanism, as opposed to ion-exchange (Morales, Gonzalez, and Troncoso, 1998).

Organic acids in dairy and other food products have commonly been analyzed by chromatographic techniques. An alternative RP-HPLC method has been optimized for analysis of multiple 11 organic acids that are metabolically of importance in dairy products (Tormo and Izco, 2004).

Solid-phase extraction (SPE) was introduced in the early 1970s to min-imize the shortcomings associated with liquid–liquid extraction, more specifically large samples or organic solvent volumes that are required. In 1998 a flow injection online SPE method was proposed, which provided high sensitivity and selectivity without interference from other concomi-tants or the solvent peak. Results are delivered within 5 min after elu-tion, as the throughput is limited only by the chromatographic conditions. The use of large amounts of expensive and hazardous organic solvents and derivatization reagents are also avoided. This method was, therefore, proposed to be used by control laboratories to identify and quantify the different preservatives in a wide variety of foods (Gonzalez, Gallego, and Valcarcel, 1998).

In one study a photochemical-CL detection system was combined with HPLC and successfully applied in analyzing organic acids in a variety of real samples, which included beer, wine, milk, fruit, and soft drinks. The HPLC method is based on selective photodecomposition of organic acids in the presence of Fe^{3+} and UO_2^{2+}, which is then combined with the sensitive determination of the Fe^{2+} produced by the CL luminol reaction in the absence of added oxidant (Perez-Ruiz et al., 2004).

An in-tube SPME-HPLC method is proposed for simultaneous determination of benzoic and sorbic acids in various types of food. This proposed method is easy, rapid, and high-throughput when used in simultaneous determination of benzoic and sorbic acids in real samples, and no cleanup procedures are required. Minimal toxic solvents are required when extracting benzoic and sorbic acids, making it environmentally friendly. All types of food samples can be easily and successfully prepared. When an in-tube SPME technique is coupled online to HPLC, using diethylamine-modified poly(GMA-co-EDMA) monolithic capillary as the extraction medium, a convenient and robust method is achieved for simultaneous determination of benzoic and sorbic acids in different food samples. This method (1) has a simple and fast pretreatment procedure, (2) low sample consumption, (3) is easy to use, (4) has good precision and accuracy, and also (5) provides high selectivity and sensitivity (Wen, Wang, and Feng, 2007).

Table 15.2 gives a concise exposition of the applicability of more contemporary detection methods.

15.4   The importance of effective detection

Analytical determination of organic acids is important for quality assurance purposes, as well as for consumer protection or even consumer interest. Limited chromatographic reports are available on the simultaneous determination of benzoic acid, sorbic acid, and the parabens, especially in food items. However, there is an increasing trend in using combinations of preservatives in the food industry, as well as in pharmaceutical formulations and cosmetic products, and such a method may become an essential tool in organic acid analysis (Saad et al., 2005).

Organic acids contribute to the flavor and aromatic properties of food and dairy products (Tormo and Izco, 2004). Their presence and relative ratio of organic acids can affect the chemical and sensorial characteristics of the food matrix (e.g., pH, total acidity, microbial stability, sweetness, and global acceptability) and can provide essential information on nutritional properties of food and means to optimize selected technological processes (Chinnici et al., 2005). Quantitative determination of organic acids is also important to monitor bacterial growth and activity. Fumaric acid is an important parameter to reveal microbial spoilage or processing of decayed fruits (Trifirò et al., 1997). For example, natural concentration in apple juices usually does not exceed 3 mg/l, whereas higher levels can arise from microbic growth or from the addition of synthetic malic acid (Kvasnicka and Voldrich, 2000).

Table 15.2 Advantages and Disadvantages of Various Modern Detection
Methods for Organic Acids

Detection method	Specific/proposed uses	Advantage	Limitations
Capillary electrophoresis (CE)	Suitable for analysis of low-molecular-mass organic acids in complex aqueous samples Determination of various organic acids in sugar refinery juices, fruit juices, foods, urine samples	Simple method Small samples required Minimum sample preparation No derivatization needed Rapid separation speed Short analysis time Highly effective Highly selective	Poor resolution for slower moving organic acids Poor solubility in aqueous solutions Not for simultaneous separation and detection of anions and cations Limited loadability
Isotachophoresis (ITP)	Separation of charged particles Suitable for separation of various anionic constituents in wine	Small samples required Good reproducibility Rapid results	Available instruments not as comfortable to use as CE instruments High electrolyte purity required
Capillary zone electrophoresis (CZE)	Popular in wine analysis	Simultaneous detection of organic acids Little sample preparation Faster method Higher resolution Higher sensitivity	Matrix interference Electrodispersion enforced by UV absorbing co-ion
Nonaqueous capillary electrophoresis with indirect absorbance (NACE-IA)	Organic acids commonly found in beverages and juices	Simple method Efficient Fast analysis High sensitivity Low cost	Baseline instability

Table 15.2 Advantages and Disadvantages of Various Modern Detection
Methods for Organic Acids (Continued)

Detection method	Specific/proposed uses	Advantage	Limitations
Solid-phase microextraction (SPME)–HPLC	All types of food samples successfully prepared Analysis of less volatile or thermally labile compounds Not yet extensively applied to food analysis	Easy method Rapid analysis High-throughput No cleanup procedures Minimal toxic solvents	Data not available

Sources: Data from Wu et al., *Journal of Chromatography A* 716:291–301, 1995; Barden et al., *Journal of Chromatography A* 785:251–261, 1997; Dashek and Micales, *Methods in Plant Biochemsitry and Molecular Biology*, pp. 107–113. Boca Raton, FL: CRC Press , 1997; González, Gallego, and Valcarcel, *Journal of Chromatography A* 823:321–329, 1998; Xiong and Li, *Journal of Chromatography A* 822:125-136, 1998; Soga and Ross, *Journal of Chromatography A* 837:231–239, 1999; Yang and Choong, *Food Chemistry* 75:101–108, 2001; Mato, Suarez-Luque, and Huidobro, *Food Research International* 38:1175–1188, 2005.

Measuring organic and inorganic levels in foods and beverages is essential for (1) monitoring a fermentation process, (2) inspecting product stability, (3) confirming authenticity of juices and concentrates, or detecting adulteration, and also (4) recording the organoleptic characteristics of fermented products, particularly wines (Barden et al., 1997; Mato, Suarez-Luque, and Huidobro, 2005; Arellano et al., 1997). The acids measured in wine come directly from the grape or from processes such as alcoholic fermentation, malolactic fermentation, oxidation of the ethanol, etc. (Masar et al., 2001; Mato, Suarez-Luque, and Huidobro, 2005).

An individual consumer may be sensitive to one or more preservatives and the type and amount must, therefore, be controlled (Tang and Wu, 2007). Benzoic and sorbic are the most commonly used additives, applied to a wide variety of foods (Gould and Jones, 1989) and the appropriate amount of benzoic or sorbic acid, for example, has been tested by many laboratories throughout the world and declared safe, but excessive use may cause adverse effects in humans, which include metabolic acidosis, convulsions, and hyperpnea (Tfouni and Toledo, 2002; Wen, Wang, and Feng, 2007).

Vitamin C is present in many beverages, juices, and medicines, and the CE method is a valuable aid when checking freshness and quality of foodstuffs in the determination of vitamin C (ascorbic acid) (Wu et al., 1995). In the process of cheesemaking, organic acids are important compounds affecting the flavors of most aged cheeses. They are produced as a result of hydrolysis of milk. Stages of hydrolysis include fat during (1) lipolysis, (2) bacterial growth, (3) normal ruminant metabolic processes, or (4) addition of acidulants during cheese making (Alkalin, Gonc, and Akbas, 2002; Izco, Tormo, and Jimenez-Florez, 2002). Quantitative analysis of organic acids is, therefore, an important tool for studying the flavor, nutritional quality, and the bacterial activity of aging cheeses (Park and Drake, 2005).

Acidity of brewed coffee is very important and has an impact on a consumer's reaction to the drink. Because acidity depends on the extent to which the coffee bean has been roasted, determining the best roasting conditions to apply to the green coffee in developing acidity and sweetness and improving sensory quality of commercialized roasted blends, it is necessary to quantify the organic acids involved. Malic and citric acids are already present in the green coffee bean, but most of the acidity is produced at the beginning of the roasting process (Rodrigues et al., 2007).

Volatile acid content is often an index to quality assurance (Yang and Choong, 2001).

15.5   Detection in specific foodstuffs

Organic constituents in wine are closely linked with the organoleptic characteristics of wine. The acids mainly responsible include the α-hydroxy acids (tartaric, malic, lactic, and citric acids). These acids are routinely determined at various stages of production and also in the final wine product. In some specific situations other acids such as acetate, ascorbate, sulfite, sulphate, phosphate, malonate, gluconate, and sorbate are also analyzed. Spectrophotometry, various column chromatography methods, and also enzymatic methods, are predominantly used in the determination of acidic wine constituents. However, capillary electrophoresis (CE) is increasingly proposed as a convenient alternative, because of the ionogenic nature of these analytes (Masar et al., 2001). Red wine normally contains high concentrations of tartrate and lactate, and tartrate is also the predominant anion in white wines. However, the lactate concentration is lower in white wine, due to the absence of malolactic fermentation, which converts malate into lactate (Arellano et al., 1997).

CE methods are increasingly used for analysis of organic acids in coffee. The primary taste sensation in coffee is acidity where low-molecular-mass organic acids contribute to both taste and flavor, due to the fact that most of them are volatile. Acids found in roasted coffee may be classified into four groups: aliphatic, chlorogenic, alicyclic, and phenolic. Organic acids in green and roasted coffee have been determined by several procedures, but mainly by GC and HPLC. However, CE has been found suitable for analysis of low-molecular-mass organic acids in complex aqueous samples, because of high efficiency and minimum sample pretreatment (Galli and Barbas, 2004).

Sorbic acid has been found to be the most popular preservative in jams, whereas benzoic acid is the most common preservative in dried fruits. A sample pretreatment procedure combined with an HPLC method was found suitable for routine determination of sorbic acid and benzoic acid in food items (Saad et al., 2005).

15.6   Characteristics of detected organic acids

Electrophoretic mobility of acids depends on the pH of the running buffer. The mobilities of tartaric, malic, and succinic acids have been found to be close to each other at alkaline pH. At pH 9.0 the migration times of various organic acids increase in order: (1) oxalate, (2) citrate, (3) tartrate, (4) malate, (5) succinate, (6) carbonate, (7) acetate, (8) lactate, (9) aspartate, (10) glutamate, (11) ascorbate, and (12) gluconate. The faster mobility of some of these acids may be attributed to the presence of hydroxyl groups, whereas the monocarboxylic acids, carbonate, acetate, and lactate migrate in ascending order in relation to their mass. Better separation of fast-moving polyacids is also obtained at pH 5.5, and at alkaline pH (9.0) a better resolution can be achieved for carbonate and the slower-moving amino acids, ascorbic, and gluconic acids. Determination of vitamin C is, however, hampered by a problem encountered as a result of oxidation. When ascorbic acid is dissolved, it is readily oxidized to dehydroascorbic acid, which is then catalyzed by air or light exposure (Wu et al., 1995).

15.7   Comparing sample preparation techniques

Conventional sample preparation techniques to achieve cleanup and preconcentration before determination of organic acids in food include: (1) liquid–liquid extraction (LLE), and (2) solid-phase extraction. However, these two methods require complex, laborious, and time-consuming procedures and the LLE technique also requires large volumes of organic solvents, rendering it a potential danger to the environment and also to human health. New miniaturized sample pretreatment techniques have, therefore, been developed, of which the: (1) solid-phase microextraction (SPME) and (2) stir-bar sorptive extraction (SBSE) are examples. These techniques have been successfully used for analysis of benzoic and sorbic acids in various foodstuffs. Both techniques have been found to provide various advantages, such as simplicity, solvent-free extraction, and high sensitivity. However, these methods are still performed offline when combined with chromatographic analysis. A method for the determination of benzoic acid based on an online pyrolytic methylation technique has been developed for which no pretreatment procedures are required (Pan et al., 2005), but this method is only applied to the analysis of soft drink samples, of which the matrices are relatively simple (Wen, Wang, and Feng, 2007).

When using a chromatographic method for analyzing complex matrices such as food samples, appropriate sample preparation techniques are important. The in-tube solid-phase microextraction (in-tube SPME) is another promising sample preparation technique, which is based on SPME. It initially utilizes an open tubular capillary of which the inner surface is coated and serves as the extraction medium. This enables online SPME coupled to HPLC analysis. Sample extraction, concentration, and introduction are hereby integrated into a single step, providing an online method capable of providing better accuracy, precision, and sensitivity than offline methods (Wen, Wang, and Feng, 2007).

Organic acids usually need to be derivatized before analysis, a process which is time-consuming (Yang and Choong, 2001). Another problem is the volatility of the derivatized compounds that is high in order to reduce recovery and to affect the accuracy and repeatability of GC quantification (McCalley, Thomas, and Leveson, 1984). These problems are addressed by the implementation of headspace analysis (Mulligan, 1995).

References

Akalin, A.S., Gonc, S., and Akbas, Y. 2002. Variation in organic acids content during ripening of pickled white cheese. *Journal of Dairy Science* 85:1670–1676.

Alur, M.D., Doke, S.N., Warrier, S.B., and Nair, P.M. 1995. Biochemical methods for determination of spoilage of foods of animal origin: A critical evaluation. *Journal of Food Science and Technology* 32:181–188.

Arellano, M., Andrianary, J., Dedieu, F., Couderc, F., and Puig, P. 1997. Method development and validation for the simultaneous determination of organic and inorganic acids by capillary zone electrophoresis. *Journal of Chromatography A* 765:321–328.

Barden, T.J., Croft, M.Y., Murby, E.J., and Wells, R.J. 1997. Gas chromatographic determination of organic acids from fruit juices by combined resin mediated methylation and extraction in supercritical carbon dioxide. *Journal of Chromatography A* 785:251–261.

Chen, M.J., Chen, H.S., Lin, C.Y., and Chang, H.T. 1999. Indirect detection of organic acids in non-aqueous capillary electrophoresis. *Journal of Chromatography A* 853:171–180.

Chinnici, F., Spinabelli, U., Riponi, C., and Amati, A. 2005. Optimization of the determination of organic acids and sugars in fruit juices by ion-exclusion liquid chromatography. *Journal of Food Composition and Analysis* 18:121–130.

Dashek, W.V. and Micales, J.A. 1997. Isolation, separation, and characterization of organic acids. In: W.V. Dashek (Ed.), *Methods in Plant Biochemsitry and Molecular Biology*, pp. 107–113. Boca Raton, FL: CRC Press.

Deng, C.R., 1997. Determination of total organic acids in wine by interfacial deviatization gas chromatographic methods. *Sepu* 15:505–507.

De Raucourt, A., Girard, D., Prigent, Y., and Boyaval, P. 1989. Lactose fermentation with cells recycled by ultra-filtration and lactate separation by electrodialysis: Modeling and simulation. *Applied Microbiology and Biotechnology* 30:521–527.

Galli, V. and Barbas, C. 2004. Capillary electrophoresis for the analysis of short-chain organic acids in coffee. *Journal of Chromatography A* 1032:299–304.

Gomis, D.B. 1992. HPLC analysis of organic acids. In: L.M.L. Nollet (Ed.), *Food Analysis by HPLC*, pp. 371–385. New York: Marcel Dekker.

González, M., Gallego, M., and Valcarcel, M. 1998. Simultaneous gas chromatographic determination of food preservatives following solid-phase extraction. *Journal of Chromatography A* 823:321–329.

Gould, G.W. and Jones, M.V. 1989. Combination and synergistic effects. In: *Mechanisms of Action of Food Preservation Procedures*. G.W. Gould (Ed.). London: Elsevier Science.

Izco, J.M., Tormo, M., and Jimenez-Florez, R. 2002. Rapid simultaneous determination of organic acids, free anion acids, and lactose in cheese by capillary electrophoresis. *Journal of Dairy Science* 85:2122–2129.

Kvasnicka, F. and Voldrich, M. 2000. Determination of fumaric acid in apple juice by on-line coupled capillary isotachophoresis-capillary zone electrophoresis with UV detection. *Journal of Chromatography A* 891:175–181.

Larsson, M. and Roos, C. 1983. Determination of C1–C4 fatty acids as p-bromoacyl esters using glass-capillary gas chromatography and electron-capture detection. *Chromatographia* 17:185–190.

Masar, M., Kaniansky, D., Bodor, R., Johnck, M., and Stanislawski, B. 2001. Determination of organic acids and inorganic anions in wine by isotachophoresis on a planar chip. *Journal of Chromatography A* 916:167–174.

Mato, I., Suarez-Luque, S., and Huidobro, J.F. 2005. A review of the analytical methods to determine organic acids in grape juices and wines. *Food Research International* 38:1175–1188.

McCalley, D.G., Thomas, C.W., and Leveson, L.L. 1984. Analysis of carboxylic acids by gas chromatography. *Chromatographia* 18:309–312.

Morales, M.L., Gonzalez, A.G., and Troncoso, A.M. 1998. Ion-exclusion chromatographic determination of organic acids in vinegars. *Journal of Chromatography A* 822:45–51.

Mulligan, K.J. 1995. Aqueous alkylation of anions for static headspace sampling with analysis by capillary gas chromatography mass spectrometry. *Journal of Microcolumn Separations* 7:567–573.

Pan, Z., Wang, L., Mo, W., Wang, C., Hu, W., and Zhang, J. 2005. Determination of benzoic acid in soft drinks by gas chromatography with on-line pyrolytic methylation technique. *Analytica Chimica Acta* 545:212–223.

Park, Y.W. and Drake, M.A. 2005. Effect of 3 months frozen-storage on organic acid contents and sensory properties, and their correlations in soft goat milk cheese. *Small Ruminant Research* 58:291–298.

Perez-Ruiz, T., Martinez-Lozano, C., Tomas, V., and Martin, J. 2004. High-performance liquid chromatographic separation and quantification of citric, lactic, malic, oxalic and tartaric acids using a post-column photochemical reaction and chemiluminescence detection. *Journal of Chromatography A* 1026:57–64.

Rodrigues, C.I., Marta, L., Maia, R., Miranda, M., Ribeirinho, M., and Maguas, C. 2007. Application of solid-phase extraction to brewed coffee caffeine and organic acid determination by UV/HPLC. *Journal of Food Composition and Analysis* 20:440–448.

Saad, B., Bari, M.F., Saleh, M.I., Ahmad, K., and Talib, M.K. 2005. Simultaneous determination of preservatives (benzoic acid, sorbic acid, methylparaben and propylparaben) in foodstuffs using high-performance liquid chromatography. *Journal of Chromatography A* 1073:393–397.

Sarkar, S.K. and Malhotra, S.S. 1979. Gas-liquid chromatographic method for separation of organic acids and its application of pine needle extracts. *Journal of Chromatography A* 171:227–232.

Schugerl, K. 2000. Integrated processing of biotechnology products. *Biotechnology Advances* 18:581–599.

Soga, T. and Ross, G.A. 1999. Simultaneous determination of inorganic anions, organic acids, amino acids and carbohydrates by capillary electrophoresis. *Journal of Chromatography A* 837:231–239.

Tang, Y. and Wu, M. 2007. The simultaneous separation and determination of five organic acids in food by capillary electrophoresis. *Food Chemistry* 103:243–248.

Tfouni, S.A.V. and Toledo, M.C.F. 2002. Determination of benzoic and sorbic acids in Brazilian food. *Food Control* 13:117–123.

Thompson, A.C. and Hedin, P.A. 1966. Separation of organic acids by thin-layer chromatography of their 2,4-dinitro phenylhydrazide derivatives and their analytical determination. *Journal of Chromatography A* 21:13–18.

Tormo, M. and Izco, J.M. 2004. Alternative reversed-phase high-performance liquid chromatography method to analyse organic acids in dairy products. *Journal of Chromatography A* 1033:305–310.

Trifirò, A., Saccani, G., Gherardi, S., et al. 1997. Use of ion chromatography for monitoring spoilage in the fruit juice industry. *Journal of Chromatography A* 770:243–252.

Weier, A.J., Glatz, B.A., and Glatz, C.E. 1992. Recovery of propionic and acetic acids from fermentation broth by electrodialysis. *Biotechnology Progress* 8:479–485.

Wen, Y., Wang, Y., and Feng, Y.Q. 2007. A simple and rapid method for simultaneous determination of benzoic and sorbic acids in food using in-tube solid-phase microextraction coupled with high-performance liquid chromatography. *Analalytical and Bioanalytical Chemistry* 388:1779–1787.

Wu, C.H., Lo, Y.S., Lee, Y.-H., and Lin, T.-I. 1995. Capillary electrophoretic determination of organic acids with indirect detection. *Journal of Chromatography A* 716:291–301.

Xiong, X. and Li, S.F.Y. 1998. Selection and optimization of background electrolytes for simultaneous detection of small cations and organic acids by capillary electrophoresis with indirect photometry. *Journal of Chromatography A* 822:125–136.

Yang, M.H. and Choong, Y.M. 2001. A rapid gas chromatographic method for direct determination of short-chain (C2–C12) volatile organic acids in foods. *Food Chemistry* 75:101–108.

Zhao, J.C., Guo, Z.A., Chang, J.H., and Wang, W. 2001. Study on reversed-phased high performance liquid chromatography separation conditions determination method of organic acids. *Sepu* 19:260–263.

Index

A

AAB. *See* Acetic acid bacteria
Acetate, 141
    resistance to, 194
Acetic acid, 25–26, 43, 78. *See also* Vinegar
    effectiveness of, 140
    production of by AAB metabolism,
        169–170
    production of from *Acetobacter* sp.,
        167–168
    resistance to, 189
    use of in combination with irradiation,
        67–68
Acetic acid bacteria (AAB), 106–109
    acid tolerance of, 214
    biopreservation and, 274–275
    organic acid production by, 167–168
    resistance of to organic acids, 109,
        186–187, 193
Acetobacter
    production of vinegar by, 167
    yeasts and, 106–107
Acetobacter acetii, acetic acid resistance of,
    109–110
Acid adaptation response, 206
    development of, 215–216
Acid resistance, 159. *See also* Natural
        resistance
    development of, 188–189
    extent of, 196–197
    inducible, 189–191
    industry strategies for, 198–199
    interacting mechanisms of, 218
    mechanisms of, 191–196
    possible advantages of, 198
    transmission of, 196
Acid stress, 205–206
    cross-resistance to secondary stresses,
        210

Acid stressing, 158
Acid tolerance response (ATR), 188, 205
    analytical procedures for determining,
        217–218
    control strategies for, 218–219
    development of, 215–216
    gastrointestinal pathogens and, 208–209
    genes involved in, 212
    implications of, 216–217
    interacting mechanisms of, 218
    mechanisms of development of, 210–213
    role of organic acids in, 206–208
Acid-tolerant organisms, 213–215
Acidic foods, organic acids in, 53
Acidified foods, 131–133, 217
    FDA regulations regarding, 254
    naturally occurring organic acids in,
        264
    use of organic acids to prevent spoilage
        of, 60
Acidity constant (pK$_a$ value), 124
Aconitase, 109
Active packaging, 71–72, 284
Active pH homeostasis, 123, 212–213
Alkyl esters, effectiveness of, 141
Allergic diseases, 151
Animal feed
    heat treatment of, 137
    optimizing organic acid application in,
        282–283
    organic acids used for preservation of,
        73–77
    preservation of, 254–255
    use of organic acids in, 55
Animal nutrition, organic acids in, 77
Antibacterial action of organic acids,
    128–129
Antibiotics
    regulation of, 55

resistance to, 185, 196 (*See also* Acid resistance)

use of organic acids as an alternative to, 74

Antifungal actions of organic acids, 129–131

Antifungal compounds, production of by lactic acid bacteria, 100

Antimicrobial agents
active packaging and, 284
resistance to, 192 (*See also* Acid resistance)

Antimicrobial films, 70–71

Antimicrobial packaging, 283–284
factors for the design of, 284–285
organic acids in, 70–73

Antimicrobial peptides, 101

Antimicrobial sprays, use of for meat decontamination, 56–59

Antiviral actions of organic acids, 131

Appert, Nicholas, 1

Application of organic acids, 79–80
methods of, 156–158

Aromatic compounds, use of as preservatives, 67

Ascorbic acid, 26, 152
color stability and, 69
effectiveness of, 141
effectiveness of combining with citric acid, 65
use of in combination with gamma irradiation, 68

Aspartic acid, use of in animal feeds, 74

Aspergillus, 10, 14

Aspergillus niger, 275
industrial production of citric acids by, 177–180
industrial production of lactic acids by, 168
organic acid production by, 110

ATP-binding cassette (ABC) transporter, 190

B

Bacteria, 13–14
mechanisms of acid resistance in, 191–193
resistance of to organic acids, 186–187
undissociated organic acids in cells of, 124

Bacterial membrane disruption, 121

Bacterial spoilage, lactic acid bacteria and, 98–99

Bactericidal actions of organic acids, 128–129

Bacteriocins, 101

Bakery products, 44
concentrations of organic acids in, 77–78
use of organic acids in preservation of, 53

Beef. *See also* Meat
acid adaptation in, 216
organic acids used against pathogens in, 78
steam washing of, 137–138
use of sodium lactate for preservation of, 64

Benzoates, 63
resistance to, 189

Benzoic acid, 26–28, 43, 55, 79. *See also* Sodium benzoate
bacterial membrane disruption by, 121
pH-dependent effects of, 128
sensory characteristics of, 155
use of as a food preservative, 2
use of as a fungicide, 130

Berries, naturally occurring organic acids in, 262

Bifidobacteria, biopreservation and, 273

Biofilms, 158
resistance of bacteria of to organic acids, 192

Biopreservation, 170, 271–272

Bioprofit, 274

Bipolar membranes, use of electrodialysis with, 171–173

Bone discolorization, 73

Buffered acids, 21–22

Buffering, 139

C

Campylobacter, 13–14, 254
acid tolerance of, 214
effect of formic acid on culturability of, 31
pH range for growth of, 127
use of acidified drinking water against, 81–82

Capillary electrophoresis (CE), use of for detection of organic acids, 295–298

Capillary zone electrophoresis (CZE), 298

Carboxylic acids, 21, 23

Carcass decontamination, 156–158. *See also*
  Meat
  acid adaptation and, 216
  control strategies for acid tolerance
    response during, 218–219
  food safety regulations regarding, 249
  policies on the use of organic acids for,
    244
  proposed amendments to regulations
    regarding, 252
  sensory properties of meat and, 70
  technologies for, 80–81
Carvacrol, 283
Cellular physiology, 227
  modeling of with genomics, 228–229
  stress models of, 236
Chemical stability, 70
Chicken. *See* Poultry
Chicken feed, addition of organic acids
    to, 75
Chill-wash, organic acid combinations
    and, 66
Chilled foods, organic acids as additives
    in, 82–83
Cinnamic acid, 28–29
Citric acid, 29–30
  color stability and, 69
  effectiveness of, 141
  effectiveness of combining with
    ascorbic acid, 65
  foodstuffs naturally containing, 42–44
  industrial production of, 177–180
  sensory characteristics of, 155–156
  use of as a fungicide, 131
  use of in animal feeds, 74
Clostridium botulinum, 83
  regulations regarding, 245
  use of organic acid combinations to
    inhibit, 65
Clostridium perfringens, 12–13, 81, 82–83
  proposed regulations regarding control
    of, 253
  use of organic acid salt combinations
    for, 66
  use of sodium salts to inhibit outgrowth
    of, 64
Codex Alimentarius Commission, 250–251
Coffee, 44
  acidity of, 302
  detection of organic acids in, 302
  naturally occurring organic acids in,
    264

Colibacillosis, control of with organic
    acids, 75
Color stability, 69
Concentrations of organic acids, 77–79
Confectionery, use of organic acids in
    preservation of, 53
Consumers
  acceptance of biopreservative
    technologies by, 276
  food safety concerns of, 7–9
  preservatives and perception of, 154
  role of general food safety regulations
    in protection of, 248–250
Cost, 156
Cranberries, naturally occurring organic
    acids in, 262
Cured meat, use of organic acids for
    preservation of, 52
Cyclospora cayetanensis, 15
Cytoplasmic pH, regulation of, 126–127

D

Daily consumption of organic acids, 78–79
Dairy, 44, 54
  fermentation of, 168
  hydrolysis of, 301–302
Detection methods
  contemporary, 295–299
  effective, 299–302
  traditional, 293–295
Dicarboxylic acids, 23–24
DIN EN ISO 9000ff, 248–249, 281–282
Dipping treatments, 60
Drinking water, use of organic acids in, 75,
    81–82
Dunrand, Peter, 1

E

Edible films, 72–73
  naturally occurring organic acids in,
    265
Effectiveness, comparisons among organic
    acids, 139–142
Electrodialysis (ED), 170–171
  production of fermented organic acids
    by, 172
Electrophoretic mobility, 303
Emulsifiers, 134
  pharmaceutical, 68
Environmental factors for microbial
    growth, 227

Escherichia coli
  acid tolerance of, 213–214
  effect of irradiation on, 67–68
  inhibition of by organic acids, 140–141
  sensitivity of to lactic acid, 34
  square root-type model for growth of,
    235
Escherichia coli O157:H45, lactic acid and,
    66–67
Escherichia coli O157:H7, 13, 53, 59, 80–81, 83,
    159, 243
  acid adaptation response of, 208–209
  control of with sodium benzoate, 61
  control strategies for acid tolerance
    response of, 218–219
  cross-resistance to secondary stresses
    of, 210
  development of acid tolerance in,
    215–216
  development of acid tolerance response
    in, 211
  effects of physical factors of organic
    acids on, 136–139
  FDA regulations regarding, 254
  inhibition of with organic acid
    combinations, 66
  resistance of to antimicrobial agents,
    192, 197
  use of liquid smoke against, 68
  use of sodium lactate to limit
    proliferation of, 64
Ethanoic acid. *See also* Acetic acid
  effectiveness of combining with lactic
    acid, 65
Ethanol
  antimicrobial effects of, 262–263
  use of in combinations, 67
Ethanol tolerance, acetic acid bacteria and,
    108
European Union
  food safety policies in, 243–244
  meat handling regulations in, 249
  regulation of meat industry in, 6
  White Paper on Food Safety, 243, 251
Eurotium, 14
Extreme acid resistance, 206

F

Farm to fork food safety, 74
FDA, 6
  food safety regulations of, 249–250
  GRAS food additives, 246–247

Fermentation
  heterolactic, 104
  homolactic, 103
  industrial, 170–173
  inhibition of, 135–136
  lactic acid bacteria, 98, 101–105
  lactic acid production from, 174–177
  substrates and yields, 168–170
Films
  antimicrobial, 70–71
  edible, 72–73
Fish
  fermentation of with lactic acid bacteria,
    105
  use of organic acids for preservation of,
    52
Flavor, 69
  organic acids and, 155–156
Food additives, FDA list of those generally
    regarded as safe, 246–247
Food allergies, definition of, 152
Food preservation
  evolution of, 1–3
  industrial fermentation for, 170
  lactic acid bacteria and, 102
  microbial resistance to, 196 (*See also*
    Acid resistance)
  significance of modeling of, 236–237
  spectra of inhibition of organic acids,
    134
  use of organic acids for, 51–55
Food production
  lactic acid bacteria and, 102
  legislative issues in, 6–7
Food safety
  alternatives for, 10–11
  consumer concerns, 7–9
  control measures, 10
  policies regarding, 243
  proposed amendments to regulations
    for, 251–253
  regulations for, 248–250
Foodborne illness
  economic costs of, 3–4
  foodstuffs associated with, 244
  global outbreaks of, 243
  proposed amendments to food safety
    regulations for, 251–253
  role of government in control of,
    253–254
Foodborne microorganisms, growth pH
    limits for, 127
Foodborne pathogens

acid tolerance of, 188, 208–209, 213–215
cross-protection of, 210
stress response capabilities of, 135
Foodstuffs
acidic, 217
adverse effects of organic acids on, 153
detection of organic acids in, 302–303
naturally occurring organic acids in, 266–267
pH of, 132
Formic acid, 30–31
effectiveness of, 140
use of in animal feeds, 76–77
Freezing, 138
Fruit
alternatives to washing techniques for, 286–288
organic acids in, 42–43
use of organic acids against pathogens in, 82
Fruit juices. *See* Juices
Fruits, use of organic acids in preservation of, 53
Fumarate, 64
Fumaric acid, 31–32, 103–104
formation of, 169
production of by *Rhizopus*, 111
Fungal spoilage, 4–5, 14–15
use of potassium sorbate to inhibit, 62
Fungi
mechanisms of organic acid resistance in, 193–196
organic acid production by, 110–111
resistance of to organic acids, 187–188
Fungicides, use of organic acids as, 129–131
Fusarium, 10, 14

G

Gallic acid, 40–41
Gamma concept of predictive microbiology, 234
Gas chromatography, use of for detection of organic acids, 293–295
Gastrointestinal pathogens, acid tolerance of, 208–209
Genetic analysis, lactic acid bacteria, 105–106
Genomic transcript profiling, 227
Genomics, modeling organic acids using, 227–229
Glacial acetic acid. *See* Acetic acid
Gluconic acid, 32–33

Gluconoacetobacter, production of vinegar by, 167
Gluconobacter, 106
production of vinegar by, 167
Glucose tolerance, acetic acid bacteria and, 108
Gram negative bacteria
acid stress responses of, 207
disruption of membrane of by organic acids, 121
Growth models, 229–230
Growth pH limits, 127

H

H_2O_2 production, 126
Haa1p, 195
Heat shock, 210
Heat treatments, 137
organic acids combined with, 75–76
Heat-shock response
effects of organic acids on, 159
influence of sorbic and benzoic acids on, 190
Heterofermentative lactic acid bacteria, 98, 100, 273
organic acid production by, 165–166
Heterolactic fermentation, 104
Homofermentative lactic acid bacteria, 98, 273
Homolactic fermentation, 103
Honey, 44
naturally occurring organic acids in, 261
Hot water washings, organic acid combinations and, 66
HPLC, use of for detection of organic acids, 293–295
Hsp30, 190
Hurdle effect, 66, 153–154
Hurdle technology, 65, 245
Hydrochloric acid, effectiveness of, 140
Hydroxybutaneoic acid. *See* Malic acid
Hydroxysuccinic acid. *See* Malic acid

I

Inducible resistance, 189–191
Industrial fermentation, 170–173
lactic acid production from, 174–177
Inhibitory effects, comparisons among organic acids, 139–142
Inorganic acids

acid stress and, 207
effectiveness of *vs.* organic acids,
133–134
Interactive effects, 138–139
Intestinal health
acid tolerance of gastrointestinal
pathogens, 208–209
use of lactic acid bacteria to improve,
105
Intracellular acidification, 199
Intrinsic resistance, 185–188
Ion-exclusion chromatography, 298
Ionizing radiation, 136
Irradiation, 136, 252–253
use of in combination with organic
acids, 67–68

J

Jerky, use of marinades in processing of, 84
Juices
organic acids in, 43
use of organic acids in preservation of,
53–54

K

K+, acid resistance and, 189–190
Kensington mangos, preservation of, 82
Kiwi, preservation of, 82
Kombucha, naturally occurring organic
acids in, 264–265

L

L-Ascorbic acid, 26. *See also* Ascorbic acid
LAB. *See* Lactic acid bacteria
Labeling, 55
Lactate, 141
use of in combinations, 67
Lactic acid, 33–35, 99–100, 168
antibacterial actions of, 129
bacterial membrane disruption by, 121
effectiveness of, 140–141
effectiveness of combining with
ethanoic acid, 65
formation of in vinegars, 169
importance of in industrial
fermentation, 170
principles of fermentation of, 101–105
production of, 174–177
production of by *Listeria monocytogenes*,
111

production of by *Saccaromyces cerevisiae*,
111
production of from *Lactobacillus* sp.,
166–167
Lactic acid bacteria (LAB), 14, 97–99
acid tolerance of, 214
antimicrobial substances produced by,
99–101
biopreservation and, 272–274
genetic and bioinformatic
characterization of, 105–106
importance of in food production and
preservation, 102
organic acid production by, 165–167
resistance of to organic acids, 109,
192–193
use of a fermentation starter cultures,
168
Lactic acid producing bacterial culture
condensate mixture (LCCM),
275–276
Lactobacilli, diseases due to, 99
Lactobacillus pontis, organic acid production
by, 166
Lactobacillus reuteri, organic acid
production by, 166
Lactobacillus sanfranciscensis, organic acid
production by, 165–166
Lactobacillus sp., 14
Lactococcus lactus, organic acid production
by, 165
LCCM, 275–276
Legislative issues in food production, 6–7
application guidelines for organic acid
preservation, 245–247
organic acid concentrations, 79
preservatives concentration and, 157
recommended daily intake, 154–155
Leuconostoc mesenteroides, 105
Liquid smoke, 79
use of as an antimicrobial agent, 68–69
Liquid-liquid extraction (LLE), 303
Listeria monocytogenes, 12–13, 52, 83, 159
acid tolerance of, 213–214
control of with organic acid salts, 61
cross-resistance to secondary stresses
of, 210
effect of citric acid on growth of, 30
effect of ethanol combinations on, 67
effective concentrations of organic acids
against, 78
efficacy of sorbate against, 80
food safety regulations regarding, 250

inhibition of with organic acid
combinations, 65–66
production of L-lactic acid by, 111
resistance of to food preservation
practices, 187
sensitivity of to lactic acid, 34–35
use of liquid smoke against, 68–69
use of potassium sorbate to inhibit
growth of, 62–63
use of sodium lactate to limit
proliferation of, 64
use of stoichiometric models for studies
of, 231
Listeriosis, 11–12, 243

M

Malic acid, 35–36, 169
effectiveness of, 140
foodstuffs naturally containing, 42–44
Mangos, preservation of, 82
MAP (modified atmosphere packaging), 73
Marinating, 83–84
Market demand for organic acids, 173–174
Mathematical modeling of organic acids,
232–233
Meat. *See also* specific types of meat
decontamination of, 56–59, 80–81, 244,
249–250, 255
effects of decontamination on sensory
properties of, 156–158
organic acids used against pathogens
in, 78
preservation of with organic acids,
51–52
proposed regulations regarding
chilling of, 252
QS label for, 249
use of organic acid combinations for
decontamination of, 66
Meat industry, regulations regarding safe
food production in, 6
Metabolic reactions, inhibition of by
organic acids, 122
Microarchitecture, modeling of, 229–230
Microbial behavior, modeling of, 225
Microbial contamination, 3–4
Microbial growth inhibition
mechanisms of action of, 117
physiological actions of organic acids
in, 119–123
Microbial growth, environmental factors
for, 227

Microbial physiology, organic acids and,
165–168
Microguard, 274
Microorganisms. *See also* specific
microorganisms
acid tolerance response of, 188
inducible resistance of, 189–191
inhibition of by organic acids, 139–142
new technologies and applications for
use of, 275–276
protective effects of organic acids on,
153–154
resistance of to preservative
compounds, 185
Minimally processed foods, 285–286
washing of, 287–288
Minimum inhibitory concentration (MIC),
232
use of in predictive modeling, 235
Modeling techniques, 226
improving on, 233–236
significance of, 236–237
Modified atmosphere packaging (MAP), 73
Molasses
use of as a substrate for citric acid
production, 177–180
use of as a substrate for lactic acid
production, 167, 174–175
Monocarboxylic acids, transporters for, 120
Monopolar membranes, use of
electrodialysis with, 171
Msn2p, 195
Msn4p, 195
Multicarboxylic acids, 23
Mycotoxins, 5, 10
food safety regulations regarding, 249

N

Natural resistance, 185–188. *See also* Acid
resistance; Resistance
Nisin, 101, 137
Nonaqueous capillary electrophoresis
(NACE), use of for detection of
organic acids, 297–298
Norovirus, 15
Nutrition, consumer perceptions, 9

O

O-polysaccharides, 192
Ochratoxin A, spoilage due to, 5
Odors, 155–156

Organic acid preservation
    adverse effects of on humans and
        animals, 151–152
    application guidelines for, 245–247
Organic acid sprays, use of for meat
    decontamination, 56–58
Organic acids. *See also* specific acids
    activity of, 118–119
    adverse effects of on foodstuffs, 153
    antibacterial action of, 128–129
    antimicrobial packaging, 70–73
    antiviral actions of, 131
    application methods, 156–158
    applications in animal feed, 282–283
        (*See also* Animal feed)
    as pro-oxidants, 152
    combinations of, 64–69
    commercial trials for the evaluation of,
        285
    comparing sample preparation
        techniques, 303–304
    comparisons of effectiveness among,
        139–142
    concentrations of, 77–79
    considerations in the selection of, 69–70
    contemporary detection methods for,
        295–299
    cost of, 156
    demand for, 173–174
    effectiveness of *vs.* inorganic acids,
        133–134
    electrodialytic production of, 172
    evolution of as food preservatives, 1–2
    factors (physical) that will enhance
        effectiveness of, 136–139
    factors that influence activity of,
        123–126
    food products naturally containing,
        42–44
    general applications of, 41–42
    general characterization of, 21–23
    improving efficacy of, 134–136
    industrial applications of, 55–60
    ineffectiveness of, 158
    inhibition by, 186
    LAB and AAB susceptibility and
        resistance to, 109–110
    microbial inhibition by, 117
    microbial physiology and, 165–168
    odors and palatability, 155–156
    oxidation of, 158
    physiological actions of, 119–123
    predictive indices for, 232–233
    protective effects of on microorganisms,
        153–154, 198
    recommended applications of, 80–81
    recommended daily intake, 154–155
    regulations on use of in
        decontamination, 252 (*See also*
        Carcass decontamination)
    role of in ATR, 206–208
    role of pH in activities of, 126–128
    salts of, 60–64
    structural description of, 23–24
    traditional detection methods for,
        293–295
    use of as additives in chilled foods,
        82–83
Oxidation, 158
Oxygen supply, effect of on production of
    lactic acid, 176
Ozone, 136

P

Palatability, 155–156
Parabenzoates, 63
Partial least squares regression (PLS)
    models, 231
Passive homeostasis, 123, 212
Pasteur, Louis, 1
Pathogens
    control of, 81–82
    new and emerging, 11–15
    organic acid concentrations and, 78
PDR12 gene, 190
PEMA (polythylene-co-methacrylic acid),
    71
Penicillium, 10, 14
Pesticides, 9–10
pH
    acid tolerance response and, 207–208
    common foodstuffs, 132
    effect of on production of lactic acid, 176
    effect of on resistance to organic acids,
        190–191
    homeostasis, 122–123, 212–213
    role of in organic acid activities, 126–128
Pharmaceutical emulsifiers, 68
Pharmaceuticals, industrial fermentation
    for, 170
Phenyllactic acid, 40, 79, 261–262
PhoPQ, regulation of *Salmonella
    typhimurium* by, 209
pK_a value, 124

Polyphenolics, recommended daily intake, 154–155

Postweaning edema disease (PWOD), control of with organic acids, 74–75

Potassium sorbate, 78
use of as an antimicrobial, 61–63

Poultry
control of common pathogens in, 81–82
organic acid decontamination of, 156–158
organic acids used against pathogens in, 78
use of marinades in processing of, 84
use of organic acid combinations for decontamination of, 66
use of organic acids for decontamination of, 59
use of organic acids for preservation of, 52
use of potassium sorbate to inhibit microbial growth in, 62–63

Predictive microbiology, 225, 235–236
Gamma concept of, 234
indices for, 232–233
models for, 230–231

Preservation, use of organic acids for, 1–3

Preservatives
chemical reactions of humans to, 151–152
combinations of, 283
concentrations of, 77–79
consumer satisfaction and the use of, 281–282
general food safety regulations and, 248–250
mechanisms of action of, 117
protective effects of on microorganisms, 153–154
sensorial effects and consumer perception, 154

Principle properties, modeling of organic acids using, 233

Pro-oxidants, organic acids as, 152

Probiotics, 105, 276–277

Processed meats, organic acids as antimicrobials in, 58–59, 80–81

Produce washing, alternative techniques to, 286–288

Production process, assessment of, 51

Propionibacteria, organic acid production by, 167

Propionibacterium freudenreichii, production of organic acids by, 111

Propionic acid, 36–37, 168
use of in animal feeds, 76–77

Proteomics, 227–228

Proton motive efflux, 193

Proton permeability, acid tolerance response development and, 211

Protozoa, 15

Pseudomonas, 52

Pure acids, 21

PWOD (postweaning edema disease), control of with organic acids, 74–75

Pyruvic acid, use of in animal feeds, 74

Q

QS label for meat, 249

Quality management, proposed regulations for, 251–253

Quantitative microbiology techniques, 226, 232–233

R

Ready-to-eat (RTE) meats, 288–289
alternative application regimes for decontamination of, 286
food safety regulations regarding, 250
levels of organic acid preservatives in, 79
pathogens in, 12
role of government in control of food poisoning due to, 253–254

Recommended daily intake, 154–155

Regulations for food safety, 248–250

Renewable materials, lactic acid production from, 175

Resistance. *See also* Natural resistance
development of, 188–189
inducible, 189–191
industry strategies and, 198–199
mechanisms of, 191–196
transmission of, 196

Response surface modeling, 236

Rhizopus oryzae, 110
use of in industrial fermentation, 173

Rhizopus sp.
organic acid production by, 110–111
production of lactic acid by, 176–177

Rhizopus stolonifer, production of fumaric acid by, 111

RpoS, acid tolerance response and, 210–211
Ruminants, organic acids in feeds for, 74

S

Saccharomyces cerevisiae, 111, 159
   acid tolerance of, 215
   modeling of organic acid responses of, 228
   resistance of to organic acids, 187–189, 194–195
   use of sorbates to inhibit growth of, 62
Salad bars
   microbial contaminants in, 12
   use of organic acids in, 54
Salmonella enterica sv. Typhimurium
   decontamination methods and, 157–158
   reaction of to stress factors, 192
Salmonella spp., 12, 135
   comparison of organic acid effectiveness against, 140
   control of with organic acids in animal feeds, 74–76
   effects of organic acids on, 129
   modeling of organic acid responses of with genomics, 229
   resistance of, 197
   use of liquid smoke against, 68
Salmonella sv. Enteritidis, effect of aromatic compounds on, 67
Salmonella typhimurium
   acid adaptation in, 215
   acid tolerance response of, 206–209
   cross-resistance to secondary stresses of, 210
   development of acid tolerance response in, 211–212
   use of sodium lactate to limit proliferation of, 64
Salmonellosis, 11–12
Salts of organic acids
   antimicrobial activity of, 60–64
   combinations of, 66–67
Seafood
   use of L-ascorbic acid for preservation of, 59
   use of organic acids for preservation of, 52
Secondary stresses, cross-resistance to, 210
Sensory properties
   organic acid selection and, 69
   preservatives and, 154

Shiga toxin-producing *E. coli* O157:H7 (STEC). *See also Escherichia coli* O157:H7
   acid tolerance of, 209
Signaling systems, modeling of, 226–227
Simultaneous saccharification and fermentation (SSF), 174
Sodium benzoate, 26–27. *See also* Benzoic acid
   chemical reactions in humans to, 151
   use of as a food preservative, 63
Sodium chloride, hurdle effect and, 66
Sodium citrate, 64, 79
   production of by *Aspergillus niger*, 179
Sodium diacetate, 64
Sodium lactate, 34, 64, 79
Sodium salts, 79, 141
   control of microbial growth with, 61
   flavor and, 69
Soft drinks, use of organic acids in, 55
Solid-phase extraction (SPE), 298, 303
Solid-state fermentation (SsF), 177–178
Sorbates
   resistance to, 189, 194–195
   use of as an antimicrobial, 61–63
Sorbic acid, 37–39, 55, 78–79, 188
   adverse effect of on bread products, 153
   bacterial membrane disruption by, 121
   chemical reactions in humans to, 151
   inhibition of *Listeria monocytogenes* by, 80
   modeling of migration between phases of, 237
   pH-dependent effects of, 128
   use of as a food preservative, 2
   use of as a fungicide, 130
Sourdough, naturally occurring organic acids in, 261–262
Spices, use of in combination with organic acid salts, 68
Spoilage microbiota, 217, 281
   organic acid production from, 169
   resistance of to organic acids, 197
Sport drinks, use of organic acids in, 55
Spraying
   development of acid tolerance response due to, 216
   use of for decontamination, 60
Starter cultures, use of lactic acid bacteria as, 98
Steam washing, 137–138
STEC O157, 192
Stoichiometric models, 231

Storage temperature, 138
Streptococcus, use of liquid smoke against, 68
Stress proteins, 189
Stress responses, 135
   evocation of by sorbic acid, 190
   gene regulation of, 195–196
   influence of organic acids on, 159
   modeling of with genomics, 228–229
Submerged fermentation (SmF), 177
Substrates, 168–170, 178
Succinic acid, 39
   use of in animal feeds, 74
Sulfite, use of as a preservative, 2
Sulfonic acids, 21

T

Tartaric acid, 39–40
Temperature
   effect of on organic acid activity, 125–126, 137
   effect of on production of lactic acid, 176
   storage, 138
Thermal stress, resistance to, 210
Thermotolerance, 190–191, 210
   effects of organic acids on, 159
Toxic anions, accumulation of, 121–122
Transdermal synergists, 134
Trehalose, 191

U

U.S. Department of Agriculture. *See* USDA
U.S. Food and Drug Administration. *See* FDA
Ultrasound, 136
Undissociated organic acids, 124, 185–186
USDA, 6
   food safety regulations of, 249–250
   Hazard Analysis and Critical Control Point (HACCP), 244

V

Vacuum packaging, 56, 138
Vegetables
   alternatives to washing techniques for, 286–288
   organic acids in, 168
   use of organic acids against pathogens in, 82

use of organic acids in preservation of, 53
Verocytotoxin-producing *Escherichia coli* (VTEC), 243
Vibrio cholerae, development of acid tolerance response in, 209, 211–212
Vinegar, 275. *See also* Acetic acid
   industrial manufacturing of from AAB, 167
   microorganisms involved in the production of, 106–107
   naturally occurring organic acids in, 264
   organic acids in, 43
   production of, 106
   use of as a food preservative, 2
Viruses, 15

W

Washing techniques, alternatives to, 286
Water activity (a_w), 125
Water, use of for meat decontamination, 56, 80
Weak acids, 23
   adaptation to, 189
   naturally occurring, 165
   pH effects of, 128
   physiological actions of, 120
   pK_a values of, 124
   resistance to, 195 (*See also* Acid resistance)
   use of as antimicrobials, 2
Wheat flour, levels of organic acid preservatives in, 79
White Paper on Food Safety, 243, 251
Wine
   detection of organic acids in, 302
   naturally occurring organic acids in, 262–263
   organic acids in, 43
   organic acids produced from, 168
   spoilage of by acetic acid bacteria, 107–108
World Trade Organization, 251

X

Xerophilic fungi, 126

Y

Yarrowia lipolytica, 14–15, 275
    acid tolerance of, 215
Yeasts
    acetic acid bacteria and, 106–107
    acid tolerance of, 214–215
    food spoilage due to, 4–5, 14–15
    inducible resistance of, 190
    inhibition of, 130
    mechanisms of organic acid resistance
        in, 194–196
    production of organic acids from, 165

    resistance of to organic acids, 187–188
Yersinia enteric, inhibition of by organic
        acids, 140
Yersinia enterocolitica, 83
Yields, 168–170

Z

Zygosaccaromyces bailii, 14, 159
    acid tolerance of, 214–215
    resistance of to organic acids, 187–189,
        194–195